Lecture Notes in Earth Sciences   59

Editors:
S. Bhattacharji, Brooklyn
G. M. Friedman, Brooklyn and Troy
H. J. Neugebauer, Bonn
A. Seilacher, Tuebingen and Yale

Springer
Berlin
Heidelberg
New York
Barcelona
Budapest
Hong Kong
London
Milan
Santa Clara
Singapore
Paris
Tokyo

Alick C. Kibblewhite
Cheng Y. Wu

Wave Interactions As a Seismo-acoustic Source

 Springer

Authors

Prof. Dr. Alick C. Kibblewhite
Dr. Cheng Y. Wu
Department of Physics, University of Auckland
Private Bag 92019, Auckland, New Zealand

Cataloging-in-Publication data applied for

Die Deutsche Bibliothek - CIP-Einheitsaufnahme

Kibblewhite, Alick C.:
Wave interactions as a seismo-acoustic source / Alick C.
Kibblewhite ; Cheng Y. Wu. - Berlin ; Heidelberg ; New York ;
Barcelona ; Budapest ; Hong Kong ; London ; Milan ; Paris ;
Tokyo : Springer, 1996
  (Lecture notes in earth sciences ; 59)
  ISBN 3-540-60721-8 (Berlin ...)
  ISBN 0-387-60721-8 (New York ...)
NE: Wu, Cheng Y.:; GT

"For all Lecture Notes in Earth Sciences published till now please see final pages of the book"

ISBN 3-540-60721-8 Springer-Verlag Berlin Heidelberg New York

This work is subject to copyright. All rights are reserved, whether the whole or part of the material is concerned, specifically the rights of translation, reprinting, re-use of illustrations, recitation, broadcasting, reproduction on microfilms or in any other way, and storage in data banks. Duplication of this publication or parts thereof is permitted only under the provisions of the German Copyright Law of September 9, 1965, in its current version, and permission for use must always be obtained from Springer-Verlag. Violations are liable for prosecution under the German Copyright Law.

© Springer-Verlag Berlin Heidelberg 1996
Printed in Germany

Typesetting: Camera ready by authors
SPIN: 10492877   32/3142-543210 - Printed on acid-free paper

Preface

The ocean has always been reluctant to reveal its secrets. Its size and the inaccessibility of its deeper regions have made their safeguard a reasonably simple matter with the result that significant misconceptions persisted for many years. Two of the most widespread of these concerned the featureless nature of the sea floor and the silence of the deep ocean. Underwater acoustics has played a key role in discrediting both and in so doing introduced new and exciting developments in oceanography and geophysics.

In the years following World War II, echosounders and subbottom profilers based on new active sonar technology, revealed the true nature of the seafloor topography and led to the major advances represented by plate tectonics. Research driven by the requirements of passive sonar, on the other hand, was to demonstrate that the sea was not silent but was characterised by a complex noise spectrum. Many individual mechanisms and sources ranging from man-made, biological and geophysical activity to the intrinsic noise of the sea itself were found to contribute to this spectrum. A major component, which is the subject of this book, was to remain unrecognised to underwater acoustics until noise measurements could be made effectively at very low frequencies, although its presence had been indicated by seismology long before these measurements were possible.

By virtue of its geographical isolation in the Southern Ocean, New Zealand has provided an ideal environment for long-range propagation and ambient noise investigations and numerous studies have been reported. Our interest in the subject of this book was aroused initially in the course of one such experiment in 1966. For the first time it had been possible to extend the recording bandwidth to 1 Hz and the improved performance of this new system was anticipated eagerly. However the main purpose of the experiment was nearly aborted by the appearance of a new and unsuspected noise component at frequencies below 10 Hz.

Due primarily to technical limitations in the equipment then available, a subsequent programme, designed to identify the properties and origin of the source more clearly, was not productive and was soon abandoned. An opportunity to revisit the problem arose some 10 years later, when the University of Auckland became involved in a major environmental study in support of the development of an offshore gas field in Cook Strait. The technology then available provided an opportunity to examine afresh the relationship between sea state and the seismo-acoustic response generated.

An initial trial demonstrated the potential of the site. Accordingly a long-term programme, involving the parallel measurement of the ocean-wave field and acoustic response, was undertaken in a series of student research theses. The data so gathered were of sufficiently high quality to ultimately establish wave-wave interactions as the source of the acoustic effects observed and to identify many of its characteristics.

This result was soon to be confirmed by other studies. As the noise data accumulated, however, it became apparent that certain refinements to the theories describing the mechanism were required. Our attempts to provide these refinements have been reported in a number of contributions in recent years. The accounts of these and similar contributions by others have unfortunately appeared in the literature in a somewhat disjointed manner, with the result that the evolution of the subject has not been easy to follow. This book attempts to present a more coherent account of the subject and its development.

Most of the early experimental and theoretical results from our group have arisen from two key Ph.D. theses, due to Dr. K.C. Ewans and Dr. C.Y. Wu. The painstaking and careful instrumentation development and data analysis provided by Dr. Ewans were critical to the definitive correlation which we were able to establish between wind field, seastate and the acoustic response so generated. Dr. Wu's thesis presented the first phase of our attempt at the resolution of certain key theoretical issues, which were identified in the course of the experimental programme. Both studies owe much to the support of Shell BP Todd Oil Services Ltd., acting for Maui Development Ltd., and to the University of Auckland. The support of the Electricity Corporation of New Zealand Ltd. during a later experimental investigation of the Southern Ocean wave field is also acknowledged.

Much of the new theoretical work described in this book is the result of work sponsored by the US Office of Naval Research, Code 324 GG, under Research Grant N00014-92-J-1037. The sympathetic support of Dr. Randall S. Jacobson of that Office throughout this phase is appreciated. We also wish to acknowledge most warmly the help and encouragement of Dr. David L. Bradley throughout. Our analysis of the effects of porosity on the reflection coefficient, in particular, is due largely to his inspiration.

Many colleagues have obviously contributed to a programme which has run for nearly 20 years and their help is also gratefully acknowledged. The technical and moral support of Mr. D.A. Jones in the early stages of the Maui Programme was critical to the success of the seismo-acoustic measurements, as was the computer software development of Dr. S.M. Tan and the field and analysis support of Mr. P.L. Pearce during the Southern Ocean programme. The contribution of Dr. M.D. Johns in the management and development of the Department's ever expanding computer facilities over the long period of these studies has also been fundamental to their success. Thanks also go to Associate Professor C.T. Tindle and Dr. D.H. Cato (of Sydney) for their helpful comments on the draft of this book and to Freda Anderson for her patience and assistance in its preparation. But the most significant contribution to the latter phases of this project has come from my student, friend and colleague, Dr. Wu. Without his theoretical ability, perseverance, patience and application this document could not have been produced in its present form.

Finally, a study of this duration experiences many ups and downs, all of which my wife and family has had to endure with me. I will be for ever grateful to them for their tolerance, understanding and love throughout it all.

A.C. Kibblewhite

Contents

1 Introduction   1

2 A Review of Earlier Theoretical Analyses   11
  2.1 Related to Microseisms 11
  2.2 Related to Underwater Acoustics 13
  2.3 Related to the Spectral Characteristics of the Pressure Field . . 18

3 The Perturbation Procedure   21
  3.1 The Environment Model 21
  3.2 Governing Equations 21
  3.3 Boundary and Interface Conditions 24
  3.4 Linearisation 26
    3.4.1 Potential as the Field Variable 26
    3.4.2 Pressure as the Field Variable 29
  3.5 Displacement Potentials in the Sediments 32
  3.6 The Effect of the Seabed on the Induced Pressure Field 33
  3.7 Notations Covering the Fourier Transforms and Spectral Representation 34
    3.7.1 Definitions 34
    3.7.2 Spectral Representation 35

4 An Exact Expression for the Homogeneous Component   37
  4.1 Introduction 37
  4.2 The Potential Field Solution 38
    4.2.1 Equations and Boundary Conditions 38
    4.2.2 Plane-Wave Analysis 39
    4.2.3 Green's Function Analysis 41
    4.2.4 Comparison of the Plane-Wave and Green's Function Analyses 44
  4.3 The Pressure Field Solution 45
    4.3.1 Equations and Boundary Conditions 45

		4.3.2 Plane-Wave Analysis	46
		4.3.3 Green's Function Analysis	48
		4.3.4 Surface and Volume Source Distributions	49
	4.4	Source Pressure Spectra	50
		4.4.1 Source Functions	50
		4.4.2 Spectral Expressions based on the Potential as the Variable	52
		4.4.3 Spectral Expressions based on the Pressure as the Variable	52
		4.4.4 Power-Density Source Spectra (Frequency-Wavenumber)	53
		4.4.5 Standing-Wave Approximation of the Source Spectrum	54
		4.4.6 Power-Density Source Spectra (Frequency)	55
5	Generalised Solution of the Potential Field		59
	5.1	First-Order Potential	59
		5.1.1 Plane-Wave Solution	59
		5.1.2 Green's Function Solution	62
		5.1.3 Potential in a Weakly Range-Dependent Wave Guide	65
	5.2	Second-Order Potential	65
		5.2.1 General Solution	65
		5.2.2 Potential in the Reference Wave Guide	68
		5.2.3 Potential in a Weakly Range-Dependent Wave Guide	69
6	The Generalised Source Spectrum		73
	6.1	Preamble	73
	6.2	Definition of the Source Spectrum (Wavenumber-Frequency)	74
	6.3	Calculation of the Source Spectrum (Wavenumber-Frequency)	77
		6.3.1 Deep-Water Case	77
		6.3.2 Shallow-Water Case	92
	6.4	Second-order Source Spectra (Frequency)	96
7	Properties of the Wave-Induced Pressure Field		99
	7.1	Coherence Function	100
		7.1.1 First-Order Field	100
		7.1.2 Second-Order Field	101
	7.2	Ocean-Wave Spectrum and Seafloor Structure	102
		7.2.1 Ocean-Wave Spectrum	102
		7.2.2 Geoacoustic Environment	105
		7.2.3 Reflection Coefficient	110
	7.3	Theoretical Pressure Spectra	112
		7.3.1 Dependence on Observation Depth	112

		7.3.2	Influence of the Seabed Structure	121
		7.3.3	Influence of the Ocean-Wave Spectrum	122
		7.3.4	Dependence on Wind Speed	123
		7.3.5	Dependence on Water Depth	124
	7.4		Comparison with Field Data	128
		7.4.1	New Zealand	128
		7.4.2	Eleuthra	129
		7.4.3	Pacific Ocean	131
		7.4.4	Woronara Dam - Australia	133
		7.4.5	Upper Levels of the Ocean	133
	7.5		The Equivalent Source Level	134
	7.6		Normalised Coherency Spectra	138
		7.6.1	The Formalism	138
		7.6.2	Vertical Coherence Function	141
		7.6.3	Horizontal Coherence Functions	142
8	**The Seismic Response**			**151**
	8.1		Introduction	151
	8.2		The Green's Functions	152
	8.3		The Coherence Function of the Seismic Field	154
		8.3.1	Source Region of Infinite Size	156
		8.3.2	Outside a Source Region of Finite Size	159
	8.4		Spectral Characteristics of the Seismic Field	161
		8.4.1	Influence of Water Depth	162
		8.4.2	Influence of Wind Speed	164
		8.4.3	Influence of Seabed Structure	164
		8.4.4	Influence of Layer Thickness	165
	8.5		Range Dependence of the Seismic Response	165
	8.6		Potential Applications of the Source Field	172
		8.6.1	Monitoring the Ocean-Wave Field	173
		8.6.2	Inversion Studies of Seabed Structure	177
9	**Summary**			**181**
A	**Equations Quoted in Chapter 2**			**183**
B	**Basic Equations Governing Wave Motion in a Viscoelastic Layer**			**193**
C	**Basic Equations Governing Wave Motion in a Water-Saturated Porous Viscoelastic Layer**			**197**

D The Calculation of the Reflection Coefficient for a Multilayered Viscoelastic Half Space — 205

E Frictional Force in a 3-D Duct — 213

F Numerical Procedure for the Calculation of the Reflection Coefficient for a Multilayered Porous Viscoelastic Medium — 219

G Spectral Representation of a Stationary Process — 235

H General Solution of Adjoint Boundary Value Problems — 239

I Solutions to the First-Order Potential Field — 247

J Calculation of the First-Order Green's Function — 253

K Solution of the Second-Order Potential Field — 257

L Concerning the Source Pressure Spectrum of the Wave-Wave Interaction Process in a Shallow-Water Environment — 265

M The Geometric Description of Wave-Wave Interactions and the Calculation of the Source Spectrum $\langle |B^{(2)}(\vec{k},\omega)|^2 \rangle$ — 267

N Coherence and Spectral-Density Functions of the First- and Second-Order Fields — 279

O Expression for the Reflection Coefficient in Terms of the Parameters of the Top Sedimentary Layer — 289

P Green's Functions Linking the Source Pressure Field and the Seismic Field on the Seafloor — 295

Q Bibliography — 301

R Subject Index — 309

List of Figures

3.1 Schematic representation of a slowly range-dependent geoacoustic model with a locally stationary random sea surface. 22

3.2 Schematic representation of a range-independent geoacoustic model with a locally stationary random sea surface. 22

6.1 Geometric representation of two interacting gravity waves inducing a pressure wave with horizontal wavenumber vector $\vec{k}$ and angular frequency ω. Propagation directions are specified relative to the wind vector OW. 78

6.2 Representation of the nonlinear interaction of two surface gravity waves inducing a plane acoustic wave with horizontal wavenumber vector $\vec{k}$: **a** for $0 \leq \hat{m} \leq 0.958$; **b** for $0.958 < \hat{m} \leq 2$; **c** for $\hat{m} = 2$; **d** for $2 < \hat{m} < \infty$. 81

6.3 Loci of the wave vectors of two interacting gravity waves producing a fixed horizontal wavenumber vector $\vec{k}$ at different frequencies: **a** for the sum-frequency component ($0 \leq \hat{m} \leq 2$); **b** for the difference-frequency component ($2 < \hat{m} < \infty$). 81

6.4 Representation of the interaction process. 84

6.5 Schematic representation of the variation of the locus of the wave interaction vector with increasing wavenumber k: **a** for $0 \leq \hat{m} \leq 0.958$; **b** for $0.958 < \hat{m} \leq 1$; **c** for $1 < \hat{m} \leq 2$; **d** for $2 < \hat{m} < \infty$. 85

6.6 The sum-frequency source spectrum, $10\log_{10}(\langle|B^{(2)}(f,\hat{m})|^2\rangle)$, presented as a function of relative wavenumber $\hat{m}$ and frequency f for a wind speed of 15 ms^{-1}. 87

6.7 The source spectral level of the sum-frequency and difference-frequency components, $10\log_{10}(\langle|B^{(2)}(f,\hat{m})|^2\rangle)$, presented as a function of $\log_{10}(f)$ and $\log_{10}(\hat{m})$: **a** at the mean surface; **b** at depth 50 m, for a wind speed of 15 ms^{-1} 91

6.8 The spectral level of the total pressure field at different depths in a gravitationally deep ocean (H_1=1000 m) overlying a multilayered seabed (MDL1), under a prevailing wind of 15 ms^{-1} and a PM sea : **a** 15 m; **b** 100 m; **c** 500 m; **d** 1000 m. 93

6.9 Loci of the wave vector of two gravity waves interacting in a shallow-water environment and inducing a sum-frequency pressure wave with horizontal wavenumber vector $\vec{k}$: **a** depth $H_1=100$ m; **b** $H_1=50$ m; **c** $H_1=10$ m. 94

6.10 Loci of the wave vector of two gravity waves interacting in a shallow-water environment and inducing a difference-frequency pressure wave with horizontal wavenumber vector $\vec{k}$: **a** depth $H_1=100$ m; **b** $H_1=50$ m; **c** $H_1=20$ m. 95

6.11 Source spectra (frequency) of the second-order pressure field (sum-frequency component); **a** and **b** the sum of the homogeneous and inhomogeneous fields at observation depths z=10, 50, 100, 500, 1000 m and at infinity, for PM and JONSWAP seas respectively and a wind speed of 15 ms^{-1}; **c** and **d** the source spectra at 10 m observation depth, for PM and JONSWAP seas and wind speeds 2.5, 5, 7.5, 10, 15, 20, and 30 ms^{-1}. 97

7.1 Ocean surface-wave spectra: **a** Pierson-Moskowitz(PM) form, **b** JONSWAP(J) form of the wave spectrum, wind speed 2.5-30 ms^{-1} in steps of 2.5 ms^{-1}. . 104

7.2 Compressional and shear-wave velocity profiles of the geoacoustic models: **a** MDL1 and MDL2 (in MDL2 the top layer is porous); **b** MDL3(1); **c** MDL3(2); **d** MDL3(3). 106

7.3 Modulus of the plane-wave reflection coefficient for model MDL1 as a function of $\log_{10}$(frequency) and $\log_{10}$(relative wavenumber): **a** emphasising the highest peaks; **b** showing the structure in more detail. 111

7.4 Modulus of the plane-wave reflection coefficient for model MDL2. 112

7.5 Modulus of the plane-wave reflection coefficient for model MDL3: **a** with no hydrate layer; **b** with a hydrate layer; and **c** with a hydrate layer in association with a thin gas layer. 113

7.6 Reflection loss for model MDL3 as a function of frequency and incident angle: **a** with no hydrate layer; **b** with a hydrate layer; **c** with a hydrate layer in association with a thin gas layer. 114

7.7 Modulus of the plane-wave reflection coefficient for model MDL3 as a function of $\log_{10}$(frequency) and incident angle: **a** with no hydrate layer; **b** with a hydrate layer; **c** with a hydrate layer in association with a thin gas layer. 115

7.8 Wave-induced pressure spectra at 10 m, 20 m, 40 m, 100 m, 500 m, 1000 m, and 4000 m below the sea surface, for an ocean 4000 m deep, model MDL1, a PM sea and a wind speed of 15 ms^{-1}. 117

7.9 A contour plot of the spectral level of the depth-dependent, wave-induced pressure field, for an ocean 4000 m deep, model MDL1, a PM sea and a wind speed of 15 ms^{-1}. 117

7.10 Depth-dependence of the wave-induced pressure spectral level at 0.2, 0.5 and 1.06 Hz, for the environment and conditions of Fig. 7.8. 118

7.11 A contour plot of the spectral level of the depth-dependent, wave-induced pressure field, for an ocean 2500 m deep, model MDL3(3), a PM sea and a wind speed of 15 ms^{-1}. 118

7.12 Wave-induced pressure spectra at 10 m, 20 m, 40 m, 80 m, and 100 m below the sea surface, for an ocean 100 m deep, a PM sea and a wind speed of 15 ms^{-1}: **a** model MDL(1); **b** model MDL0, $R_b = 0$; **c** model MDL0, $R_b = 1$. . 119

7.13 Wave-induced pressure spectra at 10 m, 50 m, 500 m 1500 m and 2500 m below the sea surface, for an ocean 2500 m deep, a PM sea and a wind speed of 15 ms^{-1}: **a** model MDL3(1); **b** model MDL3(2); **c** model MDL3(3).120

7.14 Wave-induced pressure spectra at 10 m, 20 m, 40 m, 100 m, 500 m, 1000 m, 2000 m and 4000 m below the sea surface, for an ocean 4000 m deep, a seabed with $R_b = 1$, a PM sea and a wind speed of 15 ms^{-1}. 121

7.15 Wave-induced pressure spectra at 10 m, 20 m, 50 m, 80 m and 100 m below the sea surface, for an ocean 100 m deep, a PM sea and a wind speed of 15 ms^{-1}: **a** model MDL2; **b** model MDL3(1); **c** model MDL3(3). 123

7.16 Wave-induced pressure spectra at 10 m, 20 m, 50 m, 80 m and 100 m below the sea surface, for an ocean 100 m deep, model MDL2, a JONSWAP sea and a wind speed of 15 ms^{-1}. 124

7.17 Wave-induced pressure spectra at 10 m, 20 m, 30 m, 40 m and 50 m below the sea surface, for an ocean 50 m deep, a JONSWAP sea and a wind speed of 15 ms^{-1}: **a** model MDL1; **b** model MDL2; **c** model MDL3(3). 125

7.18 Wave-induced pressure spectra in shallow water of 100 m depth, for model MDL1, and wind speeds ranging from 2.5, 5, 10, 15, 20, 25 and 30 ms^{-1}: **a** a PM sea; **b** a JONSWAP sea. 125

7.19 Wave-induced pressure spectra in shallow water of 50 m depth, for model MDL1, and wind speeds ranging from 2.5, 5, 10, 15, 20, 25 and 30 ms^{-1}: **a** a PM sea; **b** a JONSWAP sea. 126

7.20 Wave-induced pressure spectra in shallow water of 30 m depth, for model MDL1, wind speeds ranging from 2.5, 5, 10, 15, 20, 25 and 30 ms^{-1}: **a** a PM sea; **b** a JONSWAP sea. 126

7.21 Contours of the wave-induced pressure level over a continental shelf, extending 200 km from the shore with a slope of 0.002, for a JONSWAP sea and a wind speed of 15 ms^{-1}: **a** 0.2 Hz; **b** 0.5 Hz; **c** 1.05 Hz. 127

7.22 Comparison of the experimental spectra derived from the seismic data with theoretical pressure spectra in which the effect of the swell and the interaction of a local residual sea is included. 129

7.23 Comparison of predicted and experimental noise spectra measured in water off Eleuthra: **a** water depth 13 m; **b** water depth 300 m; **c** water depth 1200 m. 130

7.24 Comparison of predicted and measured noise spectra at deep-water sites: **a** 4000 m in depth; **b** over 3000 m in depth on the Pacific Rise. 133

7.25 A comparison of synthetic pressure rate (dP/dt) spectra with measured data(right) for depths 110, 150, 190, 230, 270, 290 m below the surface. The spectral curves are each spaced upward by 50 $(Pa/s)^2/$Hz. The experimental spectra, reproduced from Fig. 3 of Ref.[78], are measured at depths 110-290 m and the curves are again spaced (by $25(Pa/s)^2/$Hz) from each other for clarity. (This figure first appeared in Natural Physical Sources of Underwater Sound - Sea Surface Sound(2), Ed. B.R. Kerman, Kluwer Academic Publishers (1993) - see Ref.[36]. Reprinted by permission of Kluwer Academic Publishers). 134

7.26 The classical deep-water form of the pressure spectra generated by non-linear wave interactions under wind speeds of 2.5, 5, 7.5, 10, 15, 20, 25, and 30 ms^{-1}, calculated using **a** the PM and **b** the JONSWAP form of the ocean-wave spectrum. 136

7.27 The vertical coherence function of the wave-induced pressure field (wind speed 15 ms^{-1}) in an ocean environment 100 m in depth: **a** at 0.25 Hz; **b** at 0.5 Hz; **c** at 1.0 Hz. 143

7.28 The horizontal coherence function of the PF pressure field: in an ocean of infinite depth, **a** at 0.05 Hz; **b** at 0.1 Hz; and in water of 15 m depth, **c** at 0.05 Hz; **d** at 0.1 Hz. 144

7.29 The horizontal coherence function of the DF pressure field, for a PM sea, wind speed 15 m s^{-1}, in an ocean of infinite depth: **a** at 0 m; **b** at 100 m; and **c** at 1000 m below the surface. 146

7.30 Distribution of the DF pressure spectral levels in the horizontal wavenumber plane, for a PM sea, wind speed 15 ms^{-1}, in an ocean of infinite depth: **a** at 0 m; **b** at 100 m; and **c** at 1000 m below the surface. 147

7.31 Normalised angular distribution of the DF pressure field, for a PM sea, wind speed 15 ms^{-1}, $f = 0.2$ Hz: **a** at 0 m; **b** at 50 m; **c** at 100 m; **d** at 1000 m below the surface. 148

8.1 Spectral levels of the vertical (solid) and horizontal (dashed) components of the seabed displacement, and their ratio, for model MDL1, wind speed 15 ms^{-1}, and a JONSWAP sea: **a** $H_1 = 15$ m; **b** $H_1 = 50$ m. 161

8.2 Spectral levels of the vertical (solid) and horizontal (dashed) components of the seabed displacement, and their ratio, for model MDL1, wind speed 15 ms^{-1}, and a PM sea: **a** $H_1 = 50$ m; **b** $H_1 = 100$ m; **c** $H_1 = 500$ m; **d** $H_1 = 1000$ m. 163

8.3 Wind-dependence of the spectral levels of **a** the vertical and **b** the horizontal component of the seabed displacement, for model MDL1 and a PM sea. . . 164

8.4 Spectral levels of the horizontal and vertical components of the seabed displacement, and their ratio, for model MDL2, wind speed 15 ms^{-1} and a PM sea: **a** $H_1 = 15$ m; **b** $H_1 = 50$ m; **c** $H_1 = 100$ m; **d** $H_1 = 500$ m. 166

8.5 Spectral levels of the horizontal and vertical components of the seabed displacement, and their ratio, for a wind speed 15 ms^{-1} and a PM sea, $H_1 = 1000$ m: **a** model MDL3(1); **b** model MDL3(2); **c** model MDL3(3). . 167

8.6 Contour plots of the amplitude ratio of the horizontal (H) and vertical (V) components of the seabed displacement, H/V, for $20\log_{10}(H/V)$ ranging from -20 to 0 dB (left plots) and from 0 to 20 dB (right plots) in steps of 5 dB as a function of frequency and layer thickness, under a PM sea and wind speed 15 ms^{-1}: **a** and **b** for the elastic sediment and **c** and **d** for the porous sediment. 168

8.7 Decay of the vertical component of the seabed displacement with distance from the centre of the active region: **a** overlying an unconsolidated bottom; **b** overlying an elastic halfspace; **c** overlying a multilayered bottom characterized by low shear-wave speed; **d** overlying a multilayered bottom characterized by high shear-wave speed. 171

8.8 A comparison of short-term histories of the significant waveheight measured from waverider and inverted seismic spectra. 175

8.9 Average power density spectra of the sea-surface displacement for the months October 1990 (133 spectra), November 1990 (175), December 1990 (146) and January 1991 (184), established from 20-min records (taken every 4 hours) of the seismic response to the offshore wave field. 176

8.10 The average vertical (V) and horizontal (H,EW/NS) power-density acceleration spectra for March 1992, established from 182 20-min records (taken every 4 hours) of the seismic response to the offshore wavefield. 177

8.11 **a** The effective operating band (in the wavenumber-frequency domain) of the pressure field generated by the wave-interaction source, as a function of wind speed and water depth, for JONSWAP and PM seas; **b** Contour levels of the reflection coefficient for model MDL2 characterised by an unconsolidated, porous upper layer 100 m in thickness. The contour levels of 0.2,0.5,1,2,5,10, and 20 are the same as in Fig.7.4. 178

M.1 Representation of the nonlinear interaction of two surface gravity waves inducing a plane acoustic wave with horizontal wavenumber vector $\vec{k}$: **a** for $|\vec{k}| = 0$; **b** for $0 \leq |\vec{k}| \leq \omega/\alpha_1$; **c** $\omega/\alpha_1 \leq |\vec{k}| < \omega^2/g$; **d** for $|\vec{k}| \geq \omega/\alpha_1$. 269

M.2 Representation of the nonlinear interaction of two surface gravity waves inducing a plane acoustic wave with horizontal wavenumber vector $\vec{k}$: $|\vec{k}| = \omega^2/g$. 270

M.3 Curves L' on the $\vec{k}$-plane for different values of $\hat{m}$ 271

M.4 Contour plot of the values of Δ 271

M.5 The values of the roots of the cubic equation and $Q(\chi)$ as a function of θ_{1k} when $\hat{m} \simeq 0.958$. 274

M.6 The values of the roots of the cubic equation and $Q(\chi)$ as a function of θ_{1k} when $\hat{m}=0.98$ (**a** and **c**) and $\hat{m}=1.0$ (**b** and **d**). 275

M.7 The values of the roots of the cubic equation χ as a function of θ_{1k} when $\hat{m}=1.5$ (**a** and **b**), and $\hat{m}=10.0$ (**c**) and $\hat{m}=40.0$ (**d**). 275

List of Tables

6.1 The relationship of the horizontal wavenumber to the integration process. . 83

7.1 Parameters of MDL1. 107
7.2 Parameters of the unconsolidated porous sediment layer. 107
7.3 Parameters of MDL3(1) 108
7.4 Parameters of MDL3(2) 109
7.5 Parameters of MDL3(3) 109

8.1 Parameters of the geoacoustic models 169

C.1 Parameters for an unconsolidated porous sediment layer. 203

Chapter 1

Introduction

While the first ambient-noise spectra were published in 1948 [1], it was the growing interest in passive sonar systems in the 1960's that stimulated research into this aspect of underwater acoustics. The first published results relating to low-frequency ambient noise appeared in a paper by Wenz[2] describing data in the range 50 Hz to 10 kHz. The noise sources reported included turbulence, wave motion, earthquakes, biological activity and shipping. Typical of other studies in that period were some concerned with wind and wave noise[3]-[5], the acoustic detection and location of underwater volcanism from ambient noise measurements [6], an intriguing diversity of low-frequency activity of biological origin [7], and, relevant to the main theme of this publication, a large rise in noise level at frequencies below 4 Hz under high sea states, reports of which were beginning to appear at this time[8],[9].

Our interest in the last topic was initially aroused during a long-range propagation experiment, CHASE V, carried out in 1966 by United States agencies[8]. New Zealand participation involved developing a temporary receiving array for deployment in relatively deep water. Transistors had just become available and underwater amplifiers could be incorporated in the hydrophones for the first time. Euphoric with the new technical freedom provided by solid-state circuitry, a low-frequency cutoff of 1 Hz was specified for the underwater array. The array was deployed successfully several days before the experiment was scheduled and was performing well. However, just prior to the designated date a severe storm passed across the array site. Ambient noise levels rose dramatically and ultimately exceeded the dynamic range of the underwater electronics. Tests demonstrated that the excess noise was confined essentially to the band below 4 Hz. The array was lifted, the low-frequency response modified, and the system relaid in time, but this initial experience of very low-frequency ocean noise was not one to be forgotten quickly! While obviously related to the conditions prevailing, the noise levels observed were unexpected, given the depth at which the array was deployed, and were clearly not related to sea state in a simple way.

Due primarily to technical limitations in the equipment then available, a subsequent programme, designed to identify the properties and origin of the noise more clearly, was not productive and was soon abandoned. However, an indication of the properties of the source began to emerge as improved low-frequency data became available in the 1970's [10]-[14]. Perrone's results in particular identified several significant spectral properties, among which were a slope of -20 dB/octave between 2 Hz and 5 Hz and a predicted maximum between 1 Hz and 2 Hz.

The 1980's brought an upsurgence of interest in very-low-frequency (VLF) acoustics and the natural physical sources responsible for sound generation at the ocean surface. A research initiative sponsored by the US Office of Naval Research played a key role in fostering these activities in the northern hemisphere. An opportunity for us to mount a related investigation and revisit the problem first encountered 10 years earlier, arose when the University of Auckland became involved in an extensive environmental study in support of the development of an offshore gas field in Cook Strait[15]. A major component of this study was to involve the measurement and description of the ocean-wave field in the area. The technology then available, the long-term nature of the study, certain unique orographic properties of the region, and the high quality of the supporting wind data available, provided a promising opportunity to examine afresh the relationship between sea state and the seismo-acoustic response generated. A programme, involving the parallel measurement of the ocean-wave field and related acoustic response, was initiated and ran for several years. The data so gathered proved to be of sufficiently high quality that a set of wind-dependent, ambient-noise spectra was ultimately established. These spectra, together with the results of other studies, clearly identified wave-wave interactions as the source of the acoustic effects observed.

The growing importance of wave-related processes was recognised by four major international conferences which were convened in the years to follow, all conducted within the NATO Advanced Research Workshop series. The first, which was held at Lerici in Italy in 1985, focussed on low

1 Introduction

frequency seismo-acoustics. While concerned primarily with propagation a section was included on seismo-acoustic noise. The proceedings were published in the NATO series in 1986 as "Ocean Seismo-acoustics", edited by T. Akal and J.M. Berkson (Plenum Press). The first of two conferences concerned specifically with surface-generated noise followed in 1987. A record of this meeting, again held in Lerici, appeared in 1988 in the NATO series as "Sea Surface Sound", edited by B.R. Kerman (Kluwer Academic Publishers). A follow-up meeting in 1990 held at Cambridge, England, was reported in 1993 under the title "Natural Physical Sources of Underwater Sound - Sea Surface Sound(2)", again edited by Kerman (Kluwer Academic Publishers). The latest meeting in this series, held in 1994 at Lake Arrowhead, California, under the title "Sea Surface Sound '94", has yet to be reported.

These proceedings provide a rich source of information on all aspects of sea-surface noise. A review paper by Copeland [16] at the Cambridge meeting provides a useful summary of the status of low-frequency ambient noise at that time. He identifies the experimental contributions which establish the general form of the spectrum in the range 0.1 Hz to 4 Hz[17]-[23] and reviews the evidence which establishes nonlinear wave-wave interactions[24],[25] rather than turbulence[26],[27], as the mechanism primarily responsible for the rise in spectral levels below 4 Hz.

This mechanism was first studied by Longuet-Higgins[24],[25] in connection with the phenomenon of microseisms, a subject of intense interest to the seismic community in the 1950s. He demonstrated that the nonlinear interaction of ocean waves produces pressure fluctuations, which are not attenuated with depth and therefore capable of generating microseisms at the seafloor. In the years to follow Hasselmann[26] and many others were also to make significant contributions to the subject. The first complementary analysis in the context of underwater acoustics came from Brekhovskikh[28]. His results express the wave-induced acoustic intensity spectrum, $S(2f)$, in terms of a surface-wave spectrum, $H(f)$. The properties of $H(f)$ were not known at that time, but $S(2f)$ is predicted to be proportional to $f^3 H^2$. Later Hughes[29] and Lloyd[30] assumed the relationships $H \propto f^{-5}$ and $H \propto f^{-4}$ to establish a spectral slope above the peak of -21 dB/octave and -15 dB/octave respectively. Kibblewhite and Ewans [19],[20] interpreted their experimental data using the relation $H \propto f^{-4.5}$, actually established in parallel measurements of the ocean-wavefield (see Sect. 7.2), to obtain a spectral slope of -18 dB/octave. This result, coupled with the other spectral characteristics identified, demonstrated that the dependence of the noise field on seastate was consistent with the predictions of the wave-interaction theory as formulated by Longuet-Higgins and others. The results of Adair et al[21] and Cotaras et al[23] and later Cato[52] also supported this interpretation.

From a theoretical point of view, the ultimate understanding of wave-induced noise requires the solution of the nonlinear fluid-dynamic equation under certain boundary conditions. Longuet-Higgins' perturbation analysis and Hasselmann's spectral description of the wave-induced pressure field have provided the basis for most of the theoretical studies which were to follow [19],[28],[29],[35]–[38],[41],[42], notable exceptions being those of Lloyd[30], Cato[51],[52] and Guo[54]. In the past these treatments have proved adequate for most purposes. More recently, however, the proper interpretation of new experimental data, now becoming available from field programmes based on much improved instrumentation, has called for the resolution of a number of key theoretical issues and extensions to the theory in its original form.

The main issues are concerned with:

(i) A full-wave analysis of the wave-interaction process

The early analyses of the acoustical consequences of interacting gravity waves found it convenient to restrict consideration to the homogeneous component of the induced pressure field. This meant that in any spectral calculations the upper limit of the horizontal wavenumber could be set at the acoustical wavenumber in seawater, ω/α_1. Furthermore, because the wavenumbers involved are so small compared with those of the interacting gravity waves, it was considered reasonable to restrict consideration to interactions between waves whose directions of propagation were exactly opposed, so that $\vec{k} \simeq 0$ (the "standing-wave concept"). Though the inhomogeneous field ($k > \omega/\alpha_1$) was recognised, its neglect was justified because its decay with depth makes it negligible in deep water. Where it has been considered[34],[52],[78], the integration over wavenumber is simply extended to infinity without justification that this is appropriate.

In shallow-water environments and in the upper levels of the deep ocean the implications of these simplifications need to be examined and the relative importance of the inhomogeneous component established as a function of depth below the sea surface. Many shallow-water seabeds are characterised by sediments of low rigidity and, as Frisk[39] has demonstrated, the seismo-acoustic response of such structures can be influenced markedly by the inhomogeneous components of any source field. A proper understanding of the properties of the wave-induced pressure field throughout the whole k plane is thus important. This requires a full-wave analysis of the interaction process, which removes the approximations of earlier treatments and provides a complete description of the induced pressure field as a function of wavenumber, frequency and observation depth.

(ii) Removal of the deep-water restriction

The deep-water assumption not only justifies the neglect of the inhomogeneous component but also permits the use of the approximate dispersion relation ($\omega^2 = gk$) when describing the ocean-wave field. In recent years, however, the wave-induced pressure field has been used as a source in inversion studies of the geoacoustical properties of shallow-water sediments [43],[46],[47]. It is therefore important to have available a source description, which is applicable in environments of any depth. This description must be based on the general dispersion relation, include both the primary and double-frequency components of the induced pressure field and take the properties of the seabed into account.

(iii) The "difference-frequency" component

In all previous analyses the calculation of the pressure and seismic spectra has involved only the sum-frequency components of the wave-interaction process, those for which the frequency of the second-order pressure wave, ω, is the sum of the frequencies, σ_i, of the two interacting components of the surface-wave field; that is $\omega = \sigma_1 + \sigma_2$ with σ_1 and σ_2 both being either positive or negative. While the reality of the difference-frequency interactions, in which σ_1 and σ_2 can be of opposite sign, has been recognised, no examination of this contribution has yet been made. A comprehensive analysis of the source mechanism should incorporate the difference-frequency interactions and include an assessment of their relative importance to VLF and ULF (ultra-low-frequency) spectra.

(iv) The effects of a multilayered seabed

Because of computational limitations the early analyses of the wave-induced seismo-acoustic fields were of necessity restricted to simple models, in which the seabed was considered as either a liquid or a homogeneous elastic halfspace. With the modern facilities now available it is appropriate to consider the seismo-acoustic response of more realistic representations of the seabed.

(v) The development of a comprehensive Green's function solution

The first- and second-order equations of the perturbation analysis are based, not on the simple acoustic wave equation, but on the acoustic-gravity wave equations, $L\Phi^{(1)} = 0$ and $L\Phi^{(2)} = F(\vec{R}, t)$, where Φ is the velocity potential, $F(\vec{R}, t)$ is a source term and the operator $L \equiv c^2 \nabla^2 - g\partial/\partial z - \partial^2/\partial t^2$ - see Chap. 3. For the investigation of the propagation of the wave-induced pressure and seismic fields and their coherence properties, a comprehensive Green's function solution is required for the full acoustic-gravity equation. Since the operator is not self-adjoint there are complications in the analysis which need to be resolved.

(vi) Propagation of the wave-generated microseism field

In the early treatments of the seismic field only the response to the homogeneous component of the pressure field was considered. Furthermore calculations were restricted to the far field, for which the "residue" analysis was particularly suitable. Given the increasing sophistication of the subject, and the computational power now available, a more comprehensive analysis is required, which describes the propagation of the seismic response from the active region to the near and far fields.

(vii) Nonstationarity of the wave-induced noise process

In the interests of simplicity all theoretical treatments assume the noise-generation process to be stationary. By their nature, however, ocean waves and the pressure and seismic fields they generate are nonstationary, evolutionary processes. It is thus of importance to both theory and experiment, ultimately to incorporate nonstationarity into the analysis. Such an analysis is, however, not considered in this contribution. Along with the propagation of microseisms in range-dependent environments it is the subject of an on-going investigation.

While the pioneering works of Longuet-Higgins and Hasselmann remain as bench mark contributions, the improved computational power now available has permitted some aspects of the subject to be revisited. Extensions to theory are now available which address many of the issues outlined above and, at least in part, remove the restrictions imposed by necessity in the original treatments. Unfortunately the accounts of these modern developments have appeared in the literature in a somewhat disjointed manner, with the result that the evolution of the subject has not been easy to follow. This contribution aims to present a more coherent account of those developments which have been published in recent years and relate them to others we have not yet reported. The latter, which make up the bulk of the present contribution, are concerned primarily with the extensions to theory required for shallow-water environments. They also facilitate the use of the wave-induced pressure field as a source of low-frequency energy in the exploration of seabed properties.

Emphasis is directed throughout at developing the perturbation analysis first presented by Longuet-Higgins . While the extensions described are relevant to other theoretical treatments, the analysis presented demonstrates that certain theoretical advantages follow from using the perturbation procedure to describe the processes involved. As the analysis is self-consistent and derived without recourse to the usual simplifying assumptions, we believe predictions based on it are also more accurate than those obtained using other formulations. Exact expressions are developed, which allow the wave-induced pressure and seismic fields to be calculated for any geophysical environment that can be considered range independent . Examples of theoretical wind-dependent spectra, based on this formalism, are given for a number of typical environments and comparisons with field data are made. These are necessarily limited at this time, but the results of major programmes, based on new technologies, are expected soon. These data will permit more instructive comparisons to be made. It is hoped that the developments described here will also facilitate the interpretation of these data.

The material is presented in eight main chapters. A review of the earlier theoretical treatments is given in Chap. 2. The general environmental model to be used throughout this book is introduced in Chap. 3, along with a more detailed discussion of the basic features of the analysis first presented by Longuet-Higgins. The purpose of Chap. 4 is two-fold. The first is to extend the "standing-wave" procedure and develop an "exact" deep-water analysis of the wave-induced pressure field and its power-density spectrum, to serve as a helpful introduction before the complexities of the more general and complete solutions, appropriate to all water depths, are presented in Chap. 5. The second is to provide a quantitative comparison of analyses using the velocity potential and pressure as the field variable.

Chapter 5 removes the restrictions implicit in the deep-water and bottomless ocean assumptions of Chap. 4, develops the more general form of the dispersion relation directly from the first-order solution, and establishes both plane-wave and Green's function solutions for a general multilayered environment. Chapter 6 develops the analysis presented in Chap. 5. A geometric and analytical description of the interaction process establishes the limits of integration to be applied in calculations based on the generalised source function established in the earlier chapter. The analysis continues with an examination of the role of the difference-frequency component and its place in the total second-order source spectrum. Finally composite source spectra are presented, which include the effects of both the first- and second-order wave processes. A parallel examination of the shallow-water case completes the chapter.

In Chap. 7 the various formulations for the source field developed in the previous chapters are used to examine the properties of the wave-induced pressure field in real situations. Modelling of the real ocean environment is discussed. Several typical structures are defined and theoretical pressure and coherence spectra for a given sea state are presented for these environments. Comparisons are made with the limited field data currently available. Chapter 8 presents a similar analysis of the seismic response to the pressure field in the same environments.

A final summary is given in Chap. 9.

In an attempt to make the contribution more user-friendly most of the detail of the mathematical derivations and the key numerical procedures are presented in the form of Appendices. Appendix A presents the basic elements of the theoretical treatment followed by various authors over the past forty years. Appendices B, C and E describe the basic equations governing wave motion in solid media. While they quote extensively from the early publications of Ewing, Biot and Stoll, their inclusion is believed justified because an attempt has been made to present this material in a systematic way, with a standardised nomenclature and a free choice of the sign of the time factor $exp(\pm i\omega t)$. Appendices D and F provide detail of the numerical procedures involved in calculating the reflection coefficient (and all related transmission coefficients) for any defined seabed and includes the effects of porosity. In Appendices H, I, J and K, the Green's function solutions of the non self-adjoint gravity-acoustic wave equations for the first- and second-order fields are derived in detail. Appendix L gives detail relating to the development, in Chap. 5, of the expression for the source spectrum in a shallow-water environment. Appendix M provides a complete analysis of the wave-wave interaction process and details the calculation of the source function over the whole horizontal wavenumber plane. Calculations of the coherence and power-density spectra of the wave-induced field are detailed in Appendices G and N. Appendices O and P are concerned with the calculation of the spectra describing the seismic response to a wave-induced pressure field .

Chapter 2

A Review of Earlier Theoretical Analyses

2.1 Related to Microseisms

As is well known the prediction of the pressure field generated by wave motion at the ocean surface requires the solution of the fluid dynamic equation under specified boundary conditions. Since the dynamic equation and its associated boundary conditions are generally nonlinear, their linearisation is a critical first step in establishing an analytical solution. Most theoretical treatments achieve this by using the perturbation procedure first introduced by Longuet-Higgins in the 1950's [25] in his study of the origin of microseisms. The common feature of this procedure is the expression of the unknown field in powers of an ordering parameter of smallness ϵ and the Taylor expansion of its boundary value about the mean plane of the fluctuating surface. This establishes linearised equations of different orders, which are then solved using the standard methods of mathematical physics. Because of the complexities involved various authors have introduced additional assumptions or neglected certain terms in an attempt to simplify the procedures and yet produce acceptable solutions. Before proceeding it will be instructive to review the different procedures which have been used, being mindful of the differences in symbolism involved.

It is appropriate to begin with the pioneering work of Longuet-Higgins [25], which provides a clear understanding of the physics of the problem and defines the basic equations to be solved. He was the first to recognise that the source of microseisms was to be found in an effect originally described by Miche [31], whereby two surface waves of the same frequency, propagating in opposite directions in an incompressible fluid, generate a standing-wave pressure field at twice the frequency of the surface waves - see Eq.(A.1) of Appendix A (in the interest of the text the equations quoted in this chapter are all collated in Appendix A). The mean pressure variation of this field at depth h below the surface was found to be independent of depth.

In applying this phenomenon to microseisms, Longuet-Higgins recognised that the compressibility of water must be taken into account in a general analysis of the total pressure field. To linearise the general dynamic equation and the surface boundary condition, he introduced a perturbation series for the velocity potential ϕ, the sea-surface displacement ζ, the pressure p, and the water density ρ, and expanded the conditions at the random surface, $z = \zeta(x,y)$, in a Taylor series about the average plane $z = 0$. By collecting terms of the same order, he then established the first two orders of the dynamic equations in the form of $L\Phi_1 = 0$ and $L\Phi_2 = \partial [u_1]^2/\partial t$ where $L \equiv \partial^2/\partial t^2 - c^2\nabla^2 - g\partial/\partial z$, along with their associated boundary conditions (see Eqs.(A.2) to (A.18)). He then derived solutions of the first- and second-order potential fields (and hence the pressure) by introducing plane-wave forms with complex amplitudes, the values of which were to be determined by the corresponding surface and bottom conditions. In deriving the second-order solution, the nonhomogeneous equation of the second-order potential was made homogeneous by introducing a new potential $\Phi'_2 = \Phi_2 - f(\Phi_1)$, where $f(\Phi_1)$ is a function of the first-order solution, and making appropriate changes in the boundary conditions. The vertical displacement of the seabed (the microseism response) was calculated subsequently by combining the established source pressure distribution and the response of a water-seabed model to a point source at the sea surface. We elaborate on his procedures in Chap. 3.

In 1963 Hasselmann[26] provided a statistical description of the phenomenon. Also using the perturbation approach developed by Longuet-Higgins, but assuming water incompressibility for the first-order field, he too derived equations for the first- and second-order potentials and boundary conditions, Eqs.(A.19) to (A.25), and then used the following steps to establish the second-order pressure field.

The first-order potential was constructed in the form of a Fourier-Stjeltjes integration of plane-wave components,

$$\Phi = \int\int [d\Phi_+(\vec{k})e^{-i\sigma t} + d\Phi_-(\vec{k})e^{i\sigma t}] \exp[i\vec{k}\vec{x} + kx_3],$$

the random amplitudes of which were related to the corresponding components of the sea-surface displacement, ζ, through the surface conditions - Eqs.(A.26) and (A.27). The term, $(-1/2\alpha_1^2)\partial[(\nabla\Phi)^2]/\partial t$, on the right-hand-side of the second-order equation, Eq.(A.23), is ignored, since this (volume) contribution is small compared with the term, $-\partial[(\nabla\Phi)^2]/\partial t$, associated with the surface condition, Eq.(A.24). The product of this latter term and the density of the water is then taken as the surface value of the second-order pressure field, Eq.(A.28). Substitution of the expression for the first-order potential then leads, after some manipulation, to his power spectral-density function for the equivalent second-order pressure field, expressed in terms of the horizontal wavenumber and frequency - see Eq.(A.30). Under the approximation, $\vec{k}' = -\vec{k}''$ and $\sigma_1 = \sigma_2$, (for purposes of identification we will denote this as the "standing-wave approximation"), this expression simplifies to $F_p^{(2)}(\vec{k},\omega) = (\rho_1^2 g^2 \omega^2/2)I(\omega)$, where $I(\omega)$ relates to the power spectrum of the surface-wave field, see Eqs.(A.30) to (A.32). This pressure spectrum is then used as the source in his study of wave-induced microseisms. Using Lighthill's acoustic-analogue equation, Eq.(A.33), Hasselmann also discussed the generation of microseisms by atmospheric turbulence and concluded that this contribution is negligible in comparison.

2.2 Related to Underwater Acoustics

In a later analysis in the context of underwater acoustics rather than seismology, Brekhovskikh[28] too investigated the noise generated by wave-wave interactions. He used the perturbation procedures adopted by Longuet-Higgins to establish the basic equations and the boundary conditions for the first- and second-order velocity potentials - see Eqs.(A.34) to (A.38). After establishing the first-order potential, Eq.(A.39), he proceeds to examine its second-order equivalent, Eq.(A.37), by introducing two plane-wave solutions, Φ_2'' and Φ_2', with the coefficients again determined from the boundary conditions - see Eqs.(A.37) to (A.41). He then showed that under certain conditions the solution, Φ_2'', can become homogeneous and independent of depth, thereby contributing to underwater noise in deep water. This solution is equivalent to that produced by Hasselmann. Brekhovskikh also discussed the spectral properties of this field component in general terms.

Further evidence of the growing interest in low-frequency ambient noise was provided when Hughes[29] reexamined the contribution of nonlinear wave interactions to the ocean-noise field below 10 Hz. He too followed the perturbation procedure but chose pressure as the field variable. He also simplified the dynamic equation to an acoustic one, by neglecting the gravitational term $(g/c^2)\partial p/\partial z$, to obtain modified expressions for the first- and second-order fields and their boundary conditions - see Eqs.(A.42 to A.47). To convert the nonhomogeneous equation, (A.45), into a homogeneous one with new surface conditions, he introduced new variables by redefining the second-order pressure and sea-surface displacement - Eqs.(A.48 to A.51). Neglecting the small terms in the new surface condition (involving $g\zeta_1^2/2c^2$ and the second-order displacement), he then identified the remaining nonhomogeneous terms in Eq.(A.50) as an equivalent external source field acting at the mean surface - see Eq.(A.52). Taking advantage of new ocean-wave spectral data, which had recently become available, he could then establish an improved estimate of the spectral characteristics of the wave-induced pressure field. Apart from small differences (of the order of 3 dB) in the constant term, which remained to be reconciled, the expression obtained was essentially identical to Hasselmann's Eq.(2.15) and Brekhovskikh's Eq.(23).

All the analyses described above involve seeking plane-wave solutions of the first- and second-order equations, when these are subject to specified surface-boundary conditions. In 1981 Lloyd[30] approached the problem in a somewhat different way. By neglecting terms involving viscosity and its gradients, and replacing the water density in the stress tensor by its quiescent value at the surface, Lloyd converted his "preliminary form of Lighthill's equation" (Lloyd's Eq.(9) or Eq.(A.55)) to obtain a nonhomogeneous, nonlinear, gravity-acoustic wave equation for the pressure, see Eq.(A.57), associated with a pressure release condition at the surface. To solve the equation he introduced a reference pressure, p_0, appropriate to a dynamically incompressible ocean, so that the over pressure, $p_1 = p - p_0$, satisfies the same equation as p, but allows this to become linear if p_0 can be found, (see Eqs.(A.58 and A.60). (Lloyd calls this the Ribner-Lighthill equation. In effect this equation corresponds to the equation for the second-order potential in Longuet-Higgins' analysis, Eq.(A.13), while the reference pressure, p_0, corresponds to the first-order pressure field.) Lloyd then applied the perturbation procedure to establish the reference pressure field (through the velocity potential) up to the second order (in this sense his first-order solution is more accurate than that of others). The potentials for the first two orders are given by

Eqs.(A.61) and (A.62).

To establish the over-pressure field Lloyd first simplified the "Ribner-Lighthill" equation, by ignoring the gravity term, to obtain a nonhomogeneous, acoustic-wave equation as an approximation. He then sought a Green's function solution instead of following the well established plane-wave approach. Since both the wave equation and the surface conditions are nonhomogeneous and involve the reference field at different orders, he introduced a perturbation expansion of the Green's function, $G = G_0 + G_1 + \cdots$ and required each term to satisfy corresponding perturbation conditions at $z = 0$. In this way G_0 is the Green's function for the nonhomogeneous equation and its homogeneous boundary condition, while G_1 is that of the homogeneous equation and its nonhomogeneous boundary condition, both containing the lowest order of the nonhomogeneous term. The over pressure derived using these functions is made up of two parts, one relating to the effects of a source distribution throughout the volume and the other to one at the boundary. Lloyd showed that the pressure spectrum of this equivalent second-order source is the same as that established by Brekhovskikh and Hughes, when an error involving a factor of two in amplitude is corrected in their expressions. (It is relevant to the discussion in Sect. 4.3.4 to note that, in introducing the Green's function to construct a closed form solution of the wave-induced pressure field, Lloyd demonstrates that an equivalent surface source is an automatic consequence of replacing the random sea surface by its mean plane.)

In the late 1980's Guo[54] reported on his examination of the mechanisms involved in the noise generation by ocean waves. He used Lighthill's acoustic-analogue equation, Eq.(A.63), with water density as the field variable. He too invoked assumptions typical of the perturbation procedure, in that he adopts a known first-order solution of the surface-wave field and expands the random surface displacement about its mean plane. As a further approximation he elects to neglect the volume source distribution implicit in this procedure by evaluating the effects of the boundary source only. (We examine the implications of this approximation in Sect. 4.3.4.) Unique to his analysis is the introduction of the Green's function of the gravity-acoustic wave operator for an ocean of infinite size.

On the basis of his analysis Guo compared the underwater pressure field generated directly by air turbulence with that generated by wave-

wave interactions and concluded that air turbulence is the more important source of the ULF noise field in the ocean. This conclusion, being at variance with observation and other theoretical analyses, required explanation. This was provided by Kibblewhite and Wu[38] (see also Orcutt et al[40]) who showed that the inconsistency arises because it is tangential stress, rather than the fluctuating air pressure of turbulent air flow, that is primarily involved in supporting the wave surface in a developed sea. They demonstrated that, while Guo's assertion might be correct in the early stages of wave growth, it is not valid in the case of more developed seas.

In a more recent analysis based on the Lighthill concepts, Cato[51],[52] provides another study of sound generation in the vicinity of the sea surface. In contrast to Guo his starting point is Doak's[82] version of the Lighthill equation, in which the pressure rather than the density is used as the field variable. The gravity term is absorbed into the stress tensor, so that the expression appears as a pure acoustic-wave equation when the stress tensor is assumed known, see Eq.(A.64). By introducing a Heaviside function the whole space occupied by the air/water media is treated as a single volume with a boundary at infinity. By then invoking the field extinction condition at infinity the pressure field, $p(\vec{x}, t)$, is expressed as the sum of two volume integrations (one involving the water and the other the air medium), with the Green's function $(|\vec{x} - \vec{x}_0|)^{-1}$ calculated at the retarded time $(t - |\vec{x} - \vec{x}_0|/c_0)$ - see Eq.(A.65).

This procedure would appear to be satisfactory when calculating the noise generated by a source, such as air bubbles, distributed throughout an infinite water volume, as the small air/water interfaces in this case are not likely to invalidate the use of the simple form of the Green's function , $1/r$. Its application to calculate the wave-induced pressure field appears to consider the volume effects alone and, in contrast to previous analyses (Lloyd's in particular), neglect the surface contribution. We show in Chap. 4 that, when pressure is used as the field variable , the omission of the boundary effect can result in an error in the spectral level (of the homogeneous component) of around 3 dB. The significance of this to Cato's analysis is not immediately clear and is the subject of ongoing discussion.

To summarise, from this brief review the following points can be identified as being particularly significant: (i) All analyses of the wave-induced pressure field involve an approximation equivalent to the perturbation procedure introduced by Longuet-Higgins. (ii) The equations describing the induced field are expressed in terms of either the velocity potential or the pressure (or density) as the field parameter. In linear acoustics it is immaterial which variable is chosen. We will see in Chap. 4, however, that when nonlinear effects are involved, the selection of variable can influence the relative importance of the volume and surface components of the source field. (iii) In their general form the dynamic equations of different order are solved by using either plane-wave solutions or through the Green's function. Longuet-Higgins adopted the plane-wave approach and, without invoking any additional approximations, established an expression for the velocity potential for the problem as defined. Hasselmann used the same approach but ignored the nonhomogeneous term (volume source) and the gravity term in the second-order dynamic equation. He also established a spectral expression for the pressure field. Both analyses have had a marked influence on all subsequent theoretical developments, although later authors have introduced a number of additional approximations to deal with the complex issues involved and designed a variety of ways to simplify the solution procedures. (iv) To provide a more complete description of the effects of propagation and to examine the spatial coherence and spectral properties of the pressure field, a closed form solution based on the Green's function has been sought. Lloyd was the first to discuss Green's function solutions for the first two orders of the pressure field. Guo and Cato have also followed this approach. (v) Because of the various procedures and approximations adopted, it is often difficult to compare adequately the results of the various theoretical analyses. An examination of the literature shows, in fact, that theoretical spectral levels often differ by certain constants. Another inconsistency in level arises frequently, as the result of changing the variable from ω to f in the relevant expressions. If care is not exercised, this can lead to an anomally of 2π.

Although the differences in theoretical approach alluded to above may only change predicted source spectral levels by a few dB, increased uncertainties are inevitably involved when the resulting formalism is applied to real environments, in which some or all of the conditions implicit in the assumptions made are not fully satisfied. This situation provides the prime motivation for our attempt to develop the more definitive and rigorous account of the seismo-acoustic response to the ocean-wave field, which is presented in the following chapters.

2.3 Related to the Spectral Characteristics of the Pressure Field

An important aspect of the theoretical analysis of the wave-interaction process is concerned with the description of the second-order pressure field in the horizontal wavenumber and frequency domains, i.e., the calculation of the power-density spectrum $f_p^{(2)}(\vec{k},\omega)$. As we discussed earlier, the calculation of the pressure spectrum has traditionally been based on the "standing-wave approximation", under which $f_p^{(2)}(\vec{k},\omega) = f_p^{(2)}(0,\omega)$, and the power-density spectrum in the frequency domain takes the form,

$$F_p^{(2)}(\omega) \equiv \int f_p^{(2)}(\vec{k},\omega)d\vec{k} = \int_0^{2\pi}\int_0^{\omega/\alpha_1} f_p^{(2)}(0,\omega)k dk d\theta_k = \pi\frac{\omega^2}{\alpha_1^2}f_p^{(2)}(0,\omega) \tag{2.1}$$

From a theoretical point of view, however, nonlinear interactions can occur between any two surface wave trains of different frequency and directions of travel. The resulting pressure field will be either an ordinary acoustic wave with a horizontal wavenumber in the range $0 \leq k \leq \omega/\alpha_1$, or when $k > \omega/\alpha_1$, an evanescent wave decaying with depth according to $e^{-\gamma|z|}$, where $\gamma = \sqrt{k^2 - \omega^2/\alpha_1^2}$. (Here and later it will be convenient to follow Brekhovskikh[28] and Aki & Richards[64] and denote these as the homogeneous and inhomogeneous components of the pressure field respectively. We recognise that in the former the wave amplitude decays in the direction of propagation, while in the latter the decay is in a direction normal to it.) While the standing-wave approximation provides an adequate description of the pressure field at depths sufficiently great that only the homogeneous component is effective, modifications incorporating the inhomogeneous component are required to accurately describe the pressure field at all depths throughout the water column. The inhomogeneous component becomes especially important in shallow water when the bottom consists of low-velocity sediments.

Brekhovskikh[28] considered how much two gravity-wave trains could deviate from opposing directions of travel and still produce a homogeneous pressure wave and Lloyd[30] examined the upper limit of the frequency difference at which the generated pressure wave would still be of that form. However, both authors limited consideration to the homogeneous component alone. A significant advance was made in the mid 1980's when Schmidt and Kuperman[34] combined a full-wave solution technique for stratified elastic media[62] with a theory for distributed noise (see for example [68]), to analyse the influence on the noise field of a shallow-water wave guide bounded below by a viscoelastic medium . They demonstrated that at frequencies below water-borne propagation cut-off, interface waves provide such an effective mechanism for coupling ambient noise to the seabed, that a significant magnification of observed seismic levels results. On the basis of these findings they emphasise the need to correct for environmental propagation conditions, to ensure any noise source levels reported are independent of the location in which the data were obtained.

Applying their analysis to the data reported in the New Zealand experiment[19], they found that, by extending the integration of the product of the Green's function and the spatial wave-vector spectrum of a distributed noise source to incorporate all horizontal wave numbers (in the inhomogeneous region), the seismic levels were increased up to 40 dB in the model assumed (the pressure levels up to 25 dB). Interpreting this enhancement as the effects of "magnification" led them to question whether the noise levels reported in [19] were not simply the result of this "magnification", rather than abnormal source levels associated with the wave-wave interaction mechanism. While this was a reasonable proposal given the acknowledged limitations of the transfer functions used in [19], it was clear that a complete understanding of the seismo-acoustic response to interacting ocean waves would not be achieved, until a description of the wave-induced pressure spectrum was available over the whole k plane at the frequencies of relevance, and the transfer functions were better understood.

Kibblewhite and Wu addressed the first of these issues in a full-wave analysis of the wave-wave interaction mechanism and developed a formalism for the calculation of the source spectrum as a function of k, ω and wind speed[35],[38],[36] . It was shown that the inclusion of the inhomogeneous component of the wave-induced pressure field itself raises the pressure spectral levels in the upper parts of a deep, bottomless ocean some 40 dB above that based on the homogeneous component alone. In a shallow-water wave guide, therefore, the seismic levels will be enhanced or magnified by this component of the pressure field, irrespective of any influence of the bottom. (In Chap. 7 we demonstrate that the influence of the bottom is to modulate the average energy levels by a few dB only.) To obtain a measure of the homogeneous component (in the literature this component is defined as the source level of the wave-induced pressure field) the inhomogeneous contribution must be removed from any observation, with the subtraction based on the enhanced levels where this is appropriate. In the New Zealand experiment, however, the seismic signals were recorded onshore. A later analysis in [41], which examined the influence of the inhomogeneous component in more detail, showed that, in the model assumed, the seismic signal recorded onshore was primarily due to the homogeneous component, so that in this case the spectral levels observed differed little from the source spectral levels associated with the wave-interaction mechanism. The seismic effects observed in [19] were therefore properly interpreted, albeit fortuitously, in terms of the wave-interaction pressure field observed in deep water. The subject is, however, indebted to Schmidt and Kuperman for highlighting the issues involved. A more detailed discussion of these questions is available in [41]. We examine the question of source levels in more detail in Sect. 7.5.

In further extensions aimed at removing any restriction on the geophysical model, we incorporated the more general dispersion relation into the formalism [41],[42]. Parallel studies of the reflection coefficient [48],[49], which consider the response to both the homogeneous and inhomogeneous wave fields and include porosity in the description of a multilayered seabed structure, have provided other developments relevant to the study of the ULF noise field. The role of the difference-frequency component of the interaction process has also been examined. While this was recognised in Longuet-Higgins' original work and is referred to in others, it has never been analysed quantitatively. We will show later that this component occupies the final part of the k plane where $k > \omega^2/g$. The analysis in Chap. 6 provides a complete theoretical description of the spectral characteristics of the source field over the whole horizontal wavenumber plane.

Chapter 3

The Perturbation Procedure

3.1 The Environment Model

The motion of the sea surface and the spatial variability of the sedimentary structure are among the most significant geoacoustic features of the ocean environment. Any representation of these features for theoretical studies is at best a gross simplification of the real situation. Because of the relatively short wave lengths (less than a few hundred metres for ocean-gravity waves and a few tens of kilometers for the related acoustic waves in water) and the comparatively short observation periods involved (normal averaging times of tens of minutes), the environment is modelled here as being locally stationary and varying only slowly in space. This means we assume: (i) that the random fluctuations developed on the sea surface under the influence of a steady wind regime have reached an equilibrium state, such that in the sampling period the excitation process can be regarded as stationary; (ii) that any spatial variation in the statistical properties of the active region of the sea surface is small; and (iii) that any lateral variation in the environmental structure, including the water depth, is also sufficiently small that the area in the region of the wave activity can be described by a range-independent "reference wave-guide" — see Figs. 3.1 and 3.2.

3.2 Governing Equations

When the effects of the Earth's rotation and the viscosity of sea water are disregarded, the displacement of water particles, induced by surface-wave motion, is determined by the requirements of conservation of momentum and mass and an equation describing the thermodynamic state. If sea

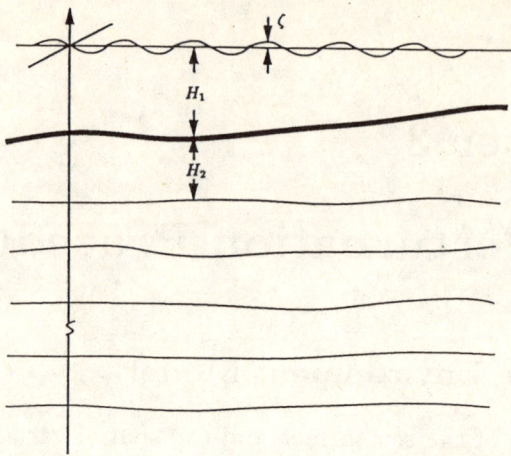

Figure 3.1: Schematic representation of a slowly range-dependent geoacoustic model with a locally stationary random sea surface.

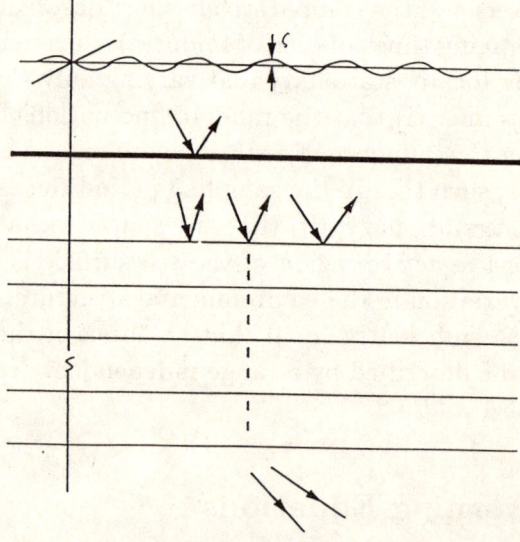

Figure 3.2: Schematic representation of a range-independent geoacoustic model with a locally stationary random sea surface.

3.2 Governing Equations

water is further regarded as a single-phase medium and certain requirements about its thermodynamic state are satisfied, then the motion can be described by the following equations[53]

$$\rho_1 \frac{d\vec{v_1}}{dt} + \nabla \tilde{p}_1 - \rho_1 \vec{g} = 0 \qquad (3.1)$$

$$\frac{d\rho_1}{dt} + \rho_1 \nabla \vec{v_1} = 0 \qquad (3.2)$$

$$\frac{d\tilde{p}_1}{d\rho_1} = \alpha_1^2 \qquad (3.3)$$

where $\vec{v}_1$ is the water-particle velocity vector, ρ_1, $\tilde{p}_1$ and α_1 are respectively the density, pressure and the adiabatic sound velocity in sea water, and g is the acceleration of gravity. In the following text the parameters p, ζ and Φ appear in two forms. The symbol marked by the tilde, eg $\tilde{p}$, is used to indicate the space/time version of the parameter, $\tilde{p}(\vec{r}, t)$, and the plain symbol that of its frequency/wavenumber equivalent, $p(\omega, \vec{k})$.

The related motion induced in the sediment is more complicated, especially when the rigidity and porosity are not negligible. In this book we invoke [48], and [49] and regard a viscoelastic material as a particular case of a porous, viscoelastic medium, and use the latter as a general model. We know two kinds of movement, the deformation of the solid frame and the flow of the pore fluid relative to the frame, can exist in a porous medium[44]-[46]. To describe this motion a pair of coupled dynamic equations are generally needed. According to Stoll[46] they are

$$\mu \nabla^2 \vec{u} + (H - \mu) \nabla \theta - C \nabla \zeta_p = \rho \ddot{\vec{u}} - \rho_f \ddot{\vec{V}} \qquad (3.4)$$

and

$$C \nabla \theta - M \nabla \zeta_p = \rho_f \ddot{\vec{u}} - m \ddot{\vec{V}} - \frac{\eta}{k_p} \dot{\vec{V}} \qquad (3.5)$$

where $\vec{u}$ and $\vec{V}$ are here respectively the displacement vector of the frame and of the pore fluid relative to the frame, $\theta = \nabla \cdot \vec{u}$ and μ, M, C, and H are moduli of the medium; ρ_f is the density of the pore fluid, while $\rho = \beta \rho_f + (1 - \beta) \rho_r$ is the bulk density of the medium and ρ_r the density of the frame; β is the porosity of the medium and $m = \alpha \rho_f / \beta$ where α is a constant related to the average pore size; η is the viscosity of the fluid, $\zeta_p = \nabla \cdot \vec{V}$, and k_p is the coefficient of permeability. The dots above a variable represent the usual time derivatives. All these quantities are specified for each of the layers. The stress in the frame and the pressure in the pore fluid relate to the displacement vectors $\vec{u}$ and $\vec{V}$ through the equations[55],[46],

$$\begin{aligned}
\tilde{p}_{xx} &= H\theta - 2\mu(\partial u_y/\partial y + \partial u_z/\partial z) - C\zeta_p \\
\tilde{p}_{xy} &= \mu(\partial u_x/\partial y + \partial u_y/\partial x) \\
\tilde{p}_{yy} &= H\theta - 2\mu(\partial u_z/\partial z + \partial u_x/\partial x) - C\zeta_p \\
\tilde{p}_{yz} &= \mu(\partial u_y/\partial z + \partial u_z/\partial y) \\
\tilde{p}_{zz} &= H\theta - 2\mu(\partial u_x/\partial x + \partial u_y/\partial y) - C\zeta_p \\
\tilde{p}_{zx} &= \mu(\partial u_z/\partial x + \partial u_x/\partial z) \\
\tilde{p}_f &= M\zeta_p - C\theta
\end{aligned} \qquad (3.6)$$

where

$$\theta = \nabla \cdot \vec{u} = e_x + e_y + e_z = \frac{\partial u_x}{\partial x} + \frac{\partial u_y}{\partial y} + \frac{\partial u_z}{\partial z}$$

If a layer is solid rather than porous then $\beta = 0$, $\vec{V} \equiv 0$, $H = \lambda + 2\mu$ and Eq.(3.4) degenerates to

$$\mu \nabla^2 \vec{u} + (\lambda + \mu)\nabla\theta = \rho \ddot{\vec{u}} \qquad (3.7)$$

where λ and μ are, here and above, the two elastic moduli (the Lamé constants), Eq.(3.5) disappears and the whole of Eq.(3.6) degenerates to the well known strain-stress relations given in Appendix B.

To find the pressure (stress) and particle velocity (or displacement) induced in the water and the sedimentary layers by a given surface sea state ($z = \zeta(x,y,t)$), Eqs.(3.1) to (3.5) (or (3.6)) must be solved, under conditions preserving field continuity at each of the interfaces and field extinction at infinity.

3.3 Boundary and Interface Conditions

The boundary conditions at the ocean surface can be expressed as

$$\tilde{p}_1(x,y,\tilde{\zeta}) = P_a - \rho_1 \gamma F(\tilde{\zeta}) \qquad (3.8)$$

$$\frac{d(z-\tilde{\zeta})}{dt} = 0 \qquad (3.9)$$

where P_a is the atmospheric pressure and γ the surface-tension coefficient. (In the interest of simplicity we will later ignore the effects of the surface

3.3 Boundary and Interface Conditions

tension force, considering this to be justified at the low frequencies involved.) While turbulent air motion is of importance to the generation of surface waves and can itself also act as an acoustical radiation source, it is not a significant source of ULF noise in the ocean[26] and is not considered in this analysis. Discussion is confined to the pressure and seismic fields generated by the wave motion at the ocean surface. This random surface is assumed to have reached a state of equilibrium so that the surface motion can be treated as a stationary process in both time and space. The air pressure is regarded as constant. (Readers are referred to Refs. [26],[51],[54] for a discussion of the effects of air turbulence as a radiating acoustic source.)

The field continuity at the interface between two different layers, n and $n+1$, establishes the other boundary conditions [55],[75]:

(i) between a liquid (n) and an elastic medium ($n+1$):

$$\left.\begin{array}{rcl} u_{n,z} &=& u_{n+1,z} \\ -\tilde{p}_{n,z} &=& \tilde{p}_{n+1,zz} \\ 0 &=& \tilde{p}_{n+1,zx} \end{array}\right\} \text{at } z = -H_n \qquad (3.10)$$

where $\tilde{p}_n$ is the pressure in the liquid

(ii) between a liquid (n) and a porous elastic medium ($n+1$):

$$\left.\begin{array}{rcl} u_{n,z} &=& u_{n+1,z} - V_{n+1,z} \\ -\tilde{p}_{n,z} &=& \tilde{p}_{n+1,zz} \\ \tilde{p}_{n,z} &=& \tilde{p}_{n+1,f} \\ 0 &=& \tilde{p}_{n+1,zx} \end{array}\right\} \text{at } z = -H_n \qquad (3.11)$$

(iii) between a porous elastic medium (n) and an adjacent elastic medium ($n+1$):

$$\left.\begin{array}{rcl} u_{n,z} &=& u_{n+1,z} \\ u_{n,x} &=& u_{n+1,x} \\ \tilde{p}_{n,zz} &=& \tilde{p}_{n+1,zz} \\ \tilde{p}_{n,zx} &=& \tilde{p}_{n+1,zx} \\ V_{n,z} &=& 0 \end{array}\right\} \text{at } z = -H_n \qquad (3.12)$$

(iv) between two porous elastic media:

$$\left.\begin{aligned}
u_{n,z} &= u_{n+1,z} \\
u_{n,x} &= u_{n+1,x} \\
u_{n,z} - V_{n,z} &= u_{n+1,z} - V_{n+1,z} \\
\tilde{p}_{n,zz} &= \tilde{p}_{n+1,zz} \\
\tilde{p}_{n,zx} &= \tilde{p}_{n+1,zx} \\
\tilde{p}_{f,n} &= \tilde{p}_{f,n+1}
\end{aligned}\right\} \text{ at } z = -H_n \qquad (3.13)$$

(v) between two elastic media:

$$\left.\begin{aligned}
u_{n,z} &= u_{n+1,z} \\
u_{n,x} &= u_{n+1,x} \\
\tilde{p}_{n,zz} &= \tilde{p}_{n+1,zz} \\
\tilde{p}_{n,zx} &= \tilde{p}_{n+1,zx}
\end{aligned}\right\} \text{ at } z = -H_n. \qquad (3.14)$$

3.4 Linearisation

3.4.1 Potential as the Field Variable

It is clear from the above that the dynamic equations and interface conditions governing the wave-induced movement of sedimentary particles are linear. They are however the linear terms of more general equations. Linearisation is justified because particles of the solid medium are bound and experience only small deviations from an equilibrium position. In contrast seawater particles can move quite freely in space and the assumption of very small displacement would, in many cases, not provide an adequate description of the processes involved. Indeed, as is well known, the higher-order terms and their interactions play a vitally important role in wave growth and the formation of the final steady sea state in which energy balance is established among all frequency components of the wave-field. These same interactions are also responsible for the double-frequency, depth-independent component of the wave-induced pressure field. The study of the seismo-acoustic effects of ocean-wave interactions must therefore start from the more general equations derived from the dynamic and kinematic conservation laws. Since these equations are nonlinear the first step in establishing quantitative predictions of the effects involved is to linearise them. A straightforward procedure of linearisation involves expanding the related quantities as a perturbation series, collecting terms of the same order and solving the resulting

3.4 Linearisation

linearised equations. In the following we essentially describe Longuet-Higgins' application of this procedure for terms up to the second order[25]. We assume the motion is irrotational, introduce the velocity potential, $\tilde{\Phi}$, satisfying $\vec{v} = \nabla\tilde{\Phi}$ (note that here $\vec{v}$ is the particle velocity vector), and derive from Eqs.(3.1).

$$\nabla\left(\frac{\partial\tilde{\Phi}_1}{\partial t} + \frac{1}{2}v_1^2 + gz\right) = \frac{-\nabla\tilde{p}_1}{\rho_1}. \tag{3.15}$$

By using Eq.(3.3) the right-hand-side of Eq.(3.15) can be written as $-\alpha_1^2\nabla ln\rho_1$ so that after integration, if ρ_1 is a function of $\tilde{p}_1$ only,

$$\frac{\partial\tilde{\Phi}_1}{\partial t} + \frac{1}{2}v_1^2 + gz = -\alpha_1^2 ln(\rho_1/\rho_s) \tag{3.16}$$

where ρ_s is the density of the sea water at the surface and the positive direction of the z-axis is set upward. (In deriving the above, the relation $\vec{v}_1\cdot\nabla\vec{v}_1 = \nabla\times\vec{v}_1\times\vec{v}_1 + \nabla(\frac{1}{2}v_1^2) = \nabla(\frac{1}{2}v_1^2)$ has been used for the irrotational motion and an arbitrary time function has been absorbed into $\tilde{\Phi}_1$.) Further, from Eq.(3.2) $\nabla^2\tilde{\Phi}_1 = -(d/dt)\ln(\rho_1/\rho_s)$. Combining these two leads to

$$\frac{\partial^2\tilde{\Phi}_1}{\partial t^2} - \alpha_1^2\nabla^2\tilde{\Phi}_1 + g\frac{\partial\tilde{\Phi}_1}{\partial z} + \frac{\partial}{\partial t}((\nabla\tilde{\Phi}_1)^2) + \nabla\tilde{\Phi}_1\cdot\nabla(\frac{1}{2}(\nabla\tilde{\Phi}_1)^2) = 0 \tag{3.17}$$

Referring now to Eq.(3.15), we can approximate the pressure field near the surface as $\tilde{p}_1 = -\rho_{s0}(\partial\tilde{\Phi}_1/\partial t + \tilde{v}_1^2/2 + gz)$ by using the average density of the surface water, ρ_{s0}, ignore the effects of the surface-tension force (meaning that $\tilde{p}_1$ always equals $P_a \equiv 0$), and write the surface dynamic condition, Eq.(3.8), as

$$\frac{d}{dt}\left[\frac{\partial\tilde{\Phi}_1}{\partial t} + \frac{1}{2}(\nabla\tilde{\Phi}_1)^2 + gz\right] = 0 \quad \text{at} \quad z = \tilde{\zeta}. \tag{3.18}$$

while the kinematic condition, Eq.(3.9), takes the form

$$\frac{\partial\tilde{\Phi}_1}{\partial z} = \frac{d\tilde{\zeta}}{dt} = \frac{\partial\tilde{\zeta}}{\partial t} + \nabla\tilde{\Phi}_1\nabla\tilde{\zeta} \quad \text{at} \quad z = \tilde{\zeta} \tag{3.19}$$

We now express $\tilde{\Phi}_1$ and $\tilde{\zeta}$ as

$$\tilde{\Phi}_1 = \varepsilon\tilde{\Phi}_1^{(1)} + \varepsilon^2\tilde{\Phi}_1^{(2)} + \cdots \tag{3.20}$$

$$\tilde{\zeta} = \varepsilon\tilde{\zeta}^{(1)} + \varepsilon^2\tilde{\zeta}^{(2)} + \cdots \tag{3.21}$$

where ε is a quantity of the order of the sea-surface slope[56], and substitute these expressions into Eqs.(3.17) to (3.19) after expanding the values of the parameters at $z = \zeta$ in a Taylor series about $z = 0$. Collecting quantities of order ε and ε^2 establishes two sets of linearised equations and surface conditions describing the first- and second-order fields respectively:

$$L\tilde{\Phi}_1^{(1)} = 0 \tag{3.22}$$

$$\left(\frac{\partial^2}{\partial t^2} + g\frac{\partial}{\partial z}\right)\tilde{\Phi}_1^{(1)} = 0, \text{ at } z = 0 \tag{3.23}$$

$$\frac{\partial \tilde{\Phi}_1^{(1)}}{\partial z} - \frac{\partial \tilde{\zeta}^{(1)}}{\partial t} = 0 \text{ at } z = 0 \tag{3.24}$$

and

$$L\tilde{\Phi}_1^{(2)} = \frac{1}{\alpha_1^2}\frac{\partial}{\partial t}(\nabla\tilde{\Phi}_1^{(1)})^2 \tag{3.25}$$

$$\left(\frac{\partial^2}{\partial t^2} + g\frac{\partial}{\partial z}\right)\tilde{\Phi}_1^{(2)} = -\frac{\partial}{\partial t}(\nabla\tilde{\Phi}_1^{(1)})^2 - \tilde{\zeta}^{(1)}\frac{\partial}{\partial z}\left(\frac{\partial^2 \tilde{\Phi}_1^{(1)}}{\partial t^2} + g\frac{\partial \tilde{\Phi}_1^{(1)}}{\partial z}\right) \text{ at } z = 0 \tag{3.26}$$

$$\frac{\partial \tilde{\Phi}_1^{(2)}}{\partial z} - \frac{\partial \tilde{\zeta}^{(2)}}{\partial t} = \nabla\tilde{\Phi}_1^{(1)} \cdot \nabla\tilde{\zeta}^{(1)} - \tilde{\zeta}^{(1)}\frac{\partial^2 \tilde{\Phi}_1^{(1)}}{\partial z^2}, \text{ at } z = 0 \tag{3.27}$$

where

$$L \equiv \nabla^2 - \frac{g}{\alpha_1^2}\frac{\partial}{\partial z} - \frac{1}{\alpha_1^2}\frac{\partial^2}{\partial t^2} \tag{3.28}$$

These equations, combined with Eqs.(3.4) and (3.5) (or (3.6)) and the corresponding interface conditions, form the sets of linear equations required to describe the first- and second-order pressure and seismic fields in the half space $z \leq \tilde{\zeta}$. The relation between the pressure and velocity potential can be derived from Eq.(3.16), by first rewriting it as

$$\rho_1 = \rho_s \exp\left\{-\frac{1}{\alpha_1^2}\left[\frac{\partial \tilde{\Phi}_1}{\partial t} + \frac{1}{2}(\nabla\tilde{\Phi}_1)^2 + gz\right]\right\} \tag{3.29}$$

or after expanding in a power series as,

$$\rho_1 = \rho_s e^{-gz/\alpha_1^2}\left\{1 - \frac{1}{\alpha_1^2}\left(\frac{\partial \tilde{\Phi}_1}{\partial t} + \frac{1}{2}(\nabla\tilde{\Phi}_1)^2\right) + \frac{1}{2\alpha_1^4}\left(\frac{\partial \tilde{\Phi}_1}{\partial t}\right)^2 + \cdots\right\} \tag{3.30}$$

3.4 Linearisation

Using the relation, $d\tilde{p}/d\rho_1 = \alpha_1^2$ (Eq.(3.3)), then establishes the zero-order density,

$$\rho_1^{(0)} = \rho_s e^{-gz/\alpha_1^2} \tag{3.31}$$

the first-order pressure,

$$\tilde{p}_1^{(1)} = -\rho_1^{(0)} \frac{\partial \tilde{\Phi}_1^{(1)}}{\partial t} \tag{3.32}$$

and the second-order pressure

$$\tilde{p}_1^{(2)} = -\rho_1^{(0)} [\frac{\partial \tilde{\Phi}_1^{(2)}}{\partial t} + \frac{1}{2}(\nabla \tilde{\Phi}_1^{(1)})^2 - \frac{1}{2\alpha_1^2}(\frac{\partial \tilde{\Phi}_1^{(1)}}{\partial t})^2 + \cdots] \tag{3.33}$$

Ignoring the last terms, the second-order field becomes

$$\tilde{p}_1^{(2)} = -\rho_1^{(0)} [\frac{\partial \tilde{\Phi}_1^{(2)}}{\partial t} + \frac{1}{2}(\nabla \tilde{\Phi}_1^{(1)})^2] \tag{3.34}$$

It is to be noted here that the relation between the pressure and potential in the case of the second-order field (Eq.(3.34)) is different from that of the first-order field (Eq.(3.32)) normally adopted in linear acoustics.

3.4.2 Pressure as the Field Variable

As was noted earlier two approaches have been used in the study of the wave-induced pressure field. In the first the solution of the relevant equations is made in terms of the velocity potential and the pressure field is derived from this. In the second pressure is used as the field variable and the solution gives the pressure directly. The two procedures have been assumed to be identical. We will demonstrate in the next chapter, however, that this assumption is not necessarily true in respect of the second-order field. Although the spectral levels of the total pressure field are the same, the relative importance of the volume and boundary effects proves to be different in the two approaches, and this has led to misunderstandings. To help later discussion we derive at this point a parallel set of equations, based on pressure as the field variable.

By applying the relation $d()/dt = \partial()/\partial t + \vec{v}\nabla()$ we first rewrite Eq.(3.1) and Eq.(3.2) in the form,

$$\rho_1 \frac{\partial \vec{v}_1}{\partial t} + \rho_1 \vec{v}_1 (\nabla \vec{v}_1) + \nabla \tilde{p}_1 + \rho_1 g \vec{z}_0 = 0 \tag{3.35}$$

where z_0 is a unit vector in the z-direction, and

$$\frac{\partial \rho_1}{\partial t} + \nabla(\rho_1 \vec{v}_1) = 0 \qquad (3.36)$$

Multiplying Eq.(3.36) by $\vec{v}_1$, adding Eq.(3.35), and applying the operator ∇ to the resulting equation, gives

$$\nabla^2 \tilde{p}_1 + \frac{\partial}{\partial t} \nabla(\rho_1 \vec{v}_1) + \nabla^2(\rho_1 \vec{v}_1^2) + g \frac{\partial \rho_1}{\partial z} = 0 \qquad (3.37)$$

Since from Eq.(3.36) we have $(\partial/\partial t)\nabla(\rho_1 \vec{v}_1) = -\partial^2 \rho_1/\partial t^2 = -\partial^2 \tilde{p}_1/(\alpha_1^2 \partial t^2)$, the equation for the pressure field becomes

$$\nabla^2 \tilde{p}_1 + \frac{g}{\alpha_1^2} \frac{\partial}{\partial z} \tilde{p}_1 - \frac{1}{\alpha_1^2} \frac{\partial^2}{\partial t^2} \tilde{p}_1 + \nabla^2(\rho_1 \vec{v}_1^2) = 0 \qquad (3.38)$$

The surface dynamic condition simply retains the form of Eq.(3.8), while the kinematic condition can be written as

$$\frac{dv_{1z}}{dt} = (\frac{\partial}{\partial t} + \vec{v}_1 \nabla)(\frac{\partial \tilde{\zeta}}{\partial t} + \vec{v}_1 \nabla \tilde{\zeta}) \quad \text{at} \quad z = \tilde{\zeta} \qquad (3.39)$$

after taking the total time derivative of Eq.(3.9). Combining this condition with the z-component of Eq.(3.1) leads to

$$\frac{\partial \tilde{p}_1}{\partial \tilde{z}} = -\rho_1 g - \rho_1 [\frac{\partial^2 \tilde{\zeta}}{\partial t^2} + \vec{v}_1 \nabla \frac{\partial \tilde{\zeta}}{\partial t} + \frac{\partial}{\partial t}(\vec{v}_1 \nabla \tilde{\zeta}) + \vec{v}_1 \nabla(\vec{v}_1 \nabla \tilde{\zeta})] \quad \text{at} \quad z = \tilde{\zeta} \qquad (3.40)$$

We now expand the values of the parameters at the random surface, $z = \tilde{\zeta}$, as a Taylor series around the average plane $z = 0$ and then introduce the two perturbation series

$$\tilde{p}_1 = \tilde{p}_1^{(0)} + \varepsilon \tilde{p}_1^{(1)} + \varepsilon^2 \tilde{p}_1^{(2)} + \cdots \qquad (3.41)$$

and

$$\rho_1 = \rho_1^{(0)} + \varepsilon \rho_1^{(1)} + \varepsilon^2 \rho_1^{(2)} + \cdots \qquad (3.42)$$

(together with those for $\tilde{\Phi}_1$ and $\tilde{\zeta}$, see Eqs.(3.20), (3.21)) into Eq.(3.38) and the boundary conditions (3.8) and (3.40). Collecting quantities of the same order, leads to the following relations: for the zero-order field,

$$\hat{L} \tilde{p}_1^{(0)} = 0 \qquad (3.43)$$

3.4 Linearisation

$$\tilde{p}_1^{(0)} = 0 \quad \text{at} \quad z = 0 \tag{3.44}$$

$$\frac{\partial \tilde{p}_1^{(0)}}{\partial z} = -\rho_1^{(0)} g \quad \text{at} \quad z = 0 \tag{3.45}$$

for the first-order field,

$$\hat{L} \tilde{p}_1^{(1)} = 0 \tag{3.46}$$

$$\tilde{p}_1^{(1)} + \tilde{\zeta}^{(1)} \frac{\partial \tilde{p}_1^{(0)}}{\partial z} = 0 \quad \text{at} \quad z = 0 \tag{3.47}$$

$$\frac{\partial \tilde{p}_1^{(1)}}{\partial z} = -\rho_1^{(0)} \frac{\partial^2 \tilde{\zeta}^{(1)}}{\partial t^2} \quad \text{at} \quad z = 0 \tag{3.48}$$

and for the second-order field

$$\hat{L} \tilde{p}_1^{(2)} = -\rho_1^{(0)} \nabla^2 [(\nabla \tilde{\Phi}_1^{(1)})^2] \tag{3.49}$$

$$\tilde{p}_1^{(2)} + \tilde{\zeta}^{(2)} \frac{\partial \tilde{p}_1^{(0)}}{\partial z} = -\tilde{\zeta}^{(1)} \frac{\partial \tilde{p}_1^{(1)}}{\partial z} \quad \text{at} \quad z = 0 \tag{3.50}$$

$$(\frac{\partial^2}{\partial t^2} + g \frac{\partial}{\partial z}) \tilde{p}_1^{(2)} = -(\frac{\partial^2}{\partial t^2} + g \frac{\partial}{\partial z})[\tilde{\zeta}^{(1)} \frac{\partial \tilde{p}_1^{(1)}}{\partial z}]$$

$$-2\rho_1^{(0)} g \nabla \tilde{\Phi}_1^{(1)} \nabla \frac{\partial}{\partial t} \tilde{\zeta}^{(1)} - \rho_1^{(0)} g \nabla \tilde{\zeta}^{(1)} \frac{\partial}{\partial t} \nabla \tilde{\Phi}_1^{(1)} \quad \text{at} \quad z = 0 \tag{3.51}$$

In the above,

$$\hat{L} \equiv \nabla^2 + \frac{g}{\alpha_1^2} \frac{\partial}{\partial z} - \frac{1}{\alpha_1^2} \frac{\partial^2}{\partial t^2} \quad \text{at} \quad z = 0 \tag{3.52}$$

is the adjoint of the operator L defined in Eq.(3.28).

The derivation of Eq.(3.51) is reasonably straight forward. Substituting the Taylor expansions of Eqs.(3.41) and (3.42) up to the first-order into Eq.(3.40) and collecting all the terms of the second-order gives

$$\frac{\partial \tilde{p}_1^{(2)}}{\partial \tilde{z}} + \tilde{\zeta}^{(1)} \frac{\partial^2 \tilde{p}_1^{(1)}}{\partial z^2} + \tilde{\zeta}^{(2)} \frac{\partial^2}{\partial z^2} \tilde{p}_1^{(0)} = -\rho_1^{(2)} g - \tilde{\zeta}^{(1)} \frac{\partial \rho_1^{(1)}}{\partial z} g - \tilde{\zeta}^{(2)} \frac{\partial \rho_1^{(0)}}{\partial z} g - \rho_1^{(0)}$$

$$\frac{\partial^2 \tilde{\zeta}^{(2)}}{\partial t^2} - \rho_1^{(1)} \frac{\partial^2 \tilde{\zeta}^{(1)}}{\partial t^2} - \tilde{\zeta}^{(1)} \frac{\partial \rho_1^{(0)}}{\partial z} \frac{\partial^2 \tilde{\zeta}^{(1)}}{\partial t^2} - 2\rho_1^{(0)} \vec{v}_1^{(1)} \frac{\partial}{\partial t} \nabla \tilde{\zeta}^{(1)} - \rho_1^{(0)} \frac{\partial \vec{v}_1}{\partial t} \nabla \tilde{\zeta}^{(1)}$$

Then using the relations $d\tilde{p}_1/d\rho_1 = \alpha_1^2$, $\tilde{p}_1^{(i)}/\rho_1^{(i)} = \alpha_1^2$, $i = 1, 2, \cdots$ and the conditions (3.45), (3.47) and (3.50) finally establishes

$$(\frac{\partial^2}{\partial t^2} + g\frac{\partial}{\partial z})\tilde{p}_1^{(2)} = -\frac{g^2}{\alpha_1^2}(\tilde{p}_1^{(2)} + \tilde{\zeta}^{(1)}\frac{\partial \tilde{p}_1^{(1)}}{\partial z}) - (\frac{\partial^2}{\partial t^2} + g\frac{\partial}{\partial z})[\tilde{\zeta}^{(1)}\frac{\partial \tilde{p}_1^{(1)}}{\partial z}]$$

$$-2\rho_1^{(0)} g \nabla \tilde{\Phi}_1^{(1)} \nabla \frac{\partial}{\partial t}\tilde{\zeta}^{(1)} - \rho_1^{(0)} g \nabla \tilde{\zeta}^{(1)} \frac{\partial}{\partial t} \nabla \tilde{\Phi}_1^{(1)} \quad \text{at} \quad z = 0$$

which leads to Eq.(3.51) when the term involving the small factor, g^2/α_1^2, is ignored.

3.5 Displacement Potentials in the Sediments

In a procedure similar to that used above displacement potentials can also be introduced to describe the motion of the sea bed. In the case of a porous-elastic layer two pairs of scalar and vector potentials are needed, the first to express the movement of the frame and the second that of the pore water relative to it[46, 48], viz.

$$\vec{u} = \nabla \Phi_s + \nabla \times \vec{\Psi}_s \tag{3.53}$$

and

$$\vec{V} = \beta(\vec{u} - \vec{U}) = \nabla \Phi_f + \nabla \times \vec{\Psi}_f \tag{3.54}$$

Performing the operations $\nabla \cdot$ and $\nabla \times$ on Eqs.(3.4) and (3.5) leads to two sets of potential equations describing the compressional- and shear-wave motions:

$$\left.\begin{array}{rcl} H\nabla^2 \Phi_s - C\nabla^2 \Phi_f &=& \rho \ddot{\Phi}_s - \rho \ddot{\Phi}_f \\ C\nabla^2 \Phi_s - M\nabla^2 \Phi_f &=& \rho_f \ddot{\Phi}_s - m\ddot{\Phi}_f - (\eta/k_p)\dot{\Phi}_f \end{array}\right\} \tag{3.55}$$

and

$$\left.\begin{array}{rcl} \mu \nabla^2 \vec{\Psi}_s &=& \rho \ddot{\vec{\Psi}}_s - \rho_f \ddot{\vec{\Psi}}_f \\ m\ddot{\vec{\Psi}}_f + (\eta/k_p)\dot{\vec{\Psi}}_f &=& \rho_f \ddot{\vec{\Psi}}_s \end{array}\right\} \tag{3.56}$$

The components of stress can be found through the stress-strain relations of Eq.(3.7). In the case of a solid layer $\vec{V} = 0$ and Eqs.(3.55) and

(3.56) degenerate to

$$\left.\begin{array}{c}(\lambda + 2\mu)\nabla^2 \Phi_s = \rho \ddot{\vec{\Phi}}_s \\ \\ \mu \nabla^2 \vec{\Psi}_s = \rho \ddot{\vec{\Psi}}_s\end{array}\right\} \quad (3.57)$$

3.6 The Effect of the Seabed on the Induced Pressure Field

Since the equations governing particle movement in sedimentary layers are all assumed to be linear, the seismic field within the seabed (and the pressure field in the water column) can be estimated by introducing a plane-wave reflection coefficient for the sea floor and a set of transmission and reflection coefficients for each sublayer. Schmidt and Jensen have successfully examined this case for a viscoelastic seabed[62] and their analysis is now the basis of powerful numerical techniques for the study of propagation in such media[63]. In the more general case of porous media, the Biot/Stoll model predicts the existence of a third type of wave, the slow compressional wave, in addition to the fast and shear waves present in a solid viscoelastic structure. In recent contributions[48, 49] we have applied the analysis in [62] to incorporate porosity in the study of the reflection coefficient. We summarise these results here.

To model a seabed comprising a combination of solid and porous (or liquid) layers, we again follow standard practice and assign to each a set of upward and downward propagating plane-waves at any relevant frequency or wavenumber, the complex amplitudes of which are to be determined from the interface conditions. The phase velocity of each of the three waves is, in this case, determined by substituting the appropriate potential into Eqs.(3.56) and (3.57) and establishing the roots of the dispersion equations. Assigning the plane wave incident from the water column unit amplitude, the amplitudes of the reflected and transmitted waves can then be found numerically as functions of frequency and horizontal wavenumber, by solving the set of linear algebraic equations established from the corresponding interface conditions[48, 49]. In establishing the pressure field in the water column and the vertical component of the displacement of the sea floor, all that is required to describe the effect of the bottom is the reflection coefficient, R_b. To determine the horizontal component of the sea-floor motion, however, the transmitted field in the first layer of sediment must also be known, since both components are not generally continuous through the interface between the water and sediment (a fact deserving appropriate attention in any calculations).

In [48, 49] the plane-wave components in the water and sedimentary layers are established as

$$\frac{i}{\omega}\Phi_1 = e^{-i\gamma_1(z+H_1)} + R_b e^{i\gamma_1(z+H_1)} \tag{3.58}$$

$$\left.\begin{array}{ll} \Phi_{js1} = A_{21}e^{-i\gamma_{21}(z+H_1)} + B_{21}e^{i\gamma_{21}(z+H_2)} & \Phi_{2f1} = \delta_{2c1}\Phi_{2s1} \\ \Phi_{js2} = A_{22}e^{-i\gamma_{22}(z+H_1)} + B_{22}e^{i\gamma_{22}(z+H_2)} & \Phi_{2f2} = \delta_{2c2}\Phi_{2s2} \\ \Psi_{js} = A_{23}e^{-i\gamma_{23}(z+H_1)} + B_{23}e^{i\gamma_{23}(z+H_2)} & \Psi_{2f} = \delta_{2c3}\Psi_{2s} \end{array}\right\} \tag{3.59}$$

where the subscripts $js1$, $js2$ and js ($j = 2, 3, \cdots, n$) stand for the fast and slow compressional waves and the shear wave in the solid frame of each layer, and $jf1$, $jf2$ and $jf3$ refer to the same waves in the pore fluid. The ratios of the waves in the two media, δ_{jc1}, δ_{jc2}, δ_{js} are determined by the modulus, frequency and phase velocity, etc in each layer. In an elastic layer the slow wave disappears and only Φ_{js1} and Ψ_{js} exist. If the basement is a solid elastic layer then:

$$\left.\begin{array}{ll} \Phi_{n+1,s1} = A_{n+1,1}e^{-i\gamma_{n+1,1}(z+H_n)} \\ \Psi_{n+1,s} = A_{n+1,3}e^{-i\gamma_{n+1,3}(z+H_n)} \end{array}\right\} \tag{3.60}$$

Because of the field extinction conditions, no upward travelling waves are present in a basement so defined. We return to these issues in later chapters.

3.7 Notations Covering the Fourier Transforms and Spectral Representation

3.7.1 Definitions

For convenience, in this contribution we use the following notations of the Fourier transform relating the time series $f(t)$ and the space function

3.7 Notations of Fourier Transforms

$f(x)$:

$$\left.\begin{array}{rcl} f(t) & = & \frac{1}{2\pi}\int_{-\infty}^{\infty} g(\omega)e^{-i\omega t}d\omega \\[6pt] g(\omega) & = & \int_{-\infty}^{\infty} f(t)e^{i\omega t}dt \end{array}\right\}$$

and

$$\left.\begin{array}{rcl} f(x) & = & \frac{1}{2\pi}\int_{-\infty}^{\infty} g(k_x)e^{ik_x x}dk_x \\[6pt] g(k_x) & = & \int_{-\infty}^{\infty} f(x)e^{-ik_x x}dx \end{array}\right\} \quad (3.61)$$

The corresponding δ-functions are then

$$\left.\begin{array}{rcl} \frac{1}{2\pi}\int_{-\infty}^{\infty} e^{-i\omega t}d\omega = \delta(t) & \quad & \int_{-\infty}^{\infty} \delta(t)e^{i\omega t}dt = 1 \\[6pt] \int_{-\infty}^{\infty} e^{i\omega t}dt = \delta(\omega) & \quad & \frac{1}{2\pi}\int_{-\infty}^{\infty} \delta(\omega)e^{-i\omega t}d\omega = 1 \end{array}\right\} \quad (3.62)$$

and

$$\left.\begin{array}{rcl} \frac{1}{2\pi}\int_{-\infty}^{\infty} e^{ik_x x}dk_x = \delta(x) & \quad & \int_{-\infty}^{\infty} \delta(x)e^{-ik_x x}dx = 1 \\[6pt] \int_{-\infty}^{\infty} e^{-ik_x x}dx = \delta(k_x) & \quad & \frac{1}{2\pi}\int_{-\infty}^{\infty} \delta(k_x)e^{ik_x x}dk_x = 1 \end{array}\right\} \quad (3.63)$$

3.7.2 Spectral Representation

With the motion of the ocean surface regarded as being locally stationary in both time and space, the first- and second-order components of the displacement, velocity potential and pressure etc., can be expressed in the form[57]

$$\left.\begin{array}{rcl} \tilde{\zeta}^{(1,2)}(\vec{r},t) & = & \int_{-\infty}^{\infty} e^{i(\vec{k}\vec{r}-\omega t)}dZ^{(1,2)}(\vec{k},\omega) \\[6pt] \tilde{\Phi}_1^{(1,2)}(\vec{R},t) & = & \int_{-\infty}^{\infty} e^{i(\vec{k}\vec{r}-\omega t)}dF^{(1,2)}(\vec{k},\omega,z) \\[6pt] \tilde{p}_1^{(1,2)}(\vec{R},t) & = & \int_{-\infty}^{\infty} e^{i(\vec{k}\vec{r}-\omega t)}dP^{(1,2)}(\vec{k},\omega,z) \end{array}\right\} \quad (3.64)$$

with the power-density spectra defined as

$$\left.\begin{array}{rcl} f_\zeta^{(1,2)}(\vec{k},\omega) & = & \frac{\langle |dZ^{(1,2)}(\vec{k},\omega)|^2 \rangle}{d\vec{k}d\omega} \\[6pt] f_\Phi^{(1,2)}(\vec{k},\omega) & = & \frac{\langle |dF^{(1,2)}(\vec{k},\omega,z)|^2 \rangle}{d\vec{k}d\omega} \\[6pt] f_p^{(1,2)}(\vec{k},\omega) & = & \frac{\langle |dP^{(1,2)}(\vec{k},\omega,z)|^2 \rangle}{d\vec{k}d\omega} \end{array}\right\} \quad (3.65)$$

where $\vec{k}$ is the horizontal wavenumber vector and ω the angular frequency. The integration covers the three-dimensional space $(\vec{k}, \omega)$. The generalised spectra, $dZ^{(1,2)}(\vec{k}, \omega)$ and $dF^{(1,2)}(\vec{k}, \omega, z)$, as functions of $\vec{k}$ and ω, are orthogonal processes satisfying (Appendix G)

$$\left. \begin{array}{l} \langle dZ^{(1,2)}(\vec{k}, \omega)\, dZ^{*(1,2)}(\vec{k}', \omega') \rangle = 0 \\[4pt] \langle dF^{(1,2)}(\vec{k}, \omega, z)\, dF^{*(1,2)}(\vec{k}', \omega', z) \rangle = 0 \\[4pt] \langle dP^{(1,2)}(\vec{k}, \omega, z)\, dP^{*(1,2)}(\vec{k}', \omega', z) \rangle = 0 \end{array} \right\} \vec{k} \neq \vec{k}' \text{ or } \omega \neq \omega'$$

Other quantities of the geoacoustic field, such as the pressure in the water column and the seismic displacement on the seabed and within the subbottom structure, have the same form of spectral expansion as the potential field.

We note that, since the integration is not absolutely convergent, a stationary process does not strictly have a Fourier transform in the form of Eq.(3.61). However, because the Fourier transformation will be very helpful to subsequent calculations, we elect to formally express the surface displacement, $\tilde{\zeta}^{(1,2)}(\vec{r}, t)$, the velocity potential, $\tilde{\Phi}^{(1,2)}(\vec{R}, t)$, and the pressure, $\tilde{p}^{(1,2)}(\vec{R}, t)$ etc., in the form,

$$\left. \begin{array}{l} \tilde{\zeta}^{(1,2)}(\vec{r}, t) = \frac{1}{(2\pi)^3} \int_{-\infty}^{\infty} \int_{-\infty}^{\infty} \zeta^{(1,2)}(\vec{k}, \omega) e^{i(\vec{k}\vec{r} - \omega t)} d\vec{k}\, d\omega \\[6pt] \tilde{\Phi}_1^{(1,2)}(\vec{R}, t) = \frac{1}{(2\pi)^3} \int_{-\infty}^{\infty} \int_{-\infty}^{\infty} \Phi_1^{(1,2)}(\vec{k}, \omega, z) e^{i(\vec{k}\vec{r} - \omega t)} d\vec{k}\, d\omega \\[6pt] \tilde{p}_1^{(1,2)}(\vec{R}, t) = \frac{1}{(2\pi)^3} \int_{-\infty}^{\infty} \int_{-\infty}^{\infty} p_1^{(1,2)}(\vec{k}, \omega, z) e^{i(\vec{k}\vec{r} - \omega t)} d\vec{k}\, d\omega \end{array} \right\} \quad (3.66)$$

The power-density spectra then become

$$\left. \begin{array}{l} f_\zeta^{(1,2)}(\vec{k}, \omega) = \frac{1}{(2\pi)^6} \langle |\zeta^{(1,2)}(\vec{k}, \omega)|^2 \rangle d\vec{k}\, d\omega \\[6pt] f_\Phi^{(1,2)}(\vec{k}, \omega) = \frac{1}{(2\pi)^6} \langle |\Phi_1^{(1,2)}(\vec{k}, \omega, z)|^2 \rangle d\vec{k}\, d\omega \\[6pt] f_p^{(1,2)}(\vec{k}, \omega) = \frac{1}{(2\pi)^6} \langle |p_1^{(1,2)}(\vec{k}, \omega, z)|^2 \rangle d\vec{k}\, d\omega \end{array} \right\} \quad (3.67)$$

with $\vec{k}$ ranging over the whole wavenumber plane and ω ranging from $-\infty$ to ∞.

Chapter 4

An Exact Expression for the Homogeneous Component

4.1 Introduction

From the previous chapters it is clear that a complete solution of the wave-induced pressure field will be complicated. To aid the development of the general case given later, we here first provide an extension of the "standing-wave" analysis. The ocean is again assumed to be infinitely deep and bottomless so that the use of the deep-water form of the ocean-wave dispersion relation is justified and only the homogeneous component of the pressure field need be considered. Further, by ignoring the compressibility of seawater, the dynamic equation governing the first-order field reduces to a Laplace equation, and by ignoring the gravity term the second-order field is described by the standard (Helmholtz) acoustic equation. Within the constraints of these assumptions and the approximations implicit in the perturbation procedure, we now develop an expression for the deep-water source field that is "exact", in that it incorporates all possible components of the homogeneous pressure field.

In the literature the power-density spectrum of the deep-water pressure field has been defined as the "equivalent source spectrum" for the wave-interaction mechanism, in the sense that it is independent of observation depth and bottom structure. In later chapters we will see that this definition is not perfect because the inhomogeneous component, which is depth dependent, is not negligible in many cases of interest. However, in situations where the observation depth is greater than 500 meters (say) and the bottom is not characterised by soft sediments, the source spectrum established in this chapter does provide a description of the wave-induced pressure field, which is adequate for most purposes. This "exact" relation is therefore of practical value in its own right, but its main role here is as the basis of the more complex analysis presented in Chap. 5.

The review in Chap. 2 showed that both the velocity potential and the pressure have been used as the field variable in previous theoretical treatments. Reference was also made to the fact that the relative importance of the "volume" and "boundary" components of the source field is markedly dependent on the choice of variable. To allow us to more readily demonstrate this difference, and in the interests of completeness, solutions are presented here in terms of both variables. Solutions are also presented in terms of both a plane-wave and Green's function analysis and their equivalence is demonstrated.

4.2 The Potential Field Solution

4.2.1 Equations and Boundary Conditions

We start with the solution of the potential field and recall from the last chapter that the simplified equations and boundary conditions for the first- and second-order potential fields take the form,

$$\nabla^2 \tilde{\Phi}_1^{(1)} = 0 \tag{4.1}$$

$$\left(\frac{\partial^2}{\partial t^2} + g\frac{\partial}{\partial z}\right) \tilde{\Phi}_1^{(1)} = 0, \quad \text{at} \quad z = 0 \tag{4.2}$$

$$\frac{\partial \tilde{\Phi}_1^{(1)}}{\partial z} - \frac{\partial \tilde{\zeta}^{(1)}}{\partial t} = 0 \quad \text{at} \quad z = 0 \tag{4.3}$$

and

$$\left(\nabla^2 - \frac{1}{\alpha_1^2}\frac{\partial^2}{\partial t^2}\right) \tilde{\Phi}_1^{(2)} = \frac{1}{\alpha_1^2}\frac{\partial}{\partial t}(\nabla \tilde{\Phi}_1^{(1)})^2 \tag{4.4}$$

$$\left(\frac{\partial^2}{\partial t^2} + g\frac{\partial}{\partial z}\right) \tilde{\Phi}_1^{(2)} = -\frac{\partial}{\partial t}(\nabla \tilde{\Phi}_1^{(1)})^2 \quad \text{at} \quad z = 0 \tag{4.5}$$

the other boundary condition relating to $\tilde{\zeta}^{(2)}$ (Eq.(3.27)) being suppressed since second-order surface displacements are not discussed here. The second term on the right-hand-side of Eq.(3.26) has also been eliminated because, from Eqs.(3.22) and (4.1), $(\partial^2/\partial t^2 + g\partial/\partial z)\Phi_1^{(1)} \equiv 0$.

4.2.2 Plane-Wave Analysis

Expressing the first-order surface displacement and the potential field in the form of the Fourier integrals,

$$\tilde{\zeta}^{(1)}(\vec{r},t) = \frac{1}{(2\pi)^3} \int \int \zeta^{(1)}(\vec{k},\omega) e^{i(\vec{k}\vec{r}-\omega t)} d\vec{k} d\omega \qquad (4.6)$$

and

$$\tilde{\Phi}_1^{(i)}(\vec{R},t) = \frac{1}{(2\pi)^3} \int \int \Phi_1^{(i)}(\vec{k},\omega,z) e^{i(\vec{k}\vec{r}-\omega t)} d\vec{k} d\omega \qquad (4.7)$$

where $i = 1, 2$, and substituting these into Eqs.(4.1) to (4.3) leads to a set of ordinary differential equations relating to the first-order field,

$$(\frac{d^2}{dz^2} - k^2)\Phi_1^{(1)}(\vec{k},\omega,z) = 0 \qquad (4.8)$$

$$(g\frac{d}{dz} - \omega^2)\Phi_1^{(1)} = 0 \quad \text{at} \quad z = 0 \qquad (4.9)$$

$$\frac{d}{dz}\Phi_1^{(1)} = -i\omega\zeta^{(1)}(\vec{k},\omega) \quad \text{at} \quad z = 0 \qquad (4.10)$$

and to two sets of equations (one involving a nonhomogeneous equation with a homogeneous boundary condition and the other a homogeneous equation with a nonhomogeneous boundary condition) describing the second-order field,

$$(\frac{d^2}{dz^2} + \gamma_1^2)\Phi_{1a}^{(2)} = S_{\phi a}^{(2)}(\vec{k},\omega,z) \qquad (4.11)$$

$$(g\frac{d}{dz} - \omega^2)\Phi_{1a}^{(2)} = 0 \quad \text{at} \quad z = 0 \qquad (4.12)$$

$$(\frac{d^2}{dz^2} + \gamma_1^2)\Phi_{1b}^{(2)} = 0 \qquad (4.13)$$

$$(g\frac{d}{dz} - \omega^2)\Phi_{1b}^{(2)} = S_{\phi b}^{(2)}(\vec{k},\omega,0) \quad \text{at} \quad z = 0 \tag{4.14}$$

where

$$S_{\phi a}^{(2)}(\vec{k},\omega,z) = \frac{1}{\alpha_1^2}\int\int\left[\frac{\partial}{\partial t}[\nabla\tilde{\Phi}_1^{(1)}(\vec{R},t)]^2\right]e^{-i(\vec{k}\vec{r}-\omega t)}d\vec{r}dt \tag{4.15}$$

and

$$S_{\phi b}^{(2)}(\vec{k},\omega,0) = -\int\int\left[\frac{\partial}{\partial t}[\nabla\tilde{\Phi}_1^{(1)}(\vec{r},t)]^2\right]_{z=0}e^{-i(\vec{k}\vec{r}-\omega t)}d\vec{r}dt \tag{4.16}$$

are source terms, $\vec{r} = \vec{R}$ when $z = 0$, and $\gamma_1 = \sqrt{\omega^2/\alpha_1^2 - k^2}$.

First-order solution

The solution of the first-order field can be found readily enough by first substituting $\Phi_1^{(1)} = A_\phi^{(1)} e^{kz}$ into Eqs.(4.9) and (4.10) to obtain,

$$\Phi_1^{(1)}(\vec{k},\omega,z) = \frac{-i\omega}{k}\zeta^{(1)}(\vec{k},\omega)e^{kz}\delta(\omega - \sqrt{gk}) \tag{4.17}$$

where $\delta(\omega - \sqrt{gk})$ is introduced to account for the dispersion of gravity waves in deep waters, described by $\omega^2 = gk$. Substituting Eq.(4.17) back into Eq.(4.7) then establishes the first-order field

$$\tilde{\Phi}_1^{(1)}(\vec{R},t) = \frac{-i}{(2\pi)^2}\int\sqrt{\frac{g}{k}}\zeta^{(1)}(\vec{k})e^{kz+i(\vec{k}\vec{r}-\omega t)}d\vec{k} \tag{4.18}$$

where $\omega = \sqrt{gk}$.

Second-order solution

The second-order solution of the nonhomogeneous equation (4.11) is the sum of the complementary function, $\exp(-i\gamma_1 z)$, (satisfying the extinction conditions at $z = -\infty$) and a particular solution, $Z_\phi(z)$, i.e.

$$\Phi_{1a}^{(2)}(\vec{k},\omega,z) = Z_\phi(z) + C_{\phi a}e^{-i\gamma_1 z} \tag{4.19}$$

where

$$Z_\phi(z) = \frac{S_{\phi a}^{(2)}}{D^2 + \gamma_1^2} = \left(\frac{1}{D - i\gamma_1} - \frac{1}{D + i\gamma_1}\right)\frac{S_{\phi a}^{(2)}}{2i\gamma_1}$$

4.2 Potential Field

and D is the operator, $D = d/dz$. According to the theory of ordinary differential equations (see, for example, [86]), the action of the inverse operator $(D+a)^{-1}$ upon a function $f(z)$ yields the solution

$$y(z) = \frac{f(z)}{D+a} = e^{-az} \int_{-\infty}^{z} e^{az_0} f(z_0) dz_0$$

so that

$$Z_\phi(z) = \frac{1}{\gamma_1} \int_{-\infty}^{z} sin\gamma_1(z-z_0) S_{\phi a}^{(2)}(\vec{k}, \omega, z_0) dz_0 \qquad (4.20)$$

Invoking the boundary condition, Eq.(4.12), then leads to

$$C_{\phi a} = -\frac{\omega^2 Z_\phi(0) - g Z_\phi'(0)}{\omega^2 + ig\gamma_1}. \qquad (4.21)$$

Some algebra then establishes the solution

$$\Phi_{1a}^{(2)}(\vec{k},\omega,z) = \frac{1}{2i\gamma_1} \left\{ \int_{-\infty}^{z} e^{i\gamma_1(z-z_0)} S_{\phi a}^{(2)}(\vec{k},\omega,z_0) dz_0 + \int_{z}^{0} e^{-i\gamma_1(z-z_0)} \cdot \right.$$
$$\left. \cdot S_{\phi a}^{(2)}(\vec{k},\omega,z_0) dz_0 - \left(\frac{\omega^2 - ig\gamma_1}{\omega^2 + ig\gamma_1}\right) \int_{-\infty}^{0} e^{-i\gamma_1(z+z_0)} S_{\phi a}^{(2)}(\vec{k},\omega,z_0) dz_0 \right\} \qquad (4.22)$$

The solution, $\Phi_{1b}^{(2)}$, is found similarly by substituting $\Phi_{1b}^{(2)} = C_{\phi b} e^{(-i\gamma_1 z)}$ into Eq.(4.14), which leads to

$$\Phi_{1b}^{(2)}(\vec{k},\omega,z) = -\frac{S_{\phi b}^{(2)}(\vec{k},\omega,0)}{\omega^2 + ig\gamma_1} e^{-i\gamma_1 z} \qquad (4.23)$$

4.2.3 Green's Function Analysis

As was mentioned before, the plane-wave solutions in the form established above are strictly only valid when the surface-wave field is regarded as uniform over an area of infinite size. For more general cases a solution based on the Green's function is needed. (Helpful presentations of the application of the Green's function in underwater acoustics are available in, for example, [63] and [89].) The required solution can be established in the following way.

Performing the Fourier transform

$$\tilde{\Phi}_1^{(2)}(\vec{R},t) = \frac{1}{2\pi} \int \Phi_1^{(2)}(\vec{R},\omega) e^{-i\omega t} d\omega \qquad (4.24)$$

on both sides of Eqs.(4.4) and (4.5) leads to the equations,

$$(\nabla^2 + \frac{\omega^2}{\alpha_1^2})\Phi_{1\omega}^{(2)} = \bar{S}_{\phi a}^{(2)}(\omega, \vec{r}, z) \tag{4.25}$$

and

$$(\frac{\partial}{\partial z} - \frac{\omega^2}{g})\Phi_{1\omega}^{(2)} = \frac{1}{g}\bar{S}_{\phi b}^{(2)}(\omega, \vec{r}, 0) \quad \text{at} \quad z = 0 \tag{4.26}$$

where

$$\bar{S}_{\phi a}^{(2)}(\omega, \vec{r}, z) = \frac{1}{\alpha_1^2} \int \frac{\partial}{\partial t}[\nabla \tilde{\Phi}_1^{(1)}(\vec{R}, t)]^2 e^{i\omega t} dt \tag{4.27}$$

and

$$\bar{S}_{\phi b}^{(2)}(\omega, \vec{r}, 0) = -\int \left\{ \frac{\partial}{\partial t}[\nabla \tilde{\Phi}_1^{(1)}(\vec{R}, t)]^2 \right\}_{z=0} e^{i\omega t} dt \tag{4.28}$$

The solution of Eq.(4.25) with the corresponding boundary condition (4.26), can be expressed in the form (see Appendix H, Eqs.(H.1) and (H.13), noting that the volume source term, $\rho(\vec{R})$, corresponds to the factor $(1/4\pi)\bar{S}_{\phi a}^{(2)}$ in Eq.(4.29))

$$\begin{aligned}\Phi_{1\omega}^{(2)} &= -\frac{1}{4\pi}\int_V G(\vec{R}, \vec{R}_0)\bar{S}_{\phi a}^{(2)}(\omega, \vec{r}_0, z_0)d\vec{R}_0 \\ &+ \frac{1}{4\pi g}\int_S G(\vec{R}, \vec{r}_0)\bar{S}_{\phi b}^{(2)}(\omega, \vec{r}_0, 0)d\vec{r}_0 \end{aligned} \tag{4.29}$$

where the Green's function $G(\vec{R}, \vec{R}_0)$ satisfies the equation [58]

$$(\nabla^2 + \frac{\omega^2}{\alpha_1^2})G(\vec{R}, \vec{R}_0) = -4\pi\delta(\vec{R} - \vec{R}_0) \tag{4.30}$$

and the homogeneous surface condition,

$$\frac{\partial}{\partial z}G(\vec{R}, \vec{R}_0) - \frac{\omega^2}{g}G(\vec{R}, \vec{R}_0) = 0 \quad \text{at} \quad z = 0 \tag{4.31}$$

Expressing the Green's function in the form

$$G(\vec{R}, \vec{R}_0) = \frac{1}{(2\pi)^2}\int g_\omega(z, z_0) e^{i\vec{k}(\vec{r}-\vec{r}_0)} d\vec{k} \tag{4.32}$$

4.2 Potential Field

and substituting back into Eqs.(4.30) and (4.31), leads to the equations required to determine the function $g_\omega(z, z_0)$,

$$(\frac{d^2}{dz^2} + \gamma_1^2) g_\omega(z, z_0) = -4\pi \delta(z - z_0) \tag{4.33}$$

and

$$(\frac{d}{dz} - \frac{\omega^2}{g}) g_\omega(z, z_0) = 0 \quad \text{at} \quad z = 0 \tag{4.34}$$

It can be shown by integration that the nonhomogeneous equation (4.33) can be replaced by its homogeneous equivalent,

$$(\frac{d^2}{dz^2} + \gamma_1^2) g_\omega(z, z_0) = 0 \tag{4.35}$$

with the following connection conditions applying at the source point, z_0,

$$\frac{d}{dz} g_\omega(z, z_0) \Big|_{z=z_0-\epsilon}^{z=z_0+\epsilon} = -4\pi \tag{4.36}$$

and

$$g_\omega(z, z_0)|_{z=z_0-\epsilon}^{z=z_0+\epsilon} = 0 \quad \text{with} \quad \epsilon \simeq 0 \tag{4.37}$$

Introducing the complementary function of the homogeneous equation,

$$g_\omega(z, z_0) = Ae^{i\gamma_1 z} + Be^{-i\gamma_1 z} \quad \text{for} \quad z_0 \leq z \leq 0 \tag{4.38}$$

$$g_\omega(z, z_0) = Ce^{-i\gamma_1 z} \quad \text{for} \quad z < z_0 \tag{4.39}$$

into Eqs.(4.34), (4.36), and (4.37) then leads, after some algebra, to the required Green's functions:

for $z_0 \leq z \leq 0$,

$$G(\vec{R}, \vec{R}_0) = i \int_0^\infty \frac{1}{\gamma_1} \left[e^{i\gamma_1(z-z_0)} - \left(\frac{\omega^2 - ig\gamma_1}{\omega^2 + ig\gamma_1} \right) e^{-i\gamma_1(z+z_0)} \right]_{z_0 \leq z \leq 0} J_0(k\rho) k dk,$$

$$\tag{4.40}$$

and for $z < z_0$,

$$G(\vec{R}, \vec{R}_0) = i \int_0^\infty \frac{1}{\gamma_1} \left[e^{-i\gamma_1(z-z_0)} - \left(\frac{\omega^2 - ig\gamma_1}{\omega^2 + ig\gamma_1} \right) e^{-i\gamma_1(z+z_0)} \right]_{z<z_0} J_0(k\rho) k dk.$$

$$\tag{4.41}$$

The potential solution can then be expressed in the form

$$\Phi_1^{(2)}(\vec{R},\omega) = \Phi_{1v}^{(2)}(\vec{R},\omega) + \Phi_{1s}^{(2)}(\vec{R},\omega) \tag{4.42}$$

where from Eq.(4.29)

$$\Phi_{1v}^{(2)}(\vec{R},\omega) = \frac{-i}{4\pi} \int\int \left\{ \int_{-\infty}^{z} [\cdot]_{z_0 \leq z \leq 0} \bar{S}_{\phi a}^{(2)}(\omega,\vec{r}_0,z_0) dz_0 + \right.$$

$$\left. \int_z^0 [\cdot]_{z<z_0} \bar{S}_{\phi a}^{(2)}(\omega,\vec{r}_0,z_0) dz_0 \right\} J_0(k\rho) \frac{k}{\gamma_1} dk d\vec{r}_0 \tag{4.43}$$

and

$$\Phi_{1s}^{(2)}(\vec{R},\omega) = \frac{-1}{2\pi} \int\int \frac{1}{\omega^2 + ig\gamma_1} \bar{S}_{\phi b}^{(2)} e^{-i\gamma_1 z} J_0(k\rho) k dk d\vec{r}_0. \tag{4.44}$$

In the above, $[\cdot]_{z_0 \leq z \leq 0}$ and $[\cdot]_{z<z_0}$ denote the corresponding expressions inside the square brackets in Eqs.(4.40) and (4.41) respectively.

4.2.4 Comparison of the Plane-Wave and Green's Function Analyses

We now demonstrate that, when the active wave surface is infinite in size, the above expressions for the second-order potential become identical in the two treatments. According to definition,

$$\tilde{\Phi}_{1v,s}^{(2)}(\vec{R},t) = \frac{1}{2\pi} \int \Phi_{1v,s}^{(2)}(\vec{R},\omega) e^{-i\omega t} d\omega \tag{4.45}$$

and

$$S_{\phi a,b}^{(2)}(\vec{k},\omega,z) = \int \bar{S}_{\phi a,b}^{(2)}(\omega,\vec{r},z) e^{-i\vec{k}\vec{r}} d\vec{r} \tag{4.46}$$

so that from Eqs.(4.43) to (4.45)

$$\tilde{\Phi}_{1v}^{(2)}(\vec{R},t) = \frac{1}{(2\pi)^3} \int\int \frac{1}{2i\gamma_1} \left[\int_{-\infty}^{z} e^{i\gamma_1(z-z_0)} S_{\phi a}^{(2)}(\vec{k},\omega,z_0) dz_0 \right.$$

$$\left. + \int_z^0 e^{-i\gamma_1(z-z_0)} S_{\phi a}^{(2)}(\vec{k},\omega,z_0) dz_0 \right] e^{i(\vec{k}\vec{r}-\omega t)} d\vec{k} d\omega$$

$$-\frac{1}{(2\pi)^3} \int\int \frac{1}{2i\gamma_1} \left(\frac{\omega^2 - ig\gamma_1}{\omega + ig\gamma_1} \right) \left[\int_{-\infty}^{0} e^{-i\gamma_1 z_0} S_{\phi a}^{(2)}(\vec{k},\omega,z_0) dz_0 \right] \cdot$$

$$\cdot e^{i(\vec{k}\vec{r}-\omega t) - i\gamma_1 z} d\vec{k} d\omega \tag{4.47}$$

and
$$\tilde{\Phi}_{1s}^{(2)}(\vec{R},t) = \frac{-1}{(2\pi)^3} \int\int \frac{1}{\omega^2 + ig\gamma_1} S_{\phi b}^{(2)}(\vec{k},\omega,0) e^{i(\vec{k}\vec{r}-\omega t)-i\gamma_1 z} d\vec{k} d\omega \quad (4.48)$$

These expressions are the same as those established in the plane-wave analysis when Eqs.(4.22) and (4.23) are substituted into Eq.(4.7). The two procedures are therefore mutually consistent and produce identical results when the area of the active region tends to infinity. This demonstrated equivalence will be helpful to us later on.

4.3 The Pressure Field Solution

4.3.1 Equations and Boundary Conditions

Again referring to Chap. 3 the simplified equations and boundary conditions for the first- and second-order pressure fields are

$$\nabla^2 \tilde{p}_1^{(1)} = 0 \quad (4.49)$$

$$\tilde{p}_1^{(1)} + \tilde{\zeta}^{(1)} \frac{\partial \tilde{p}_1^{(0)}}{\partial z} = 0 \quad \text{at} \quad z = 0 \quad (4.50)$$

$$\frac{\partial \tilde{p}_1^{(1)}}{\partial z} + \rho_1^{(0)} \frac{\partial^2 \tilde{\zeta}^{(1)}}{\partial t^2} = 0 \quad \text{at} \quad z = 0 \quad (4.51)$$

and

$$\left(\nabla^2 - \frac{1}{\alpha_1^2}\frac{\partial^2}{\partial t^2}\right) \tilde{p}_1^{(2)} = -\rho_1^{(0)} \nabla^2 [(\nabla \tilde{\Phi}_1^{(1)})^2] \quad (4.52)$$

$$\left(\frac{\partial^2}{\partial t^2} + g\frac{\partial}{\partial z}\right) \tilde{p}_1^{(2)} \equiv \tilde{h}_b(\vec{r},t) = -(g\frac{\partial}{\partial z} + \frac{\partial^2}{\partial t^2})[\tilde{\zeta}^{(1)}\frac{\partial \tilde{p}_1^{(1)}}{\partial z}]$$
$$-2\rho^{(0)} g \nabla \tilde{\Phi}_1^{(1)} \nabla \frac{\partial}{\partial t}\tilde{\zeta}^{(1)} - \rho_1^{(0)} g \nabla \tilde{\zeta}^{(1)} \frac{\partial}{\partial t} \nabla \tilde{\Phi}_1^{(1)} \quad \text{at} \quad z=0 \quad (4.53)$$

The zero-order pressure field can be readily established as

$$\tilde{p}_1^{(0)} = \rho_1^{(0)} g z \quad (4.54)$$

It is relevant to the discussion in Sect. 4.3.4 and later to note that the dynamic equation for the second-order pressure field, Eq.(4.52), becomes identical to the Doak version of Lighthill's equation adopted by Cato[51] (see Eq.(A.64)), when other volume (pressure) forces and stresses are ignored, the relation $p = c_0^2 \rho$ is regarded as always valid, and the density ρ is replaced by its average value ρ_0.

4.3.2 Plane-Wave Analysis

We first express the pressure field in terms of the Fourier integral,

$$\tilde{p}_1^{(i)}(\vec{R},t) = \frac{1}{(2\pi)^3} \int \int p_1^{(i)}(\vec{k},\omega,z) e^{i(\vec{k}\vec{r}-\omega t)} d\vec{k}d\omega \qquad (4.55)$$

where $i = 1,2$. Substitution into Eqs.(4.49) to (4.53), along with the Fourier transform of the displacement field, Eq.(4.6), leads to the following differential equations for the first-order field,

$$(\frac{d^2}{dz^2} - k^2)p_1^{(1)}(\vec{k},\omega,z) = 0 \qquad (4.56)$$

$$p_1^{(1)} = \rho_1^{(0)} g \zeta^{(1)}(\vec{k},\omega) \text{ at } z = 0 \qquad (4.57)$$

$$\frac{dp_1^{(1)}}{dz} = \rho_1^{(0)} \omega^2 \zeta^{(1)}(\vec{k},\omega) \text{ at } z = 0 \qquad (4.58)$$

and to two comparable sets for the second-order field,

$$(\frac{d^2}{dz^2} + \gamma_1^2)p_{1a}^{(2)} = S_{pa}^{(2)}(\vec{k},\omega,z) \qquad (4.59)$$

$$(g\frac{d}{dz} - \omega^2)p_{1a}^{(2)} = 0 \text{ at } z = 0 \qquad (4.60)$$

and

$$(\frac{d^2}{dz^2} + \gamma_1^2)p_{1b}^{(2)} = 0 \qquad (4.61)$$

$$(g\frac{d}{dz} - \omega^2)p_{1b}^{(2)} = S_{pb}^{(2)}(\vec{k},\omega,0) \text{ at } z = 0 \qquad (4.62)$$

where

$$S_{pa}^{(2)}(\vec{k},\omega,0) = -\rho_1^{(0)} \int \int \nabla^2 \left\{[\nabla \tilde{\Phi}_1^{(1)}(\vec{r},t)]^2\right\} e^{-i(\vec{k}\vec{r}-\omega t)} d\vec{r}dt \qquad (4.63)$$

4.3 Pressure Field

and

$$S_{pb}^{(2)}(\vec{k},\omega,0) = -\int\int\left[g\tilde{\zeta}^{(1)}\frac{\partial^2\tilde{p}_1^{(1)}}{\partial z^2} + \frac{\partial^2}{\partial t^2}(\tilde{\zeta}_1\frac{\partial\tilde{p}_1^{(1)}}{\partial z}) + 2\rho_1^{(0)}g\nabla\tilde{\Phi}_1^{(1)}\nabla\frac{\partial}{\partial t}\tilde{\zeta}^{(1)} \right.$$
$$\left. +\rho_1^{(0)}g\nabla\tilde{\zeta}^{(1)}\frac{\partial}{\partial t}\nabla\tilde{\Phi}_1^{(1)}\right]_{z=0} e^{-i(\vec{k}\vec{r}-\omega t)}d\vec{r}dt \tag{4.64}$$

In a development, which parallels that for the potential field, we can now use a plane-wave substitution to establish solutions for the first- and second-order pressure fields.

First-order field

For the first-order field this solution is

$$p_1^{(1)}(\vec{k},\omega,z) = \rho_1^{(0)}g\zeta^{(1)}(\vec{k},\omega)e^{kz}\delta(\omega - \sqrt{gk}) \tag{4.65}$$

and by referring to Eq.(4.55)

$$\tilde{p}_1^{(1)}(\vec{R},t) = \frac{\rho_1^{(0)}g}{(2\pi)^2}\int\zeta^{(1)}(\vec{k})e^{i\vec{k}\vec{r}}d\vec{k}$$

Second-order field

In the case of the second-order field (a bar is used to differentiate from the expression derived from the potential solutions)

$$\bar{p}_{1a}^{(2)}(\vec{k},\omega,z) = \frac{1}{2i\gamma_1}\left\{\int_{-\infty}^z e^{i\gamma_1(z-z_0)}S_{pa}^{(2)}(\vec{k},\omega,z_0)dz_0 + \int_z^0 e^{-i\gamma_1(z-z_0)}\right.$$
$$\left. \cdot S_{pa}^{(2)}(\vec{k},\omega,z_0)dz_0 - \left(\frac{\omega^2 - ig\gamma_1}{\omega^2 + ig\gamma_1}\right)\int_{-\infty}^0 e^{-i\gamma_1(z+z_0)}S_{pa}^{(2)}(\vec{k},\omega,z_0)dz_0\right\}$$
$$\tag{4.66}$$

The solution, $\bar{p}_{1b}^{(2)}$, is found by substituting $\bar{p}_{1b}^{(2)} = C_{\phi b}\exp(-i\gamma_1 z)$ for $p_{1b}^{(2)}$ in Eq.(4.62), to obtain

$$\bar{p}_{1b}^{(2)}(\vec{k},\omega,z) = -\frac{S_{pb}^{(2)}(\vec{k},\omega,0)}{\omega^2 + ig\gamma_1}e^{-i\gamma_1 z} \tag{4.67}$$

which corresponds to its potential equivalent, Eq.(4.23).

4.3.3 Green's Function Analysis

In an analysis which again parallels that for the potential (Sect. 4.2.3), we can take the Fourier transform (time domain) of both sides of the two Eqs.(4.52) and (4.53), to obtain

$$(\nabla^2 + \frac{\omega^2}{\alpha_1^2})p_1^{(2)} = \bar{S}_{pa}^{(2)}(\omega, \vec{r}, z) \tag{4.68}$$

$$(\frac{\partial}{\partial z} - \frac{\omega^2}{g})p_1^{(2)} = \frac{1}{g}\bar{S}_{pb}^{(2)}(\omega, \vec{r}, 0) \quad \text{at} \quad z = 0 \tag{4.69}$$

where

$$\bar{S}_{pa}^{(2)}(\omega, \vec{r}, z) = -\rho_1^{(0)} \int \nabla^2 [\nabla \tilde{\Phi}_1^{(1)}(\vec{R}, t)]^2 e^{i\omega t} dt \tag{4.70}$$

and then use Eq.(4.53)

$$\bar{S}_{pb}^{(2)}(\omega, \vec{r}, 0) = \int \tilde{h}_b(\vec{r}, t) e^{i\omega t} dt \tag{4.71}$$

to establish the solution,

$$\begin{aligned} p_1^{(2)} &= -\frac{1}{4\pi} \int_V G(\vec{R}, \vec{R}_0) \bar{S}_{pa}^{(2)}(\omega, \vec{r}_0, z_0) d\vec{R}_0 \\ &+ \frac{1}{4\pi g} \int_S G(\vec{R}, \vec{r}_0) \bar{S}_{pb}^{(2)}(\omega, \vec{r}_0, 0) d\vec{r}_0 \end{aligned} \tag{4.72}$$

where the Green's function, $G(\vec{R}, \vec{R}_0)$, satisfies the same equation and boundary condition that applied in the case of the potential field. Thus, using Eqs.(4.40) and (4.41) we establish the required solution,

$$p_1^{(2)}(\vec{R}, \omega) = p_{1v}^{(2)}(\vec{R}, \omega) + p_{1s}^{(2)}(\vec{R}, \omega) \tag{4.73}$$

where

$$p_{1v}^{(2)}(\vec{R}, \omega) = \frac{-i}{4\pi} \int \int \left\{ \int_{-\infty}^{z} [\cdot]_{z_0 \leq z \leq 0} \bar{S}_{pa}^{(2)}(\omega, \vec{r}_0, z_0) dz_0 + \int_{z}^{0} [\cdot]_{z < z_0} \bar{S}_{pa}^{(2)}(\omega, \vec{r}_0, z_0) dz_0 \right\} J_0(k\rho) \frac{k}{\gamma_1} dk d\vec{r}_0 \tag{4.74}$$

with $[\cdot]_{z_0 \leq z \leq 0}$ and $[\cdot]_{z<z_0}$ again denoting the expressions in the square brackets of Eqs.(4.40) and (4.41), and

$$p_{1s}^{(2)}(\vec{R}, \omega) = \frac{-1}{2\pi} \int \int \frac{1}{\omega^2 + ig\gamma_1} \bar{S}_{pb}^{(2)} e^{-i\gamma_1 z} J_0(k\rho) k dk d\vec{r}_0 \tag{4.75}$$

4.3 Pressure Field

As in the case of the potential (see Eqs.(4.45) and (4.46)) we can use the relations

$$\tilde{p}^{(2)}_{1v,s}(\vec{R},t) = \frac{1}{2\pi} \int p^{(2)}_{1v,s}(\vec{R},\omega) e^{-i\omega t} d\omega \qquad (4.76)$$

and

$$S^{(2)}_{pa,b}(\vec{k},\omega,z) = \int \bar{S}^{(2)}_{pa,b}(\omega,\vec{r},z) e^{-i\vec{k}\vec{r}} d\vec{r} \qquad (4.77)$$

to show that, when the active surface is infinite in size, the Green's function solutions become identical with the plane-wave solutions, Eqs.(4.66) and (4.67), derived above.

4.3.4 Surface and Volume Source Distributions

From the foregoing it is clear that the source responsible for the first-order pressure field is the surface displacement (represented by the term, $\partial \tilde{\zeta}_1/\partial t$, in the surface condition Eq.(4.3) and $\tilde{\zeta}_1$ in Eq.(4.50)). On the other hand two source distributions apparently contribute to the second-order field, one represented by the nonhomogeneous term of the dynamic equation, Eq.(4.4) (or Eq.(4.52)), and the other by that of the surface boundary condition, Eq.(4.5) (or Eq.(4.53)). The nature of these sources is apparent enough in mathematical terms, but it is clear from the early literature that the assessment of their individual contributions to the total pressure field has created some difficulty. Because these two terms have been central to all theoretical developments it will be instructive to discuss them and their physical significance before proceeding.

The preceding analyses, especially that based on the Green's function , clearly establish the two distinct components of the second-order field. The integral expressions in Eq.(4.29) (or Eq.(4.72)) identify these with a source distributed throughout the volume (arising from the non-linear terms of Eq.(4.25) (or Eq.(4.68)) and a surface-source distribution (associated with the nonlinear term of the boundary condition at $z = 0$, Eq.(4.26) (or Eq.(4.69)). The nature of the surface source is clear enough. In physical terms the volume contribution can be visualised as arising from the nonlinear interaction throughout the water column of components of the exponentially decaying primary-frequency wave fields. The source level of the resulting second-order pressure field is thus depth-dependent, but the homogeneous component of the induced field propagates from its zone of generation without decay, in a manner similar to that arising from the source distribution at the surface, associated with the vertical displacement about the mean plane, $z = 0$.

The acoustical effects of these two source distributions have been handled in different ways in earlier analyses - see Chap. 2. Longuet-Higgins and Brekhovskikh, for instance, incorporated both distributions into their plane-wave solutions but the resulting formalism is complicated and difficult to use. Hasselmann neglected the volume source in his analysis based on the potential field. Hughes invoked a strategy to convert the (pressure) dynamic equation to a homogeneous one and effectively combined the two source distributions without resolving their individual effects. Lloyd, on the other hand, calculated the contribution of both sources under specified conditions. Guo's analysis ignores the volume effects. Cato uses a strategy which appears to result in neglect of the boundary contribution.

To facilitate a meaningful examination of all treatments we have here taken both distributions into account and established, without invoking any additional approximations, exact solutions (in terms of both the potential and pressure) using both plane-wave and Green's function procedures. The acoustical effects of the two distributions are defined specifically in Eqs.(4.22), (4.23) and their equivalents. These allow a comprehensive comparison of the different theoretical analyses to be made.

4.4 Source Pressure Spectra

With solutions available in terms of both the potential and pressure, we now proceed to establish the power-density spectral functions of the second-order pressure field. Functions are established in terms of both variables, with the volume and surface effects considered as independent processes. As a final step we also consider the function obtained when the additional assumptions of the "standing-wave approximation" are invoked.

4.4.1 Source Functions

We start by calculating the source terms, $S^{(2)}_{\phi a}(\vec{k}, \omega, z)$, $S^{(2)}_{\phi b}(\vec{k}, \omega, 0)$, $S^{(2)}_{pa}(\vec{k}, \omega, z)$, and $S^{(2)}_{pb}(\vec{k}, \omega, 0)$.

With the angular frequency and wavenumber vector of the ocean-wave field now denoted by σ and $\vec{q}$ respectively, the first-order potential field

4.4 Source Pressure Spectra

becomes (see Eq.(4.18)),

$$\tilde{\Phi}_1^{(1)}(\vec{R},t) = \frac{-i}{(2\pi)^2}\int \frac{\sigma}{q}\zeta^{(1)}(\vec{q})e^{i(\vec{q}\vec{r}-\sigma t)+qz}d\vec{q} \quad (4.78)$$

so that

$$\left[\nabla\tilde{\Phi}_1^{(1)}(\vec{R},t)\right]^2 = \frac{-1}{(2\pi)^4}\int\int \sigma_1\sigma_2(1-cos\theta_{12})\zeta^{(1)}(\vec{q}_1)\zeta^{(1)}(\vec{q}_2)\cdot$$

$$\cdot e^{i[(\vec{q}_1+\vec{q}_2)\vec{r}-(\sigma_1+\sigma_2)t]+(q_1+q_2)z}d\vec{q}_1 d\vec{q}_2 \quad (4.79)$$

where $cos\theta_{12} \equiv \vec{q}_1\vec{q}_2/q_1q_2$. Substituting Eq.(4.79) into Eqs.(4.15), (4.16) and (4.63), and Eqs.(4.78) and (4.65) into (4.64), and completing the integrations with respect to $d\vec{r}$, dt and $d\vec{q}_2$, establishes the source potential functions,

$$S^{(2)}_{\phi a}(\vec{k},\omega,z) = \frac{-1}{(2\pi)^2\alpha_1^2}\int F[\vec{q}_1,\vec{q}_2]e^{(q_1+q_2)z}d\vec{q}_1 \quad (4.80)$$

$$S^{(2)}_{\phi b}(\vec{k},\omega,0) = \frac{1}{(2\pi)^2}\int F[\vec{q}_1,\vec{q}_2]d\vec{q}_1 \quad (4.81)$$

and their pressure equivalents

$$S^{(2)}_{pa}(\vec{k},\omega,z) = i\frac{\rho_1^{(0)}}{2\pi^2}\int F[\vec{q}_1,\vec{q}_2]\frac{q_1q_2}{(\sigma_1+\sigma_2)}(1-cos\theta_{12})e^{(q_1+q_2)z}d\vec{q}_1 \quad (4.82)$$

$$S^{(2)}_{pb}(\vec{k},\omega,0) = i\frac{\rho_1^{(0)}g}{(2\pi)^2}\int F[\vec{q}_1,\vec{q}_2]\cdot$$

$$\cdot\frac{q_2\sigma_1(2\sigma_2+\sigma_1)-q_1\sigma_2(2\sigma_1+\sigma_2)cos\theta_{12}}{\sigma_1\sigma_2(\sigma_1+\sigma_2)(1-cos\theta_{12})}d\vec{q}_1 \quad (4.83)$$

In the above we note that,

$$F[\vec{q}_1,\vec{q}_2] = (-i)\sigma_1\sigma_2(\sigma_1+\sigma_2)(1-cos\theta_{12})\zeta^{(1)}(\vec{q}_1)\zeta^{(1)}(\vec{q}_2)\delta(\sigma_1+\sigma_2-\omega) \quad (4.84)$$

and

$$\sigma_1 = \sqrt{gq_1}, \quad \sigma_2 = \sqrt{gq_2}, \quad \vec{q}_2 = \vec{k}-\vec{q}_1.$$

4.4.2 Spectral Expressions based on the Potential as the Variable

By now introducing Eqs.(4.80) and (4.81) into Eqs.(4.22) and (4.23) we establish the potential solutions

$$\Phi_{1a}^{(2)}(\vec{k},\omega,z) = \frac{1}{(2\pi)^2 \alpha_1^2} \int F[\vec{q}_1,\vec{q}_2] \frac{1}{(q_1+q_2)^2 + \gamma_1^2} \cdot$$

$$\cdot \left[\frac{\omega^2 - g(q_1+q_2)}{\omega^2 + ig\gamma_1} e^{-i\gamma_1 z} - e^{(q_1+q_2)z} \right] d\vec{q}_1 \quad (4.85)$$

$$\Phi_{1b}^{(2)}(\vec{k},\omega,z) = \frac{-1}{(2\pi)^2} \frac{e^{-i\gamma_1 z}}{\omega^2 + ig\gamma_1} \int F[\vec{q}_1,\vec{q}_2] d\vec{q}_1. \quad (4.86)$$

These can be converted, using Eq.(3.34) and the Fourier transform (3.66), to obtain corresponding spectral expressions for the pressure field

$$p_1^{(2)}(\vec{k},\omega,z) = p_{1a1}^{(2)}(\vec{k},\omega,z) + p_{1a2}^{(2)}(\vec{k},\omega,z) + p_{1b}^{(2)}(\vec{k},\omega,z) + p_{1c}^{(2)}(\vec{k},\omega,z) \quad (4.87)$$

where

$$p_{1a1}^{(2)} = \frac{i\rho_1^{(0)}\omega}{(2\pi)^2 \alpha_1^2} \int \frac{1}{[(q_1+q_2)^2 + \gamma_1^2]} \left(\frac{\omega^2 - g(q_1+q_2)}{\omega^2 + ig\gamma_1} \right) F[\vec{q}_1,\vec{q}_2] e^{-i\gamma_1 z} d\vec{q}_1 \quad (4.88)$$

$$p_{1a2}^{(2)} = \frac{-i\rho_1^{(0)}\omega}{(2\pi)^2 \alpha_1^2} \int \frac{1}{[(q_1+q_2)^2 + \gamma_1^2]} F[\vec{q}_1,\vec{q}_2] e^{(q_1+q_2)z} d\vec{q}_1 \quad (4.89)$$

$$p_{1b}^{(2)} = \frac{-i\rho_1^{(0)}\omega}{(2\pi)^2} \int \frac{1}{\omega^2 + ig\gamma_1} F[\vec{q}_1,\vec{q}_2] e^{-i\gamma_1 z} d\vec{q}_1 \quad (4.90)$$

$$p_{1c}^{(2)} = \frac{i\rho_1^{(0)}}{2(2\pi)^2} \int \frac{1}{(\sigma_1+\sigma_2)} F[\vec{q}_1,\vec{q}_2] e^{(q_1+q_2)z} d\vec{q}_1 \quad (4.91)$$

4.4.3 Spectral Expressions based on the Pressure as the Variable

Substituting the source functions, Eqs.(4.82) and (4.83), into Eqs.(4.66) and (4.67) establishes an alternative set of pressure spectra, $\bar{p}^{(2)}(\vec{k},\omega,z)$,

4.4 Source Pressure Spectra

based on pressure as the field variable and the direct solution of the pressure-wave equation. In this case

$$\bar{p}_1^{(2)}(\vec{k},\omega,z) = \bar{p}_{1a1}^{(2)}(\vec{k},\omega,z) + \bar{p}_{1a2}^{(2)}(\vec{k},\omega,z) + \bar{p}_{1b}^{(2)}(\vec{k},\omega,z) \quad (4.92)$$

where

$$\bar{p}_{1a1}^{(2)} = \frac{-i2\rho_1^{(0)}}{(2\pi)^2} \int \frac{(1-\cos\theta_{12})}{(\sigma_1+\sigma_2)} \frac{q_1 q_2}{(q_1+q_2)^2 + \gamma_1^2} \cdot$$

$$\cdot \frac{\omega^2 - g(q_1+q_2)}{\omega^2 + ig\gamma_1} F[\vec{q}_1,\vec{q}_2] e^{-i\gamma_1 z} d\vec{q}_1 \quad (4.93)$$

$$\bar{p}_{1a2}^{(2)} = \frac{i2\rho_1^{(0)}}{(2\pi)^2} \int \frac{(1-\cos\theta_{12})}{(\sigma_1+\sigma_2)} \frac{q_1 q_2}{(q_1+q_2)^2 + \gamma_1^2} F[\vec{q}_1,\vec{q}_2] e^{(q_1+q_2)z} d\vec{q}_1 \quad (4.94)$$

$$\bar{p}_{1b}^{(2)} = \frac{-i\rho_1^{(0)} g}{(2\pi)^2} \int \frac{q_2\sigma_1(2\sigma_2+\sigma_1) - q_1\sigma_2(2\sigma_1+\sigma_2)\cos\theta_{12}}{\sigma_1\sigma_2(\sigma_1+\sigma_2)(1-\cos\theta_{12})} \cdot$$

$$\cdot \frac{1}{(\omega^2+ig\gamma_1)} F[\vec{q}_1,\vec{q}_2] d\vec{q}_1 \quad (4.95)$$

4.4.4 Power-Density Source Spectra (Frequency-Wavenumber)

Whatever approach is followed, the power-density spectra of the second-order pressure field in deep water can now be expressed in a general form

$$p(\vec{k},\omega,z) = \int A(q_1,q_2) F[\vec{q}_1,\vec{q}_2] d\vec{q}_1$$

Referring to Eq.(3.67), the frequency-wavenumber spectrum can be written,

$$f_p(\vec{k},\omega) = <|p(\vec{k},\omega,z)|^2> \frac{d\omega d\vec{k}}{(2\pi)^6} = \frac{1}{(2\pi)^6} \int\int A(q_1,q_2) A^*(q_1',q_2')$$

$$< F[\vec{q}_1,\vec{q}_2] F^*[\vec{q}'_1,\vec{q}'_2] > d\vec{q}_1 d\vec{q}'_1 d\vec{k} d\omega$$

Further, by invoking the definition of $F[\vec{q}_1,\vec{q}_2]$ given in Eq.(4.84), noting that because of the orthogonality of a random process

$$<\zeta^{(1)}(\vec{q}_i)\zeta^{(1)*}(\vec{q}'_i)> = \begin{cases} 0 & \vec{q}_i \neq \vec{q}'_i \\ \frac{(2\pi)^4}{d\vec{q}_i} f_\zeta(\vec{q}_i) & \vec{q}_i = \vec{q}'_i \end{cases} \quad (4.96)$$

introducing the relations, $d\vec{k} = d\vec{q_2}$, $d\vec{q_1} = q_1(\partial q_1/\partial\omega)d\omega d\theta_1$, and carrying out the two integrations of the delta functions, $\delta(\sigma_1 + \sigma_2 - \omega)$, establishes the "exact" expression for the homogeneous component of the pressure field,

$$f_p(\vec{k},\omega) = (2\pi)^4 \int |A(q_1,q_2)|^2 f_\zeta(\vec{q_1}) f_\zeta(\vec{q_2}) \sigma_1^2 \sigma_2^2 (\sigma_1+\sigma_2)^2 (1-\cos\theta_{12})^2$$
$$q_1(\frac{\partial q_1}{\partial \omega})d\theta_1 \qquad (4.97)$$

The value of the analysis, which leads to Eq.(4.97), will become more apparent in Chaps. 5 and 6, when we consider a generalised source spectrum applicable in all environments. For our immediate purpose of examining the relative contributions of the volume- and surface- sources, when the velocity potential and the pressure respectively are used as the variable of the field equation, it is more convenient to make comparison in terms of the simpler standing-wave approximation.

4.4.5 Standing-Wave Approximation of the Source Spectrum

By recognising the deep-water dispersion relation, $\sigma_i^2 = gq_i$, and the relations (which apply under the standing-wave approximation),

$$\vec{q_1} + \vec{q_2} = \vec{k} = 0 \quad \text{(hence} \quad \theta = \pi) \qquad (4.98)$$

$$\sigma_1 = \sigma_2 = \omega/2. \qquad (4.99)$$

and

$$\frac{\partial q_1}{\partial \omega} = \frac{\omega}{2g}$$

Eq.(4.97) reduces to

$$f_p(\vec{k},\omega) = \frac{(2\pi)^4 \omega^9}{32g^2} \int |A(q_1,q_1)|^2 f_\zeta(\vec{q_1}) f_\zeta(-\vec{q_1}) d\theta_1 \qquad (4.100)$$

Further, introducing the expression for the surface-wave spectrum [26],[56],

$$f_\zeta(\vec{q_1}) \equiv F_a(\sigma_1) H(\theta_1) \frac{d\sigma_1 d\theta_1}{d\vec{q_1}} = \frac{g^2}{2\sigma^3} F_a(\sigma_1) H(\theta_1) \qquad (4.101)$$

4.4 Source Pressure Spectra

where $H(\theta)$ is the normalized directional distribution function of the spectral energy of the surface-wave field, satisfying

$$\int_0^{2\pi} H(\theta)d\theta = 1, \qquad (4.102)$$

and using the definition

$$I_\omega = \int_0^{2\pi} H(\theta_1)H(\theta_1 + \pi)d\theta_1 \qquad (4.103)$$

finally establishes the familiar expression for the power-density spectrum of the pressure field developed in the earlier treatments,

$$f_p(\vec{k},\omega) = \frac{(2\pi)^4 g^2 \omega^3}{2} |A(q_1, q_1)|^2 F_a^2(\frac{\omega}{2}) I_\omega \qquad (4.104)$$

4.4.6 Power-Density Source Spectra (Frequency)

The power-density spectrum of the second-order field can now be established as a function of frequency only, by the integration of Eq.(4.104) on the $\vec{k}$-plane. We recall that in the deep-water analysis the observation point is considered to be sufficiently deep that the inhomogeneous component of the pressure field can be neglected. In this case the integration with respect to k is restricted to the range 0 to ω/α_1. Further since the acoustical wavenumber, ω/α_1, is much smaller then the wavenumber of wind-driven gravity waves, the integration can be simply replaced by the multiplying factor, $\pi(\omega/\alpha_1)^2$ [26]. (A full discussion of the inhomogeneous component is given in later chapters.) Completing the calculation of the factor $|A(q_1, q_1)|^2$ for each component of the pressure field, Eqs.(4.87) to (4.95), finally establishes the following expressions for the power-density spectra (note the symbols $F_x^{(2)}$ and $\bar{F}_x^{(2)}$ differentiate the expressions based on the potential and pressure):

Based on the potential:

$$\begin{aligned}F_{p1a1}^{(2)} &= \frac{8\pi \rho_1^2 g^4 \omega^5}{\alpha_1^2(\omega^2 + 2ag)(\omega^2 + 8ag)^2} a^2 F_a^2\left(\frac{\omega}{2}\right) I_\omega(\omega) \\ &\simeq \frac{2\pi \rho_1^2 g^2 \omega^3}{\alpha_1^2}\left(\frac{2ga}{\omega^2}\right)^2 F_a^2\left(\frac{\omega}{2}\right) I_\omega(\omega) \qquad (4.105)\end{aligned}$$

$$F^{(2)}_{p1a2}(\omega) = \frac{32\pi\rho_1^2 g^4 \omega^3}{\alpha_1^2(\omega^2+8ag)^2} a^2 \exp(\frac{\omega^2}{g}z) F_a^2\left(\frac{\omega}{2}\right) I_\omega$$

$$\simeq \frac{8\pi\rho_1^2 g^2 \omega^3}{\alpha_1^2}(\frac{2ga}{\omega^2})^2 \exp(\frac{\omega^2}{g}z) F_a^2\left(\frac{\omega}{2}\right) I_\omega \qquad (4.106)$$

$$F^{(2)}_{p1b}(\omega) = \frac{\pi\rho_1^2 g^2 \omega^3}{2\alpha_1^2} F_a^2\left(\frac{\omega}{2}\right) I_\omega, \qquad (4.107)$$

$$F^{(2)}_{p1c}(\omega) = \frac{\pi\rho_1^2 g^2 \omega^3}{8\alpha_1^2} \exp(\frac{\omega^2}{g}z) F_a^2\left(\frac{\omega}{2}\right) I_\omega. \qquad (4.108)$$

Based on the pressure:

$$\bar{F}^{(2)}_{p1a1}(\omega) = \frac{\pi\rho_1^2 g^2 \omega^3}{8\alpha_1^2} F_a^2\left(\frac{\omega}{2}\right) I_\omega, \qquad (4.109)$$

$$\bar{F}^{(2)}_{p1a2}(\omega) = \frac{\pi\rho_1^2 g^2 \omega^3}{2\alpha_1^2} \exp(\frac{\omega^2}{g}z) F_a^2\left(\frac{\omega}{2}\right) I_\omega, \qquad (4.110)$$

$$\bar{F}^{(2)}_{p1b}(\omega) = \frac{9\pi\rho_1^2 g^2 \omega^3}{32\alpha_1^2} F_a^2\left(\frac{\omega}{2}\right) I_\omega. \qquad (4.111)$$

Several features of the above relations are noteworthy. First of all in the above $a = g/2\alpha_1^2$ is such a small quantity that $2ga/\omega^2 \ll 1$. As a result the components represented by Eqs.(4.105) and (4.106) can be neglected in comparison with Eq.(4.107), which we recognise as the classical expression of the second-order field established in earlier analyses. Secondly, a comparison of all expressions shows that those established through the potential field are somewhat different from those derived directly using pressure as the field variable. In the case of the potential field, the spectral components $F^{(2)}_{p1a1}$, $F^{(2)}_{p1a2}$ and $F^{(2)}_{p1c}$ are all associated with the volume source, and only $F^{(2)}_{p1b}$ with the boundary source. It is clear that in this case the pressure field in deep water is dominated by the boundary source, (a feature that was recognised in the early stages of the subject's development [26],[25]), the effects of the volume source either being small or negligible at depth. The situation is, however, quite different in the analysis based on pressure as the variable. As Eqs.(4.109)

4.4 Source Pressure Spectra

to (4.111) show, the contribution of the volume source, mainly because of $\bar{F}^{(2)}_{p1a1}$, is now the same order of magnitude as that of the boundary source, $\bar{F}^{(2)}_{p1b}$. This rather surprising result emphasises the unusual nature of the nonlinearity of the wave-interaction process. Thirdly, in the analysis based on the potential, the dominant spectral component, Eq.(4.107), is approximately twice the level of its equivalent in the pressure analysis, Eq.(4.111). The spectral levels in the two analyses only become (very nearly) identical when the contributions represented by Eqs.(4.109) and (4.111) are added. (The small residual difference is attributable to the approximation inherent in the perturbation procedure.) This comparison identifies a distinct advantage in selecting the potential as the field variable, as in any calculations of the pressure spectrum only the boundary contribution needs to be taken into account. When the pressure is used as the variable on the other hand, it is clear that both the boundary and volume contributions need to be considered and the calculations obviously become more complicated. As we noted in Chap. 2 and Sect. 4.3.4, this requirement has not always been recognised.

Chapter 5

Generalised Solution of the Potential Field

In the last chapter qualified solutions for the first two orders of the potential and pressure fields were derived by assuming that (i) the dynamic equations of the first- and second-order fields can be simplified respectively to a Laplace equation and an acoustic wave equation ; (ii) the effects of the ocean bottom can be ignored; and (iii) the observation point is deep. In this chapter more complete solutions are established by removing these restrictions. The development here (and throughout the remainder of the book) is carried out using the potential as the field variable, to exploit the fact demonstrated in the last chapter, that in so doing the volume term can be neglected and the analysis carried out in terms of the boundary source alone. The pressure field is obtained, as before, in a subsequent conversion.

5.1 First-Order Potential

5.1.1 Plane-Wave Solution

We start by considering the plane-wave solution of the first-order potential field in the reference wave guide defined in Sect. 3.1. Since all the interfaces are considered parallel to the horizontal plane and the displacement and potential fields are assumed stationary in time and space, the required potential field can be expressed as a linear combination of plane waves, with amplitudes (complex valued) to be determined by the boundary conditions. To establish this field we perform the Fourier transformation of Eq.(3.66) and substitute the plane-wave components, $\Phi_1^{(1)}(\vec{k},\omega,z)$ and $\zeta_1^{(1)}(\vec{k},\omega)$, back into the first-order field equations, (3.22) to (3.24).

This leads to

$$\left(\frac{d^2}{dz^2} - \frac{g}{\alpha_1^2}\frac{d}{dz} + \gamma_1^2\right)\Phi_1^{(1)}(\vec{k},\omega,z) = 0 \qquad (5.1)$$

$$\left(g\frac{d}{dz} - \omega^2\right)\Phi_1^{(1)} = 0 \qquad \text{at} \quad z = 0 \qquad (5.2)$$

$$\frac{d}{dz}\Phi_1^{(1)} + i\omega\zeta^{(1)} = 0 \qquad \text{at} \quad z = 0 \qquad (5.3)$$

and

$$L_b\Phi_1 = 0 \qquad \text{at} \quad z = -H_1 \qquad (5.4)$$

where $\gamma_1^2 = (\omega/\alpha_1)^2 - k^2$ and the operator L_b, here and later, defines the boundary conditions at the seafloor. The solution of the second-order differential equation, Eq.(5.1), has the form

$$\Phi_1^{(1)}(\vec{k},\omega,z) = A_0^{(1)}\left[e^{\gamma_+(z+H_1)} + R_b e^{\gamma_-(z+H_1)}\right] \qquad (5.5)$$

where

$$\left.\begin{array}{l}\gamma_\pm = a \mp \sqrt{a^2 - \gamma_1^2} \text{ for } \gamma_1^2 - a^2 \leq 0 \\[6pt] \gamma_\pm = \mp i\left(\sqrt{\gamma_1^2 - a^2} \pm ia\right) \text{ for } \gamma_1^2 - a^2 > 0\end{array}\right\} \qquad (5.6)$$

are the roots of its characteristic equation, $\chi^2 - 2a\chi + \gamma_1^2 = 0$, and $a \equiv g/2\alpha_1^2$. The parameter $A_0^{(1)}$ represents the unknown amplitude of the velocity potential and R_b the plane-wave reflection coefficient for the seafloor. Both parameters are functions of frequency and horizontal wavenumber. They are constant in the reference wave guide but can be slowly varying functions of time and space in a real environment.

Substitution of Eq.(5.5) into Eq.(5.3) establishes $A_0^{(1)}$ as

$$A_0^{(1)} = \frac{-i\omega}{\gamma_+}\frac{e^{-\gamma_+ H_1}}{1 + \epsilon R_b e^{(\gamma_- - \gamma_+)H_1}}\zeta^{(1)}(\vec{k},\omega) \qquad (5.7)$$

so that the first-order potential becomes

$$\Phi_1^{(1)}(\vec{k},\omega,z) = \frac{-i\omega}{\gamma_+}\frac{e^{\gamma_+(z+H_1)} + R_b e^{\gamma_-(z+H_1)}}{e^{\gamma_+ H_1} + \epsilon R_b e^{\gamma_- H_1}}\zeta^{(1)}(\vec{k},\omega) \qquad (5.8)$$

where

$$\epsilon = \frac{\gamma_-}{\gamma_+} \simeq -1 \qquad (5.9)$$

5.0 First-Order Potential

The dispersion relation

At this point we examine the more general form of the dispersion relation for the ocean-wave field. Substituting the solution (5.8) into condition (5.2) establishes the relation

$$\omega^2 = g\gamma_+ \frac{1 + \epsilon R_b e^{(\gamma_- - \gamma_+)H_1}}{1 + R_b e^{(\gamma_- - \gamma_+)H_1}}. \tag{5.10}$$

If seawater is considered incompressible (i.e. $\alpha_1 \to \infty$) it follows that $a = 0$ and, from Eq.(5.6), that $\gamma_\pm = \pm k$, whereupon Eq.(5.10) reduces to

$$\omega^2 = gk \frac{1 - R_b e^{-2kH_1}}{1 + R_b e^{-2kH_1}} \tag{5.11}$$

Further, when the bottom is "hard" ($R_b(k,\omega) = 1$) Eq.(5.11) takes on its more familiar form

$$\omega^2 = gk \tanh(kH_1) \tag{5.12}$$

As is well known Eq.(5.12) simplifies to

$$\omega^2 = gk \tag{5.13}$$

in deep-ocean environments, the form used in the deep-water analysis of Chap. 4.

Equation (5.10) describes the most general form of the dispersion relation, but because the non-dimensional quantity, ag/ω^2, is small and the reflection coefficient $R_b(\vec{k},\omega)$ is very close to 1 for ocean gravity waves, Eq.(5.12) is in general a good enough approximation for any real environment. This is not so, however, when the water is shallow and the bottom is very soft (low rigidity). In this case the full relation is required.

The potential function

We now return to the expression for $\tilde{\Phi}_1^{(1)}$. Substituting Eq.(5.8) into Eq.(3.66) establishes the first-order potential in the reference waveguide,

$$\tilde{\Phi}_1^{(1)}(\vec{R},t) = \frac{-i}{(2\pi)^3} \int \int_{-\infty}^{\infty} \frac{\omega}{\gamma_+} \frac{e^{\gamma_+(z+H_1)} + R_b e^{\gamma_-(z+H_1)}}{e^{\gamma_+ H_1} + \epsilon R_b e^{\gamma_- H_1}} \zeta^{(1)}(\vec{k},\omega) \cdot$$

$$\cdot e^{i(\vec{k}\vec{r}-\omega t)} \delta(\omega - \omega_k) d\vec{k}\, d\omega =$$

$$\frac{-i}{(2\pi)^2} \int \frac{\omega}{\gamma_+} \frac{e^{\gamma_+(z+H_1)} + R_b e^{\gamma_-(z+H_1)}}{e^{\gamma_+ H_1} + \epsilon R_b e^{\gamma_- H_1}} \zeta^{(1)}(\vec{k}) e^{i(\vec{k}\vec{r}-\omega t)} \Big|_{\omega=\omega_k} d\vec{k} \tag{5.14}$$

As in the last chapter the function, $\delta(\omega - \omega_k)$, has been introduced to account for the dispersion relation. When $\alpha_1 \to \infty$ and $R_b = 0$ (or $H \to \infty$), $\gamma_\pm = \pm k$, $\omega = \sqrt{gk}$ and Eq.(5.14) degenerates, as it should, to the simplified solution given by Eq.(4.18).

5.1.2 Green's Function Solution

Following the development in Chap. 4 we now derive the Green's function solution of the first-order field. By performing the Fourier transformation on both the surface displacement and velocity potential,

$$\tilde{\zeta}^{(1)}(\vec{r},t) = \frac{1}{2\pi} \int_{-\infty}^{\infty} \zeta_\omega^{(1)}(\vec{r}) e^{-i\omega t} d\omega \qquad (5.15)$$

$$\tilde{\Phi}_1^{(1)}(\vec{R},t) = \frac{1}{2\pi} \int_{-\infty}^{\infty} \Phi_{1\omega}^{(1)}(\vec{R}) e^{-i\omega t} d\omega \qquad (5.16)$$

and invoking Eqs.(3.22) to (3.24), the required equation and boundary conditions for $\Phi_{1\omega}^{(1)}(\vec{R})$ become

$$L\Phi_{1\omega}^{(1)}(\vec{R}) = 0 \qquad (5.17)$$

$$\left(\omega^2 - g\frac{\partial}{\partial z}\right)\Phi_{1\omega}^{(1)} = 0 \qquad \text{at} \quad z = 0 \qquad (5.18)$$

$$\frac{\partial}{\partial z}\Phi_{1\omega}^{(1)} = -i\omega\zeta_\omega^{(1)} \qquad \text{at} \quad z = 0 \qquad (5.19)$$

$$L_b\Phi_{1\omega}^{(1)} = 0 \qquad \text{at} \quad z = -H_1 \qquad (5.20)$$

where

$$L = \nabla^2 - \frac{g}{\alpha_1^2}\frac{\partial}{\partial z} + \frac{\omega^2}{\alpha_1^2} \qquad (5.21)$$

The form of the differential operator L is here slightly different from the operator used in the last chapter. As is well known, for the Helmholtz operator, $L_H = \nabla^2 + \omega^2/\alpha_1^2$, the Green's theorem [58] takes the form (see Appendix H)

$$uL_H v - vL_H u = \nabla \cdot (u\nabla v - v\nabla u)$$

where u and v are two arbitrary scalar functions. With certain boundary conditions applying, the solution of the wave equation,

$$L_H \psi(\vec{R}) = -4\pi\rho(\vec{R})$$

5.0 First-Order Potential

can then be expressed in the general form (see Eq.(H.10))

$$\psi(\vec{R}) = \int_V \rho(\vec{R}_0) G(\vec{R}, \vec{R}_0) d\vec{R}_0$$

$$+ \frac{1}{4\pi} \int_S \left[G(\vec{R}, \vec{R}_0) \frac{\partial}{\partial n_0} \psi(\vec{R}_0) - \psi(\vec{R}_0) \frac{\partial}{\partial n_0} G(\vec{R}, \vec{R}_0) \right] d\vec{R}_0$$

by using the Green's function G satisfying

$$L_H G(\vec{R}, \vec{R}_0) = -4\pi \delta(\vec{R} - \vec{R}_0)$$

and noting the reciprocity

$$G(\vec{R}, \vec{R}_0) = G(\vec{R}_0, \vec{R})$$

For the more general operator, L, the simple form of the Green's theorem given above is no longer valid. However as is shown in Appendix H, we can always find an "adjoint operator", $\hat{L}$, such that the following generalised Green's theorem applies,

$$uLv - v\hat{L}u = \nabla \cdot \vec{P}(u, v)$$

(where $\vec{P}$ is the bilinear concomitant vector) and the adjoint Green's function, $\hat{G}$, satisfying the equation

$$\hat{L}\hat{G}(\vec{R}, \vec{R}_0) = -4\pi \delta(\vec{R} - \vec{R}_0)$$

(here $\hat{L} = \nabla^2 + (g/\alpha_1^2)\partial/\partial z + \omega^2/\alpha_1^2$) under certain boundary conditions, is reciprocal to the Green's function $G(\vec{R}, \vec{R}_0)$ satisfying the equation

$$LG(\vec{R}, \vec{R}_0) = -4\pi \delta(\vec{R} - \vec{R}_0).$$

under the same boundary conditions, i.e.,

$$\hat{G}(\vec{R}_0, \vec{R}) = G(\vec{R}, \vec{R}_0),$$

The general solution of the equation $L\psi(\vec{R}) = -4\pi\rho(\vec{R})$ (Eq.(H.20)) can then be expressed through the adjoint Green's function as

$$\psi(\vec{R}_0) = \int_V \rho(\vec{R}) \hat{G}(\vec{R}, \vec{R}_0) d\vec{R} + \frac{1}{4\pi} \int_S \vec{n} \cdot \vec{P}\left[\hat{G}(\vec{R}, \vec{R}_0), \psi(\vec{R})\right] d\vec{R}$$

which can also be expressed as

$$\psi(\vec{R}) = \int_V \rho(\vec{R}_0) G(\vec{R}, \vec{R}_0) d\vec{R}_0 + \frac{1}{4\pi} \int_S \vec{n}_0 \cdot \vec{P}\left[G(\vec{R}, \vec{R}_0), \psi(\vec{R}_0)\right] d\vec{R}_0$$

by interchanging the symbols $\vec{R}$ and $\vec{R}_0$ and applying the reciprocal relation.

Based on the analysis in Appendix I and noting that Eq.(5.17) is homogeneous, i.e., $\rho(\vec{R}_0) = 0$, we can, by assigning appropriate boundary conditions in the establishment of the adjoint Green's function, we finally establish the required solution of the first order field, $\Phi^{(1)}_{1\omega}(\vec{R})$, as (Eqs.(I.11) and (I.26))

$$\Phi^{(1)}_{1\omega}(\vec{R}) = \frac{1}{4\pi} \int_S G^{(1)}(\vec{R}, \vec{R}_0) \frac{\partial}{\partial z_0} \Phi^{(1)}_{1\omega}(\vec{r}_0) d\vec{r}_0$$

$$= \frac{-i\omega}{4\pi} \int_S G^{(1)}(\vec{R}, \vec{R}_0) \zeta^{(1)}_\omega(\vec{r}_0) d\vec{r}_0 \qquad (5.22)$$

where the Green's function for the reference waveguide is

$$G^{(1)}(\vec{R}, \vec{R}_0) = \frac{1}{\pi} \int_{\vec{k}} \frac{1}{\gamma_+} \frac{e^{\gamma_+(z+H_1)} + R_b e^{\gamma_-(z+H_1)}}{e^{\gamma_+ H_1} + \epsilon R_b e^{\gamma_- H_1}} e^{i\vec{k}(\vec{r}-\vec{r}_0)} d\vec{k} \qquad (5.23)$$

By now recalling the definition,

$$\zeta^{(1)}_\omega(\vec{r}_0) = \int \zeta^{(1)}(\vec{r}_0, t) e^{i\omega t} dt$$

and introducing the dispersion relation, $\omega = \omega_k$, determined by the surface kinematic condition, Eq.(5.18), we can write

$$\zeta^{(1)}(\vec{r}_0, t) = \frac{1}{(2\pi)^3} \int_{\vec{k}'} \int_{\omega'} \zeta^{(1)}_{\omega'}(\vec{k}') \delta(\omega' - \omega'_k) e^{i(\vec{k}'\vec{r}_0 - \omega' t)} d\vec{k}' d\omega'$$

and

$$\zeta^{(1)}_\omega(\vec{r}_0) = \frac{1}{(2\pi)^2} \int_{\vec{k}'} \zeta^{(1)}_\omega(\vec{k}') \delta(\omega - \omega_k) e^{i\vec{k}'\vec{r}_0} d\vec{k}'.$$

where $\zeta^{(1)}_\omega(\vec{k}') \equiv \zeta^{(1)}(\vec{k}', \omega)$. When $S \to \infty$, completing the integration upon $d\vec{r}_0$ in Eq.(5.22), leads to

$$\Phi^{(1)}_{1\omega}(\vec{R}) = \frac{1}{(2\pi)^2} \int S^{(1)}(\vec{k}, \omega) \frac{e^{\gamma_+(z+H_1)} + R_b e^{\gamma_-(z+H_1)}}{e^{\gamma_+ H_1} + \epsilon R_b e^{\gamma_- H_1}} e^{i\vec{k}\vec{r}} d\vec{k} \qquad (5.24)$$

where

$$S^{(1)}(\vec{k}, \omega) = -\frac{i\omega}{\gamma_+} \zeta^{(1)}_\omega(\vec{k}) \delta(\omega - \omega_k)$$

We now note that, by taking an additional Fourier transform of Eq.(5.24) about ω, $\int(\cdot)e^{-i\omega t}d\omega/2\pi$, results in an expression for $\tilde{\Phi}^{(1)}_1$, which is exactly the same as that derived directly from the plane-wave solution - see Eq.(5.14). This demonstrates the equivalence of the two procedures.

5.1.3 Potential in a Weakly Range-Dependent Wave Guide

A more general model, allowing weak range-dependence, can be considered by introducing as an approximation

$$G^{(1)}(\vec{R}, \vec{R}_0) = \frac{1}{\pi} \int_{\vec{k}} \frac{F(\vec{r}, \vec{r}_0, k)}{\gamma_+} \frac{e^{\gamma_+(z+H_1)} + R_b e^{\gamma_-(z+H_1)}}{e^{\gamma_+ H_1} + \epsilon R_b e^{\gamma_- H_1}} e^{i\vec{k}(\vec{r}-\vec{r}_0)} d\vec{k}$$

where $F(\vec{r}, \vec{r}_0, k)$ is a function describing the (slow) range-dependence of the model. The corresponding potential then becomes

$$\Phi^{(1)}_{1\omega}(\vec{R}) = \frac{1}{(2\pi)^2} \int_{\vec{k}} S^{(1)}(\vec{k}, \omega, \vec{r}) \frac{e^{\gamma_+(z+H_1)} + R_b e^{\gamma_-(z+H_1)}}{e^{\gamma_+ H_1} + \epsilon R_b e^{\gamma_- H_1}} e^{i\vec{k}\vec{r}} d\vec{k} \quad (5.25)$$

where

$$S^{(1)}(\vec{k}, \omega, \vec{r}) = \frac{-i\omega}{\gamma_+} \int_S \zeta^{(1)}_\omega(\vec{r}_0) F(\vec{r}, \vec{r}_0, \vec{k}) e^{-i\vec{k}\vec{r}_0} d\vec{r}_0 \quad (5.26)$$

and the area S is the effective active wave region. When $S \to \infty$ and $F(\vec{r}, \vec{r}_0, \vec{k}) = 1$ (no range-dependence), $S^{(1)}(\vec{k}, \omega, \vec{r})$ becomes $S^{(1)}(\vec{k}, \omega)$, and Eq.(5.25) reverts to the same form as Eq.(5.24), as it should.

5.2 Second-Order Potential

5.2.1 General Solution

To determine the second-order velocity potential we turn to Eqs.(3.25) and (3.26) and a corresponding equation defining the boundary condition at the sea-floor interface, i.e.,

$$L\tilde{\Phi}^{(2)}_1(\vec{R}, t) = \frac{1}{\alpha_1^2} \frac{\partial}{\partial t}[\nabla \tilde{\Phi}^{(1)}_1]^2$$

$$\left(\frac{\partial^2}{\partial t^2} + g\frac{\partial}{\partial z}\right)\tilde{\Phi}^{(2)}_1 = -\frac{\partial}{\partial t}[\nabla \tilde{\Phi}^{(1)}_1]^2 \quad \text{at} \quad z = 0$$

$$L_b \tilde{\Phi}^{(2)}_1 = 0 \quad \text{at} \quad z = -H_1$$

while Eq.(3.27) can be used to determine the second-order surface displacement, $\tilde{\zeta}^{(2)}$, when this is of interest. In the above we have deliberately ignored the second term on the right-hand-side of the surface condition, Eq.(3.26). We can justify this because, in later calculations of the second-order field, we assume the first-order field satisfies the Laplace equation, i.e., $\nabla^2 \tilde{\Phi}^{(1)}_1 = 0$. This means that $(\partial^2/\partial t^2 + g\partial/\partial z)\Phi^{(1)}_1 \equiv 0$.

Because of the linearity of the equations, the solution of $\tilde{\Phi}_1^{(2)}$ can be regarded as the sum of two parts, $\tilde{\Phi}_{1a}^{(2)}$ and $\tilde{\Phi}_{1b}^{(2)}$, the first of which satisfies an inhomogeneous equation with a homogeneous surface condition, while the other satisfies a homogeneous equation and inhomogeneous surface condition, viz.

$$L\tilde{\Phi}_{1a}^{(2)} = \frac{1}{\alpha_1^2}\frac{\partial}{\partial t}[\nabla\tilde{\Phi}_1^{(1)}]^2$$

$$\left(\frac{\partial^2}{\partial t^2} + g\frac{\partial}{\partial z}\right)\tilde{\Phi}_{1a}^{(2)} = 0 \quad \text{at} \quad z = 0$$

$$L_b\tilde{\Phi}_{1a}^{(2)} = 0 \quad \text{at} \quad z = -H \quad (5.27)$$

and

$$L\tilde{\Phi}_{1b}^{(2)} = 0$$

$$\left(\frac{\partial^2}{\partial t^2} + g\frac{\partial}{\partial z}\right)\tilde{\Phi}_{1b}^{(2)} = -\frac{\partial}{\partial t}[\nabla\tilde{\Phi}_1^{(1)}]^2 \quad \text{at} \quad z = 0$$

$$L_b\tilde{\Phi}_{1b}^{(2)} = 0 \quad \text{at} \quad z = -H \quad (5.28)$$

Physically $\tilde{\Phi}_{1a}^{(2)}$ accounts for sound generated by a source distribution throughout the water column, while $\tilde{\Phi}_{1b}^{(2)}$ describes that generated by a distribution at the ocean surface - see Sect. 4.3.4. In that discussion it was shown that the boundary source dominates in the analysis based on the velocity potential as the field variable. We therefore concentrate here and in later chapters on the solution $\tilde{\Phi}_{1b}^{(2)}$. (If necessary readers can refer to Sect. 4.4 for an account of the volume component calculated under certain simplifying assumptions.)

We elect to use the more general Green's function analysis. To establish $\tilde{\Phi}_{1b}^{(2)}$ we first perform the Fourier transform

$$\tilde{\Phi}_{1b}^{(2)}(\vec{R}, t) = \frac{1}{2\pi}\int \Phi_{1b\omega}^{(2)}(\vec{R})e^{-i\omega t}d\omega$$

to establish the following new relations (see Appendix K):

$$L\Phi_{1b\omega}^{(2)}(\vec{R}) = 0 \quad (5.29)$$

5.2 Second-Order Potential

$$\left(\frac{\partial}{\partial z} - \frac{\omega^2}{g}\right)\Phi^{(2)}_{1b\omega}(\vec{R}) = \frac{1}{g}\bar{S}^{(2)}_b(\omega,\vec{r},0) \quad \text{at} \quad z=0 \qquad (5.30)$$

$$L_b \Phi^{(2)}_{1b\omega}(\vec{R}) = 0 \quad \text{at} \quad z=-H_1 \qquad (5.31)$$

where

$$\bar{S}^{(2)}_b(\omega,\vec{r},0) = -\int_{-\infty}^{\infty}\left\{\frac{\partial}{\partial t}[\nabla\tilde{\Phi}^{(1)}_1(\vec{R},t)]^2\right\}_{z=0} e^{i\omega t}dt. \qquad (5.32)$$

(The form of the operator L here is defined in Eq.(5.33), while L_b again defines the corresponding boundary conditions at the seafloor.) We now introduce the Green's functions, $G^{(2)}_b(\vec{R},\vec{R}_0)$ and $\hat{G}^{(2)}_b(\vec{R},\vec{R}_0)$, satisfying

$$LG^{(2)}_b(\vec{R},\vec{R}_0) = -4\pi\delta(\vec{R}-\vec{R}_0), \quad L = \nabla^2 - 2a\frac{\partial}{\partial z} + \frac{\omega^2}{\alpha_1^2} \qquad (5.33)$$

$$\hat{L}\hat{G}^{(2)}_b(\vec{R},\vec{R}_0) = -4\pi\delta(\vec{R}-\vec{R}_0), \quad \hat{L} = \nabla^2 + 2a\frac{\partial}{\partial z} + \frac{\omega^2}{\alpha_1^2} \qquad (5.34)$$

where $\hat{L}$ is the adjoint operator of L, and note from Appendix H that $G^{(2)}_b$ and $\hat{G}^{(2)}_b$ are reciprocal relations, (Eqs.(K.18) to (K.20)).

Multiplying Eq.(5.34) by $\Phi^{(2)}_{1b\omega}$ and Eq.(5.29) by $\hat{G}^{(2)}_b(\vec{R},\vec{R}_0)$ and invoking Gauss' theorem,

$$\int_V \nabla \cdot \vec{F} dv = \int_S \vec{F} \cdot d\vec{s}$$

establishes (see Appendix K for details)

$$\Phi^{(2)}_{1b\omega}(\vec{R}_0) = \frac{1}{4\pi}\int_S \vec{n}\cdot\vec{P}\left[\hat{G}^{(2)}_b(\vec{R},\vec{R}_0),\Phi^{(2)}_{1b\omega}(\vec{R})\right]d\vec{R} \qquad (5.35)$$

where

$$\vec{P}(u,v) = u\nabla v - v\nabla u - 2auv\vec{z}_0 \qquad (5.36)$$

and

$$\nabla \cdot \vec{P}(u,v) = uLv - v\hat{L}u \qquad (5.37)$$

Interchanging the symbols $\vec{R}$ and $\vec{R}_0$ gives

$$\Phi^{(2)}_{1b\omega}(\vec{R}) = \frac{1}{4\pi}\int_S \vec{n}_0\cdot\vec{P}\left[\hat{G}^{(2)}_b(\vec{R}_0,\vec{R}),\Phi^{(2)}_{1b\omega}(\vec{R}_0)\right]d\vec{R}_0 \qquad (5.38)$$

Then replacing $\hat{G}^{(2)}_b(\vec{R}_0,\vec{R})$ by $G^{(2)}_b(\vec{R},\vec{R}_0)$ leads to the required solution

$$\Phi^{(2)}_{1b\omega}(\vec{R}) = \frac{1}{4\pi g} \int_S G_b^{(2)}(\vec{R}, \vec{R}_0) \bar{S}_b^{(2)}(\omega, \vec{r}_0, 0) d\vec{r}_0, \qquad (5.39)$$

where (see Appendix K)

$$G_b^{(2)}(\vec{R}, \vec{R}_0)\Big|_{z_0=0} = 2 \int_0^\infty \frac{1}{\gamma_+} \left[\frac{e^{\gamma_+(z+H_1)} + R_b e^{\gamma_-(z+H_1)}}{e^{\gamma_+ H_1} + \frac{\epsilon}{b} R_b e^{\gamma_- H_1}} \right] \frac{1}{1 - \frac{\omega^2}{g\gamma_+}} J_0(k\rho) k\, dk \qquad (5.40)$$

and

$$b = \frac{(1 - \frac{\omega^2}{g\gamma_+})}{(1 - \frac{\omega^2}{g\gamma_-})}, \qquad \epsilon = \frac{\gamma_-}{\gamma_+} \qquad (5.41)$$

5.2.2 Potential in the Reference Wave Guide

To establish the factor $\bar{S}_b^{(2)}(\omega, \vec{r}_0, 0)$ in Eq.(5.39) we express $\tilde{\Phi}_1^{(1)}(\vec{R}, t)$ as

$$\tilde{\Phi}_1^{(1)}(\vec{R}, t) = \frac{1}{(2\pi)^2} \int A_0^{(1)}(q) \zeta^{(1)}(\vec{q}) \left[e^{\gamma_+(z+H_1)} + R_b e^{\gamma_-(z+H_1)} \right] e^{i(\vec{q}\vec{r} - \sigma t)} d\vec{q} \qquad (5.42)$$

where from Eq.(5.14)

$$A_0^{(1)}(q) = \frac{-i\sigma}{\gamma_+} \frac{e^{-\gamma_+ H_1}}{1 + \epsilon R_b e^{(\gamma_- - \gamma_+) H_1}} \qquad (5.43)$$

If, as an approximation, we take

$$\gamma_\pm = a \mp i \sqrt{\frac{\omega^2}{\alpha_1^2} - q^2} \cong \pm q$$

it follows that (Appendix L)

$$\bar{S}_b^{(2)}(\omega, \vec{r}_0, 0) = \frac{-i}{(2\pi)^4} \int\int (\sigma_1 + \sigma_2) \sigma_1 \sigma_2 \zeta^{(1)}(\vec{q}_1) \zeta^{(1)}(\vec{q}_2) \{\vec{q}_1, \vec{q}_2\} \cdot$$

$$\cdot \delta(\omega - \sigma_1 - \sigma_2) e^{i(\vec{q}_1 + \vec{q}_2)\vec{r}_0} d\vec{q}_1\, d\vec{q}_2 \qquad (5.44)$$

with

$$\{\vec{q}_1, \vec{q}_2\} = \frac{C_- \left[1 + R_{b1} R_{b2} e^{-2(q_1+q_2)H_1} \right] - C_+ \left[R_{b2} e^{-2q_2 H_1} + R_{b1} e^{-2q_1 H_1} \right]}{(1 - R_{b1} e^{-2q_1 H_1})(1 - R_{b2} e^{-2q_2 H_1})}$$

$$(5.45)$$

5.2 Second-Order Potential

where

$$C_- = 1 - \cos\theta_{12}, \quad C_+ = 1 + \cos\theta_{12}, \quad R_{b1,2} \equiv R_b(q_{1,2}), \quad \cos\theta_{12} = \frac{\vec{q}_1 \vec{q}_2}{q_1 q_2},$$

and σ_i ($i = 1, 2$) satisfies the dispersion relation (see Eq.(5.11))

$$\sigma_i^2 = gq_i \frac{1 - R_{bi} e^{-2q_i H_1}}{1 + R_{bi} e^{-2q_i H_1}}$$

Equation (5.44) extends the relationship we established previously for deep-water[38], to cover environments of any water depth. Substituting Eq.(5.44) into Eq.(5.39) gives an expression for the second-order potential in the reference waveguide. When the effects of the bottom are ignored, i.e., $R_{b1,2} = 0$ (or $H_1 \to \infty$), Eq.(5.45) reduces to the form $(1 - \cos\theta_{12})$ used in [38] and Chap. 4. The additional terms only become important when the water is shallow.

5.2.3 Potential in a Weakly Range-Dependent Wave Guide

In a weakly range-dependent ocean $\Phi_{1b\omega}^{(2)}(\vec{R})$ can still be expressed in the form given in Eq.(5.39), except that the Green's function, Eq.(5.40), is now replaced by

$$G_b^{(2)}(\vec{R}, \vec{R}_0)\Big|_{z_0=0} = 2\int_0^\infty F(\vec{r}, \vec{r}_0, k) \frac{1}{\gamma_+} \left[\frac{e^{\gamma_+(z+H_1)} + R_b e^{\gamma_-(z+H_1)}}{e^{\gamma_+ H_1} + \frac{\epsilon}{b} R_b e^{\gamma_- H_1}} \right]$$

$$\cdot \frac{1}{1 - \frac{\omega^2}{g\gamma_+}} J_0(k\rho) k\, dk \quad (5.46)$$

with $F(\vec{r}, \vec{r}_0, k)$ describing the range dependence. Replacing Eq.(5.46) by

$$G_b^{(2)}(\vec{R}, \vec{R}_0)\Big|_{z_0=0} = \frac{1}{\pi} \int_{\vec{k}} F(\vec{r}, \vec{r}_0, k) \frac{1}{\gamma_+} \left[\frac{e^{\gamma_+(z+H_1)} + R_b e^{\gamma_-(z+H_1)}}{e^{\gamma_+ H_1} + \frac{\epsilon}{b} R_b e^{\gamma_- H_1}} \right]$$

$$\cdot \frac{1}{1 - \frac{\omega^2}{g\gamma_+}} e^{i\vec{k}(\vec{r}-\vec{r}_0)} d\vec{k} \quad (5.47)$$

and altering the sequence of integration in Eq.(5.39), we can express $\Phi_{1b\omega}^{(2)}(\vec{R})$ in the alternative form,

$$\Phi_{1b\omega}^{(2)}(\vec{R}) = \frac{1}{(2\pi)^2} \int_{\vec{k}} S_b^{(2)}(\vec{k}, \omega) \left[\frac{e^{\gamma_+(z+H_1)} + R_b e^{\gamma_-(z+H_1)}}{e^{\gamma_+ H_1} + \frac{\epsilon}{b} R_b e^{\gamma_- H_1}} \right] e^{i\vec{k}\vec{r}} d\vec{k} \quad (5.48)$$

where

$$S_b^{(2)}(\vec{k},\omega) = \frac{-1}{\omega^2 - g\gamma_+} \int_S \bar{S}_b^{(2)}(\omega,\vec{r}_0,0) F(\vec{r},\vec{r}_0,k) e^{-i\vec{k}\vec{r}_0} d\vec{r}_0$$

$$= \frac{i}{(2\pi)^4(\omega^2 - g\gamma_+)} \int\int \sigma_1\sigma_2(\sigma_1+\sigma_2)\zeta^{(1)}(\vec{q}_1)\zeta^{(1)}(\vec{q}_2)\{\vec{q}_1,\vec{q}_2\} \cdot$$
$$\cdot D_S(\vec{q}_1,\vec{q}_2,\vec{k},\vec{r})\delta(\omega-\sigma_1-\sigma_2)\,d\vec{q}_1\,d\vec{q}_2 \qquad (5.49)$$

with $\{\vec{q}_1,\vec{q}_2\}$ again defined by Eq.(5.44) and D_S by:

$$D_S(\vec{q}_1,\vec{q}_2,\vec{k},\vec{r}) = \int_S F(\vec{r},\vec{r}_0,k) e^{-i(\vec{k}-\vec{q}_1-\vec{q}_2)\vec{r}_0} d\vec{r}_0 \qquad (5.50)$$

In Eq.(5.48) the term in [·] describes the plane-wave response of the medium to the surface source distribution, $S_b^{(2)}(\vec{k},\omega)$, associated with the wave-wave interaction process. In the ideal case when $S \to \infty$ and $F(\vec{r},\vec{r}_0,k) = 1$ the function D_S tends to the δ-function $\delta(\vec{k}-\vec{q}_1-\vec{q}_2)$.

Comparison with the plane wave solution

Even though the equivalence between the plane-wave and Green's function analysis (for the second-order field) in the ideal case of an infinite source region and range-independent environmental model was demonstrated in Chap. 4, it would still be comforting to confirm this equivalence for the present case. This can be done by assuming plane-wave expressions for the potential

$$\Phi_{1b\omega}^{(2)}(\vec{R}) = \frac{1}{(2\pi)^2} \int_{\vec{k}} \Phi_{1b}^{(2)}(\vec{k},\omega,z) e^{i\vec{k}\vec{r}} d\vec{k} \qquad (5.51)$$

where

$$\Phi_{1b}^{(2)}(\vec{k},\omega,z) = A_0(e^{\gamma_+(z+H_1)} + R_b e^{\gamma_-(z+H_1)}) \qquad (5.52)$$

and the source term

$$\frac{1}{g}\bar{S}_b^{(2)}(\omega,\vec{r},0) = \frac{1}{(2\pi)^2 g} \int_{\vec{k}} u(\vec{k},\omega) e^{i\vec{k}\vec{r}} d\vec{k} \qquad (5.53)$$

and introducing both into the surface condition, Eq.(5.30). Noting the definitions of the symbols ϵ and b we can, after some manipulation, establish the amplitude A_0 as

$$A_0 = -\frac{u(\vec{k},\omega)}{(\omega^2 - g\gamma_+)(e^{\gamma_+ H_1} + \frac{\epsilon}{b} R_b e^{\gamma_- H_1})} \qquad (5.54)$$

5.2 Second-Order Potential

where

$$u(\vec{k},\omega) = \int \bar{S}_b^{(2)}(\omega,\vec{r},0)e^{-i\vec{k}\vec{r}_0}d\vec{r}_0 \qquad (5.55)$$

Equation (5.51) then becomes identical to Eq.(5.48), as expected, when $F(\vec{r},\vec{r}_0,k) = 1$.

In this chapter we have removed the restrictions of earlier treatments, demonstrated the equivalence of the plane-wave and Green's function analyses and exploited the advantages of describing the wave-induced pressure field in terms of the potential. We are now in a position to examine the properties of the wave-induced pressure field and its associated seismic response. This is done in the following chapters. To define the appropriate integration limits clearly we need first to examine the interaction mechanism in more detail.

Chapter 6

The Generalised Source Spectrum

6.1 Preamble

In the previous chapter expressions were established for the first- and second-order solutions of the potential field associated with ocean-gravity waves, when the assumptions basic to the deep-water analysis are removed. To develop useable relations for the spectral and coherence functions of the related pressure and seismic fields, based on this formalism, it is necessary to be able to calculate the first- and second-order source functions , $\langle |S^{(1)}(\vec{k},\omega)|\rangle^2$ and $\langle |S_b^{(2)}(\vec{k},\omega)|\rangle^2$, as functions of frequency and wave vector. The source function for the first-order field can be readily established from Eq.(5.24) and consideration of it is deferred till later. Discussion in this chapter is focussed on the second-order field and the calculation of Eq.(5.49). To proceed we need a full-wave analysis of the interaction process.

As an acoustic source the wave-wave interaction mechanism differs in a number of significant ways from a distribution of ordinary point sources[50]. First, by virtue of the dispersion relation governing the gravity-wave field, it is environment dependent. Secondly, its source spectral function is not only a function of frequency but also of wavenumber. As we shall see later, the whole horizontal wavenumber plane is divided into three segments by two characteristic wavenumbers, the acoustic wavenumber, $k_{\alpha_1} = \omega/\alpha_1$, and the gravity-wave wavenumber, $k_g = \omega^2/g$. The source components for which $0 \leq k \leq k_{\alpha_1}$ (the homogeneous component of the pressure field) propagate energy to large depths without significant decay; in contrast, those for which $k_{\alpha_1} \leq k \leq k_g$ (the inhomogeneous component) decay exponentially with depth. The components of both groups are generated by sum-frequency interactions - see Sect. 6.3.

A third group for which $k_g < k < \infty$, arises from difference-frequency interactions. We will see that it, too, is depth dependent and is only of significance at frequencies below the band in which the other two are dominant. Thirdly, the relative importance of all components to the total pressure field is dependent not only on frequency, but also on the water depth and the geoacoustic parameters of the seabed structure.

As we have seen consideration of such complexities has been avoided in earlier analyses, by restricting discussion to deep environments. The calculation of the source spectrum has then been restricted to the homogeneous component alone. While the decay with depth of both the inhomogeneous component of the sum-frequency process and the difference-frequency component justifies such an approximation in deep-ocean observations, their neglect could clearly introduce significant errors in the upper levels of the ocean and in shallow-water applications. A comprehensive investigation, which considers all components of the source spectrum, is therefore important to the development of the subject.

An analysis, which considers the general case where the water and observation point can be of any depth, is presented in the sections which follow. Furthermore, whereas only the sum-frequency component of the wave-interaction process has been considered in the past, the investigation here also includes the difference-frequency component of the pressure field - see Sect. 6.3.1. These extensions allow the total source spectrum to be established for any situation and over the whole frequency-wavenumber domain. By way of example, specific calculations of deep- and shallow-water spectra, based on the generalised formalism, are described, but a detailed discussion of the spectral and coherency properties of the pressure and seismic fields in typical ocean environments is deferred to the next chapter.

6.2 Definition of the Source Spectrum (Wavenumber-Frequency)

The total acoustic effect of wave-wave interactions in a given sea state is embodied in the source term, $S_b^{(2)}(\vec{k}, \omega)$, given by Eq.(5.49). In the ideal case, when the source distribution is regarded as stationary and infinite in extent and the geoacoustic model is range independent, the

6.2 Definition of the Source Spectrum

term $D_S(\vec{q}_1, \vec{q}_2, \vec{k}, \vec{r})$ in $S_b^{(2)}(\vec{k}, \omega)$ becomes a δ-function

$$D_S(\vec{q}_1, \vec{q}_2, \vec{k}, \vec{r}) = \delta(\vec{k} - \vec{q}_1 - \vec{q}_2)$$

so that one fold of the integration in Eq.(5.49) can be completed giving

$$S_b^{(2)}(\vec{k}, \omega) = \frac{i}{(2\pi)^2(\omega^2 - g\gamma_+)} \int \sigma_1 \sigma_2 (\sigma_1 + \sigma_2) \zeta^{(1)}(\vec{q}_1) \zeta^{(1)}(\vec{q}_2) \{\vec{q}_1, \vec{q}_2\} \cdot$$
$$\cdot \delta(\omega - \sigma_1 - \sigma_2) \, d\vec{q}_1 \qquad (6.1)$$

with

$$\vec{q}_1 + \vec{q}_2 = \vec{k}$$

The δ-function in Eq.(6.1) can be written as a function of q_1 by changing the variable[60] so that

$$\delta(\omega - \sigma_1 - \sigma_2) = \delta[f(q_1)] = \frac{\delta(q_1 - \bar{q}_1)}{|f'(\bar{q}_1)|} = \frac{\delta(q_1 - \bar{q}_1)}{|\frac{\partial(\sigma_1+\sigma_2)}{\partial q_1}|_{\bar{q}_1}}$$

where the value of $\bar{q}_1$ satisfies the equation

$$f(\bar{q}_1) = \omega - \sigma_1(\bar{q}_1) - \sigma_2(\bar{q}_1) = 0$$

With some manipulation it can be shown that the mean square of $S_b^{(2)}(\vec{k}, \omega)$ then takes the form,

$$\langle |S_b^{(2)}(\vec{k}, \omega)|^2 \rangle = \frac{(2\pi)^6 \omega^2}{d\omega d\vec{k} |\omega^2 - g\gamma_+|^2} \int_0^{2\pi} \sigma_1^2 \sigma_2^2 f_\zeta^{(1)}(\vec{q}_1) f_\zeta^{(1)}(\vec{q}_2) |\{\vec{q}_1, \vec{q}_2\}|^2 \cdot$$
$$\cdot \frac{q_1}{|\frac{\partial(\sigma_1+\sigma_2)}{\partial q_1}|} d\theta_1 \qquad (6.2)$$

In establishing the above use has been made of the orthogonality of the spectral components of the stationary process[57],

$$\langle \zeta^{(1)}(\vec{q}_1) \zeta^{(1)*}(\vec{q}_1') \zeta^{(1)}(\vec{k} - \vec{q}_1) \zeta^{(1)*}(\vec{k} - \vec{q}_1') \rangle =$$

$$\begin{cases} \langle |\zeta^{(1)}(\vec{q}_1)|^2 \rangle \langle |\zeta^{(1)}(\vec{k} - \vec{q}_1)|^2 \rangle & \text{for } \vec{q}_1 = \vec{q}_1' \\ 0 & \text{for } \vec{q}_1 \neq \vec{q}_1' \end{cases}$$

the relations

$$\vec{q}_1 + \vec{q}_2 = \vec{k}, \ d\vec{q}_i = q_i \, dq_i \, d\theta_i, \qquad f_\zeta^{(1)}(\vec{q}_i) \equiv \frac{d\vec{q}_i}{(2\pi)^4} \langle |\zeta^{(1)}(\vec{q}_i)|^2 \rangle,$$

and the expression

$$\frac{1}{2\pi}\int f(\omega)\delta(\omega-\omega_0)d\omega = f(\omega_0).$$

For convenience we can now define a generalised source spectral-density, $\langle|B^{(2)}(\vec{k},\omega)|\rangle$, as

$$\langle|B^{(2)}(\vec{k},\omega)|^2\rangle = \frac{d\omega d\vec{k}}{(2\pi)^6}\langle|S_b^{(2)}(\vec{k},\omega)|^2\rangle \tag{6.3}$$

We recall (see Sect. 4.4.4) that the surface-wave spectrum in Eq.(6.2), $f_\zeta^{(1)}(\vec{q}_i)$, is usually expressed as the product of the frequency spectrum, $F_a(\sigma_i)$, and the angular-distribution function, $H(\theta)$, through the relation

$$\int f_\zeta^{(1)}(\vec{q})\,d\vec{q} = \int\int F_a(\sigma)H(\theta)\,d\sigma\,d\theta \tag{6.4}$$

where $H(\theta)$ is the normalised directional distribution function satisfying

$$\int_0^{2\pi} H(\theta)d\theta = 1$$

By using the differential relation $f_\zeta^{(1)}(\vec{q})d\vec{q} = F_a(\sigma)d\sigma$ we can write

$$\langle|B^{(2)}(\vec{k},\omega)|^2\rangle = \frac{\omega^2}{|\omega^2 - g\gamma_+|^2}\int_0^{2\pi} F_a(\sigma_1)F_a(\sigma_2)H(\theta_1)H(\theta_2)\cdot$$

$$\cdot|\{\vec{q}_1,\vec{q}_2\}|^2\frac{\sigma_1^2\sigma_2^2}{q_2}\left(\frac{\partial\sigma_1}{\partial q_1}\right)\left(\frac{\partial\sigma_2}{\partial q_2}\right)\frac{1}{\left[\frac{\partial(\sigma_1+\sigma_2)}{\partial q_1}\right]}d\theta_1 \tag{6.5}$$

with the conditions

$$\left.\begin{array}{l} \sigma_1 + \sigma_2 = \omega \\[6pt] q_{1x} + q_{2x} = k_x \\[6pt] q_{1y} + q_{2y} = k_y \\[6pt] F(q_{1x},q_{1y},\sigma_1) = 0 \\[6pt] F(q_{2x},q_{2y},\sigma_2) = 0 \\[6pt] \tan^{-1}(q_{1y}/q_{1x}) = \theta_1 \end{array}\right\} \tag{6.6}$$

where $F(q_{ix}, q_{iy}, \sigma_i)$ is the relevant dispersion relation and the term $\{\vec{q}_1, \vec{q}_2\}$ is defined by Eq.(5.45). When $\vec{k}$ and ω are given, the values of $\vec{q}_1$, $\vec{q}_2$, σ_1 and σ_2 can be found for each value of θ_1, by solving the set of equations (6.6). A geometric analysis, which provides insight into the interaction process helpful to the evaluation of Eq.(6.5), is described in the next section. The generalised relation, Eq.(6.5), is in effect an extension of the "exact" deep-water relation developed earlier. This is made clear from the discussion in the next section and a comparison of the spectral expressions for the pressure field given by Eqs.(4.97) and (6.28).

6.3 Calculation of the Source Spectrum (Wavenumber-Frequency)

6.3.1 Deep-Water Case

The cubic equation

The calculation of the generalised source function is somewhat involved. To develop the discussion we again start with the simpler case in which the ocean and observation depth are deep and the effect of the bottom is negligible. In this situation the dispersion relation governing the interacting surface waves takes the well known form, $\sigma_i^2 = gq_i$, and the conditions (6.6) become

$$\left.\begin{aligned} \sigma_1 + \sigma_2 &= \omega \\[4pt] q_{1x} + q_{2x} &= k_x \\[4pt] q_{1y} + q_{2y} &= k_y \\[4pt] \sigma_1^2 &= gq_1 \\[4pt] \sigma_2^2 &= gq_2 \\[4pt] \tan^{-1}(q_{1y}/q_{1x}) &= \theta_1 \end{aligned}\right\} \quad (6.7)$$

Referring to Fig. 6.1, the geometric relation between wave vectors $\vec{q}_1$, $\vec{q}_2$ and $\vec{k}$ can be expressed as

$$q_2^2 = q_1^2 + k^2 - 2kq_1 \cos\theta_{1k} \qquad (6.8)$$

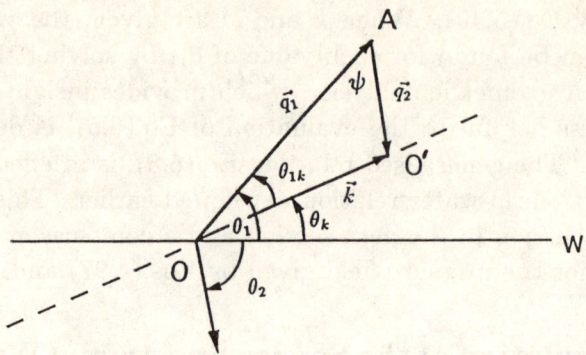

Figure 6.1: Geometric representation of two interacting gravity waves inducing a pressure wave with horizontal wavenumber vector $\vec{k}$ and angular frequency ω. Propagation directions are specified relative to the wind vector OW.

where θ_{1k} is the angle between $\vec{q}_1$ and $\vec{k}$. If the ratio of the frequencies of two interacting wavetrains is then defined as

$$\chi = \frac{\sigma_2}{\sigma_1} \qquad (6.9)$$

it follows from Eqs.(6.7) that

$$\left.\begin{array}{l} \sigma_1 = \omega/(1+\chi) \\[6pt] \sigma_2 = \sigma_1 \chi \\[6pt] q_1 = \omega^2/g^2(1+\chi)^2 \\[6pt] q_2 = q_1 \chi^2 \end{array}\right\} \qquad (6.10)$$

Substituting Eq.(6.10) into (6.8) now leads to the relation

$$\chi^4 = 1 + \frac{k^2}{q_1^2} - \frac{2k\hat{t}}{q_1} \qquad (6.11)$$

where $\hat{t} = \cos\theta_{1k}$, and hence to a cubic equation

$$(\chi^2+1)(\chi-1) - \frac{\hat{m}^2}{4}(1+\chi)^3 + \hat{m}\hat{t}(1+\chi) = 0 \qquad (6.12)$$

6.3 Calculation of the Source Spectrum

where $\hat{m}$ is defined as

$$\hat{m} = \frac{2kg}{\omega^2} \qquad (6.13)$$

Solving Eq.(6.11) allows values of σ_i and q_i to be readily established from Eq.(6.10). The angle, θ_2, of the wave vector $\vec{q}_2$ can be defined (referring to Fig. 6.1) as

$$\theta_2 = \theta_1 - \pi + \psi \qquad \text{for} \quad \theta_k \leq \theta_1 < \theta_k + \pi$$

$$\theta_2 = \theta_1 + \pi - \psi \qquad \text{for} \quad \theta_k - \pi \leq \theta_1 < \theta_k$$

or alternatively as

$$\theta_2 = \theta_1 - (\pi - \psi)\text{sgn}[\sin(\theta_1 - \theta_k)] \qquad (6.14)$$

where $\text{sgn}(x) = x/|x|$. The angles involved lie in the range, $0 \leq |\theta_i| \leq \pi$, with $(i = 1, 2, k)$, and θ_1 and θ_2 are measured from the reference axis, OW, (chosen as the prevailing wind direction) with positive values in the counter-clockwise sense. The angle ψ, the apex angle of the triangle in Fig. 6.1, can be found from the relations

$$q_1^2 + q_2^2 - 2q_1 q_2 \cos\psi = k^2 \quad \text{and} \quad q_1^2 + k^2 - 2q_1 k \cos\theta_{1k} = q_2^2$$

as

$$\cos\psi = \frac{[1 - \hat{m}\hat{t}(1 + \chi)^2]}{\chi^2} \qquad (6.15)$$

or

$$\cos\psi = \frac{1 + \chi^4 - k^2 g^2(1 + \chi)^4}{2\chi^2} \qquad (6.16)$$

The term $\partial q_1/\partial \omega$ involved in the integral of Eq.(6.5) now takes the form[1]

---

[1]
$$\frac{\partial q_1}{\partial(\sigma_1 + \sigma_2)} = \left[\frac{\partial \sigma_1(1+\chi)}{\partial q_1}\right]^{-1} = \left[(1+\chi)\frac{\partial \sigma_1}{\partial q_1} + \sigma_1 \frac{\partial \chi}{\partial q_1}\right]^{-1}$$

$$= \left[\frac{1}{2}(1+\chi)\sqrt{\frac{g}{q_1}} + \frac{\omega}{1+\chi}\frac{\partial \chi}{\partial q_1}\right]^{-1} = \frac{2\omega/g(1+\chi)^2}{1 + [2\omega^2/g(1+\chi)^3]\frac{\partial \chi}{\partial q_1}}$$

$$4\chi^3 \frac{\partial \chi}{\partial q_1} = \frac{2k}{q_1^2}\left(\hat{t} - \frac{k}{q_1}\right)$$

$$\frac{\partial \chi}{\partial q_1} = \frac{k}{2\chi^3 q_1^2}\left(\hat{t} - \frac{k}{q_1}\right)$$

$$\frac{\sigma_1^2 \sigma_2^2}{q_2}\frac{\partial \sigma_1}{\partial q_1}\frac{\partial \sigma_2}{\partial q_2} = \frac{\sigma_1 \sigma_2}{4q_2}g^2 = \frac{g^2 \sigma_2^2}{4\chi q_2} = \frac{g^3}{4\chi}$$

$$\frac{\partial q_1}{\partial(\sigma_1+\sigma_2)} = \frac{2\omega/g(1+\chi)^2}{1+[2\omega^2/g(1+\chi)^3]\frac{\partial\chi}{\partial q_1}}$$

$$= \frac{2\omega/g}{(1+\chi)^2 + [2\omega^2/(1+\chi)](k/2\chi^3 q_1^2)(\hat{t}-k/q_1)}$$

which, by using Eq.(6.12) and the relations

$$\frac{k}{q_1} = \frac{\hat{m}}{2}(1+\chi)^2, \quad \frac{\omega^2}{q_1} = g(1+\chi)^2$$

becomes

$$\frac{\partial q_1}{\partial(\sigma_1+\sigma_2)} = \frac{2\omega\chi^3}{g(1+\chi)^2[\chi^2 - (\frac{1}{2}\hat{m}\hat{t}+1)\chi + (1-\frac{1}{2}\hat{m}\hat{t})]}. \tag{6.17}$$

(Equation (6.17) is identical to the expression developed in the Appendix of [38], the only difference being the replacement of ω by $\sigma_1 + \sigma_2$.)

Since the equation determining the value of χ is cubic, it can have either one real root or three real roots depending on circumstances. The source spectrum is influenced accordingly. A geometric description of these circumstances and their implications regarding the source spectrum were reported in [38]. The essential features of this analysis are reproduced in Appendix M, as reference to this material will be helpful in understanding what follows. In [38], however, consideration was restricted to the sum-frequency components of the interaction process. In response to the discussion in Sect. 5.5, the analysis is extended here to include the difference-frequency component.

Geometric description and the integration limits in the k-plane

The main features of the geometric representation of the wave interaction process given in Appendix M are summarised in Fig. 6.2. In each of Figs. 6.2a to d two sets of curves are shown, curve (1) (the solid line) centered at O and curve (2) (the dashed one) centered at O'. They represent the dispersion relations $\sigma_1^2 = gq_1$ and $\sigma_2^2 = gq_2$ respectively. We assign the value $\sigma_1 = +\sqrt{gq_1}$ to the branch of curve (1) above O and $\sigma_1 = -\sqrt{gq_1}$ to that below O. Conversely, we assign the value $\sigma_2 = \sqrt{gq_2}$ to the branch of curve (2) below O' and $\sigma_2 = -\sqrt{gq_2}$ to that above O'.

6.3 Calculation of the Source Spectrum

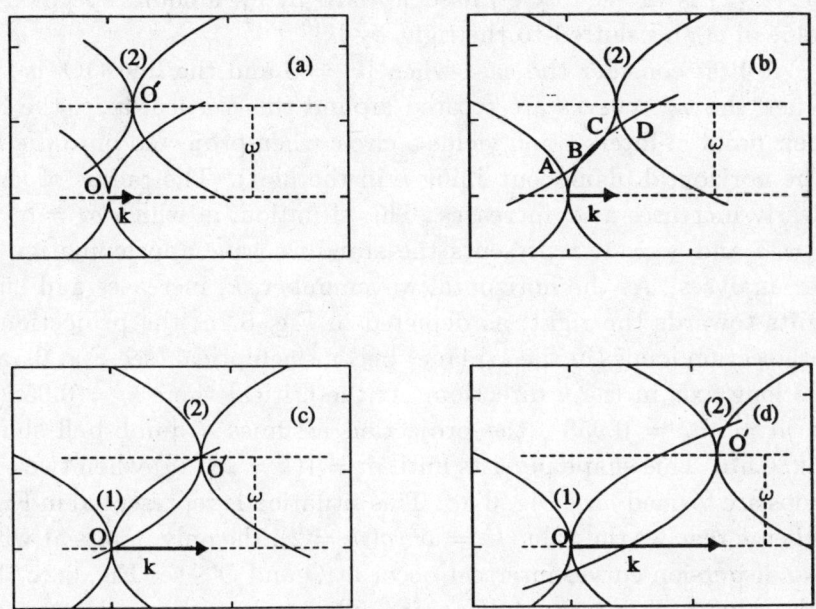

Figure 6.2: Representation of the nonlinear interaction of two surface gravity waves inducing a plane acoustic wave with horizontal wavenumber vector $\vec{k}$: **a** for $0 \leq \hat{m} \leq 0.958$; **b** for $0.958 < \hat{m} \leq 2$; **c** for $\hat{m} = 2$; **d** for $2 < \hat{m} < \infty$.

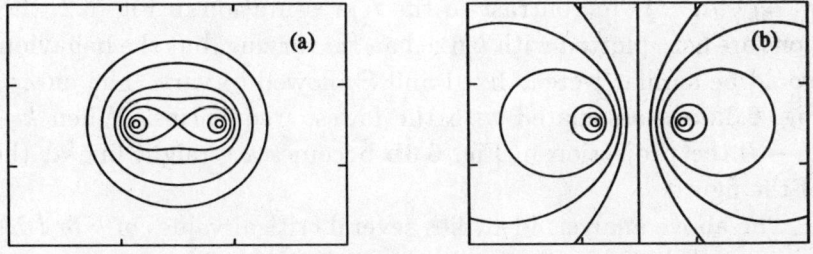

Figure 6.3: Loci of the wave vectors of two interacting gravity waves producing a fixed horizontal wavenumber vector $\vec{k}$ at different frequencies: **a** for the sum-frequency component ($0 \leq \hat{m} \leq 2$); **b** for the difference-frequency component ($2 < \hat{m} < \infty$).

Curve (2) is in each case raised upward by an amount specified by the value of ω and shifted to the right by $|\vec{k}|$.

We first consider the case when $|\vec{k}| = 0$ and the line OO' is vertical. When the two curves are rotated around this vertical axis, the locus of their point of intersection yields a circle when projected onto the $\vec{k}$-plane (the horizontal plane containing $\vec{k}$ in the plot). The radius of the circle clearly increases as ω increases. This situation, in which $\sigma_1 = \sigma_2 = \omega/2$, $\vec{k} = 0$ and $\chi = 1$, represents the standing-wave approximation of earlier analyses. As the horizontal wavenumber, k, increases and curve (2) shifts towards the right, as depicted in Fig. 6.2a, the projection of the intersection locus (in the $\vec{k}$-plane) becomes elliptical (see Fig. 6.5a), with the long axis in the $\vec{k}$ direction. At the critical value $k_c = 0.958(\omega^2/2g)$ or $\hat{m} = \hat{m}_c = 0.958$, the projection assumes a dumb-bell shape -see Fig. 6.5b. This shape persists until $\hat{m} = 1(k = \omega^2/2g)$ when two separate loops are formed - see Fig. 6.5c. This situation is represented in Fig. 6.2b. When k reaches the value $k_g = \omega^2/g(\hat{m} = 2)$, the only points at which the two dispersion curves intersect occur at O and O' - see Fig. 6.2c. Beyond this value of k intersection occurs only in the region where σ_1 and σ_2 differ in sign. All interaction associated with the sum-frequency component ceases and only the difference-frequency interactions are physically relevant - see Fig. 6.2d.

The projections of the intersection curve for the cases $0 \leq k \leq k_g$ and $k_g < k < \infty$ are shown in Figs. 6.3a and b respectively, for η=0.6, 0.65, 0.707, 0.75, 0.8, 0.9 and 1.0 where $\eta = (\sqrt{\sigma_1} + \sqrt{\sigma_2})/\sqrt{\omega} (\equiv (q_1^{1/4} + q_2^{1/4})g^{1/4}/\omega^{1/2})$. In contrast to the representation in Fig. 6.2, the projections are here plotted with k fixed and ω varying, but the behaviour shown would be similar were ω fixed and k allowed to vary. The outer loops of Fig. 6.3a are associated with the lowest frequencies. When $k \to \infty$ or $\omega \to 0$ the projection in Fig. 6.3b becomes a straight line at the centre of the figure.

The above analysis identifies several critical values of k or ($\hat{m}$). These values and their significance to the evaluation of the source spectrum are indicated in Table 6.1 and Fig. 6.4. The second column from the right in Table 6.1 lists the integration limits of θ_{1k} from which the corresponding limits of θ_1 can be established through the relation $\theta_1 = \theta_k + \theta_{1k}$, with θ_k fixed relative to the prevailing wind direction, $\theta_1 = 0$ - see Fig. (6.1). These situations are shown in an exaggerated form in Figs. 6.5a - d.

6.3 Calculation of the Source Spectrum

Table 6.1: The relationship of the horizontal wavenumber to the integration process.

Range of k	Pressure field	Interaction type	Roots of χ	Range of θ_{1k}	Range of $\hat{m}$
$0 \leq k \leq k_{\alpha_1}$	Homogeneous wave region		Single real root	$0 \leq |\theta_{1k}| \leq \pi$	$0 \leq \hat{m} \leq \frac{2g}{\omega\alpha_1}$
$k_{\alpha_1} < k \leq k_c$					$\frac{2g}{\omega\alpha_1} < \hat{m} \leq 0.958$
$k_c < k \leq k'_g$		Sum-frequency interaction		$0 \leq |\theta^{(1)}_{1k}| \leq \pi$ $\theta_{c1} \leq |\theta^{(2)}_{1k}| \leq \theta_{c2}$ $\theta_{c1} \leq |\theta^{(3)}_{1k}| \leq \theta_{c2}$	$0.958 < \hat{m} \leq 1.01$
$k'_g < k \leq k_g$	Inhomogeneous wave region		Three real roots	$0 \leq |\theta^{(1)}_{1k}| \leq 2\pi$	$1.01 < \hat{m} \leq \hat{m}_g$
$k_g < k < \infty$		Difference-frequency interaction		$0 \leq |\theta^{(2)}_{1k}| \leq \theta_c$ $0 \leq |\theta^{(3)}_{1k}| \leq \theta_c$	$\hat{m}_g < \hat{m} < \infty$

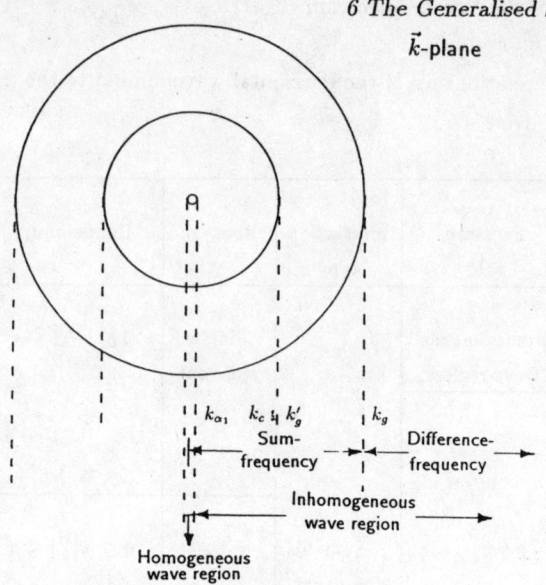

Figure 6.4: Representation of the interaction process.

Sum-frequency source component

As was described above, there can be either one or three values of $\chi^{(i)}$ for a given angle, θ_{1k}, angular frequency ω and wavenumber k. These values of $\chi^{(i)}$ ($i = 1, 2, 3$) are defined by Eq.(6.12) - see Appendix M. The general form of the source spectrum can then be expressed as (M.7)

$$\langle |B^{(2)}(\vec{k},\omega)|^2 \rangle = \frac{g^2}{2\omega} \frac{1}{V(k,\omega)} \sum_{i=1}^{3} \int_{\theta_{di}}^{\theta_{ui}} \frac{(1-\cos\theta_{12})^2 \chi_i^2}{(1+\chi_i)^2 Q(\chi_i)} F_a\left(\frac{\omega}{1+\chi_i}\right) F_a\left(\frac{\omega\chi}{1+\chi_i}\right) \cdot$$

$$\cdot H(\theta_1) H(\theta_2) d\theta_1 \qquad (6.18)$$

where

$$Q(\chi_i) = \chi_i^2 - \left(\frac{1}{2}\hat{m}\hat{t} + 1\right)\chi_i + \left(1 - \frac{1}{2}\hat{m}\hat{t}\right),$$

$V(k,\omega) = \left|1 - g\gamma_+/\omega^2\right|^2$ and $\gamma_+ = -a + \sqrt{a^2 - (\omega/\alpha_1)^2 + k^2}$, i.e.,

$$V(k,\omega) = \begin{cases} 1 - \frac{\hat{m}^2}{4} & k < \sqrt{\frac{\omega^2}{\alpha_1^2} - a^2} \\ \left[1 - \frac{\hat{m}}{2}\left(\sqrt{1 - \omega^2/\alpha_1^2 k^2 + a^2/k^2} + a/k\right)\right]^2 & k > \sqrt{\frac{\omega^2}{\alpha_1^2} - a^2} \end{cases}$$

$$(6.19)$$

6.3 Calculation of the Source Spectrum

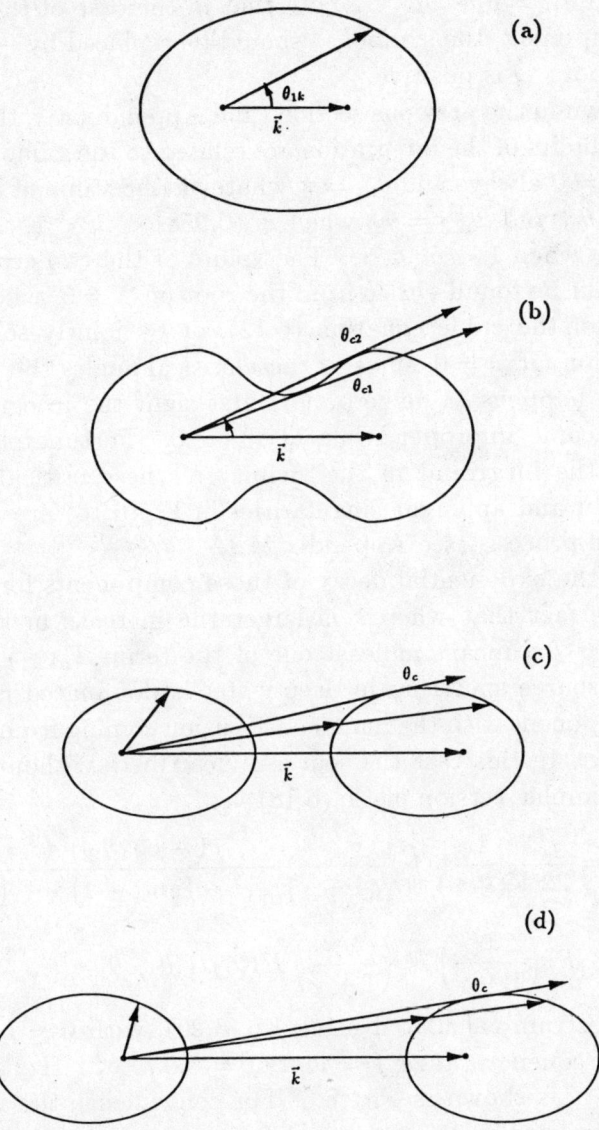

Figure 6.5: Schematic representation of the variation of the locus of the wave interaction vector with increasing wavenumber k: **a** for $0 \leq \hat{m} \leq 0.958$; **b** for $0.958 < \hat{m} \leq 1$; **c** for $1 < \hat{m} \leq 2$; **d** for $2 < \hat{m} < \infty$.

$a = g/2\alpha_1^2$ and $\theta_{1k} = \theta_1 - \theta_k$. We note that in the case of the difference-frequency component, the symbol χ_i should be replaced by $-\chi_i$ since the value of the roots χ_i is positive.

As was shown in the previous section (and Appendix M), the upper (u) and lower (d) limits of the integration are related to the value of k (or $\hat{m}$). For $i = 1$, $\theta_{d1} = 0$ always and $\theta_{u1} = \pi$ whatever the value of $\hat{m}$; for $i = 2$ and 3, $\theta_{d2,3} = \theta_{c1}$ and $\theta_{u2,3} = \theta_{c2}$ when $\hat{m}_c(0.958) < \hat{m} < 1$, and $\theta_{d2,3} = 0$ and $\theta_{u2,3} = \theta_{c2}$ when $1 < \hat{m} < \infty$. The values of the two critical angles, θ_{c1} and θ_{c2}, can be found either from the root of $\Delta = 0$, where Δ is the discriminator of the cubic equation (6.12), or by jointly solving (6.12) and the equation $Q(\chi) = 0$, since at these critical angles the value of the function $Q(\chi)$ happens to be zero. (At first sight the integral in (6.18) appears to become improper when $Q(\chi) = 0$. Fortunately, however, the values of the integrand in the vicinity of these critical angles are different in sign and apparent singularities in Eq.(6.18) are removed by the summation process - see Appendix M.)

Because of the exponential decay of those components for which $k > \omega/\alpha_1$ (and the fact that when k is larger the increase in the value of the ratio $\chi = \sigma_2/\sigma_1$ means at least one of the terms $F_a(\sigma_1)$ and $F_a(\sigma_2)$ is small), the source spectrum in deep-water is dominated by the sum-frequency component with the main contribution coming from the branch $\chi^{(1)}$ with $k \leq k_c$. In this case the source is described, without significant error, by the simpler version of Eq.(6.18) viz.

$$\langle |B^{(2)}(\vec{k},\omega)|^2 \rangle = \frac{g^2}{2\omega} \frac{1}{V(k,\omega)} \int_0^{2\pi} \frac{(1-\cos\theta_{12})^2 \chi^2}{(1+\chi)^2 \left[\chi^2 - \left(\frac{1}{2}\hat{m}\hat{t}+1\right)\chi + \left(1-\frac{1}{2}\hat{m}\hat{t}\right)\right]}$$

$$\cdot F_a\left(\frac{\omega}{1+\chi}\right) F_a\left(\frac{\omega\chi}{1+\chi}\right) H(\theta_1) H(\theta_2) d\theta_1 \qquad (6.20)$$

The source spectrum calculated using Eq.(6.20), with $0 \leq k \leq \hat{m}$ ($0 \leq \hat{m} \leq \omega^2/2g$), frequency range $f(=\omega/2\pi) = 0.05$ to 1 Hz, and a wind speed of $15 \mathrm{ms}^{-1}$, is shown in Fig. 6.6. For convenience the variables $\hat{m}$ and f have been used in place of k and ω. The figure is plotted in terms of $10 log_{10}(\langle |B^{(2)}(\hat{m},f)|^2 \rangle)$. It is of interest to observe the characteristics of the whole spectrum in the $(\hat{m}, f)$ plane, even though in practice only that part associated with small $\hat{m}$ is important. The function is smooth but clearly not constant.

The self consistency of the extended analysis is demonstrated by now introducing the additional approximations relevant to the standing-wave

6.3 Calculation of the Source Spectrum

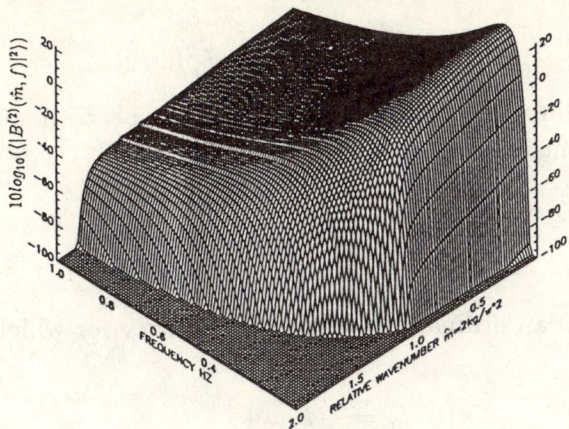

Figure 6.6: The sum-frequency source spectrum, $10\log_{10}(\langle|B^{(2)}(f,\hat{m})|^2\rangle)$, presented as a function of relative wavenumber $\hat{m}$ and frequency f for a wind speed of 15 m^{-1}.

analysis of Chap. 4, $\chi = 1(\hat{m} = 0)$ and $\theta_{12} = \pi$. In this case Eq.(6.20) simplifies further to

$$\langle|B^{(2)}(\vec{k},\omega)|^2\rangle = \frac{g^2}{2\omega}F_a^2(\omega/2)\int_0^{2\pi} H(\theta_1)H(\theta_1+\pi)d\theta_1$$

which after integration upon $\vec{k}$ leads to the familiar spectral expression

$$F_p^2(\omega) = \frac{\pi\rho^2 g^2 \omega^3}{2\alpha_1}F_a^2(\omega/2)I_\omega$$

where $I_\omega = \int_0^{2\pi} H(\theta_1)H(\theta_1+\pi)d\theta_1$.

This is the form of the spectrum established in the earlier analyses[25],[38] and developed in Chap. 4 - see Eq.(4.107). It is to be noted that this classical formula, as presented in the literature, often differs by a multiplying constant. This inconsistency is due to differences in the time-space transformation used. This fact has motivated us to define very clearly the Fourier transform used in this book.

As an aside it is also worth emphasising here that some confusion exists in the literature regarding the multiplication factor, 2π, relating the power-density spectrum of the second-order pressure field expressed in terms of angular frequency, ω, and frequency f. To clarify this point consider the source spectrum $\langle|B^{(2)}(\vec{k},\omega)|^2\rangle$. Under the standing-wave approximation it takes the form

$$\langle |B^{(2)}(\vec{k},\omega)|^2 \rangle = A(\vec{k},\omega) \cdot F_a^2(\omega/2) \tag{6.21}$$

with $\vec{k} = 0$. Since $F_a(\omega/2)$ (the power-density spectrum of the surface-wave field) satisfies the relation

$$F_a\left(\frac{f}{2}\right) = 2\pi F_a(\omega/2)\Big|_{\omega=2\pi f} \tag{6.22}$$

and $A(\vec{k},\omega)$ is an ordinary function of frequency for which

$$A(\vec{k},f) = A(\vec{k},\omega)\Big|_{\omega=2\pi f}$$

it is often concluded that

$$\begin{aligned}\langle |B^{(2)}(\vec{k},f)|^2 \rangle &= (2\pi)^2 A(\vec{k},f) \cdot F_a^2(f/2) \\ &= (2\pi)^2 \langle |B^{(2)}(\vec{k},\omega)|^2 \rangle|_{\omega=2\pi f}\end{aligned} \tag{6.23}$$

However, this expression contradicts the fact that $\langle |B^{(2)}(\vec{k},\omega)|^2 \rangle$ and $\langle |B^{(2)}(\vec{k},f)|^2 \rangle$ are themselves both power-density spectral functions. The correct relation between them is given by

$$\begin{aligned}\langle |B^{(2)}(\vec{k},f)|^2 \rangle &= (2\pi)\langle |B^{(2)}(\vec{k},\omega)|^2 \rangle|_{\omega=2\pi f} \\ &= 2\pi A(\vec{k},f) \cdot F_a^2(f/2)\end{aligned} \tag{6.24}$$

The same argument applies to other spectral functions of the second-order field, such as the pressure frequency spectrum, $F_p^{(2)}(\omega)$, and the seismic spectra $F_{v,h}^{(2)}(\omega)$.

Difference-frequency component

The previous section was concerned with the sum-frequency component of the pressure field, which results when the frequencies of the two interacting wave components, σ_1 and σ_2, are assumed to be both positive or both negative, so that in the interaction equations

$$\left.\begin{aligned}\omega &= \sigma_1 + \sigma_2 \\ \vec{k} &= \vec{q}_1 + \vec{q}_2\end{aligned}\right\}$$

6.3 Calculation of the Source Spectrum

either

$$\sigma_1 \geq 0 \quad \text{and} \quad \sigma_2 \geq 0 \qquad (6.25)$$

or

$$\sigma_1 < 0 \quad \text{and} \quad \sigma_2 < 0 \qquad (6.26)$$

(Because of the symmetry involved, the contribution from the interaction of two negative-frequency components has been taken into account by the multiplication factor 2 in the calculation of the spectrum). For completeness we now examine the contribution of the difference-frequency components, for which σ_1 and σ_2 are opposite in sign. Without loss of generality we specify $\sigma_1 \geq 0$ and $\sigma_2 < 0$.

For simplicity we again elect to discuss this component under the assumptions of the deep-water approximation. The geometric description of the equation $\omega = \sigma_1 + \sigma_2$, i.e., $\omega = \sigma_1 - |\sigma_2|$, for this case is shown in Fig. 6.2d. We recall that for a given frequency, ω, and horizontal wavenumber, $k(= |\vec{k}|)$, the dispersion relations, $\sigma_1^2 = gq_1$ and $\sigma_2^2 = gq_2$, describing the two sets of interacting waves, are shown in Figs. 6.2c and d by curve (1) centered at the point O and curve(2) at O'. We also recall that the difference-frequency interaction process only begins to contribute to the pressure field when the horizontal wavenumber, k, (represented by the distance between O and O') exceeds the value of ω^2/g, the point at which the sum-frequency interaction ceases - see Fig. 6.2d. Beyond this value of k the projection of the intersection of the two dispersion surfaces takes the form of the two loops shown in Fig. 6.2d and Fig. 6.3b. The size of the loops increases with increasing k when ω is fixed, or with decreasing ω when k is fixed. For convenience, in Fig. 6.3 the loops are plotted with k fixed. As is shown later in Fig. 6.10, the form of the loops changes with water depth because the dispersion relation is depth dependent.

The source expression, $\langle |B^{(2)}(\vec{k}, \omega)|^2 \rangle$, can still be calculated using Eq.(6.18), but with three real roots now involved it takes the form,

$$\langle |B^{(2)}(\vec{k},\omega)|^2\rangle = \frac{g^2}{2\omega V(k,\omega)} \sum_{i=1}^{3} \int_0^{\theta_{ci}} \frac{(1-\cos\theta_{12})^2 \chi_i^2}{(1+\chi_i)^2 Q(\chi_i)} F_a\left(\frac{\omega}{1+\chi_i}\right) F_a\left(\frac{\omega}{1+\chi_i}\right) \cdot$$

$$\cdot H(\theta_1) H(\theta_2) d\theta_1 \qquad (6.27)$$

Since $\theta_1 = \theta_{1k} + \theta_k$, θ_{1k} can be used as the variable for each given θ_k. From the geometry of Figs. 6.1 and 6.5a and the symmetry involved, it is clear that the integration for the first branch, $\chi^{(1)}$, should be carried out from $\theta_{1k} = 0$ to $\theta_{1k} = \pi$, while for the second and third branches the limits are $\theta_{1k} = 0$ to $\theta_{1k} = \theta_c$, where θ_c is the angle at which the branches $\chi^{(2)}$ and $\chi^{(3)}$ meet - see Fig. 6.5d. We recall that no singularity occurs when $Q(\chi) = 0$. Difference-frequency spectra calculated on this basis are presented in Fig. 6.7 and discussed in the next section.

Total second-order source spectrum

Theoretical source spectra, which include both the sum- and difference-frequency components of the induced pressure field, are presented in Fig. 6.7. The wind speed is again 15ms^{-1} and spectra are shown for the mean surface and at a depth of 50 m. For convenience both the $\hat{m}$ and f axes have been plotted in a logarithmic scale.

As expected, the sum-frequency component occupies the region of high frequencies and $\hat{m}$ less than $\hat{m}_g = 2$, while the difference-frequency component is associated with the region of very low frequencies and $\hat{m} > \hat{m}_g$. It is of interest to note that, while the inhomogeneous component of the sum-frequency spectrum decays rapidly at high values of k, the difference-frequency component remains significant in this region. This difference in behaviour arises mainly because the two interacting frequencies, $|\sigma_1|$ and $|\sigma_2|$, are much closer in the latter case, so that the product $F_a^2(\sigma_1)F_a^2(\sigma_2)$ is much greater than it is for the sum-frequency component.

Composite ULF pressure spectrum

To establish some appreciation for the relative levels of the wave-induced components of the pressure field, Fig. 6.8 presents the primary frequency (PF) spectrum and the two second-order (DF) spectra for an environment in which an ocean 1000 m deep overlies the seabed structure

6.3 Calculation of the Source Spectrum

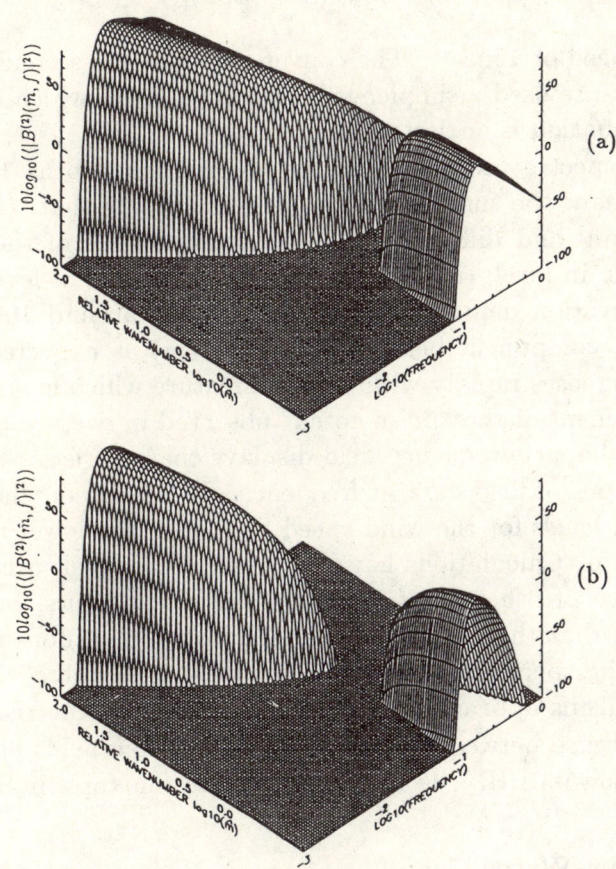

Figure 6.7: The source spectral level of the sum-frequency and difference-frequency components, $10\log_{10}(\langle|B^{(2)}(f,\hat{m})|^2\rangle)$, presented as a function of $\log_{10}(f)$ and $\log_{10}(\hat{m})$: **a** at the mean surface; **b** at depth 50 m, for a wind speed of 15 ms^{-1}

MDL1, described in detail in Chap. 7. The calculations of the second-order spectra were based on the relation

$$f_{p0}^{(2)}(\vec{k},\omega) = \rho_1^2 \omega^2 \langle |B^{(2)}(\vec{k},\omega)|^2 \rangle \tag{6.28}$$

and a wind speed of 15ms^{-1}. The conversion from the second-order potential to pressure used a simpler form of Eq.(3.34) in which the second term of the equation is neglected.

The three spectra are presented for observation depths $10, 100, 500$ and 1000 m below the surface. The plots demonstrate that: (i) near the surface the sum- and difference-frequency components of the DF field are comparable in level; (ii) both components decrease in level with increasing observation depth but stabilise in value around 1000 m, the sum-frequency component being the greater; (iii) as expected, the PF component decreases rapidly with depth, a feature which is fundamental to the development of the "noise-notch" observed in deep-water spectra [40]; (iv) the difference-frequency field displays characteristics which mirror the properties of field data at frequencies below the notch[33],[40], but the theoretical levels for the wind speed of 15ms^{-1} are lower than those reported to date. Calculations have shown that the levels are enhanced by the presence of other low-frequency swell components, but present indications are that the difference-frequency component does not, on its own, explain the spectral behaviour below the "noise-notch". Other reported mechanisms appear more likely to account for the rise in levels and high coherence between the pressure and seismic fields observed at frequencies below 0.01 Hz[20],[33],[40]. We return to this topic in Sect. 7.4.3.

6.3.2 Shallow-Water Case

The general dispersion relation

In this section we consider the source spectrum in shallow-water environments. The acoustic effects of the wave-wave interaction process are influenced by the water depth in two ways; (i) it changes the form of the dispersion relation governing the ocean-wave field and (ii) it changes the conditions under which the induced acoustic field propagates. In the present section we are only concerned with the first effect. As we discussed in Chap. 5 this is embodied in the general form of the dispersion

6.3 Calculation of the Source Spectrum

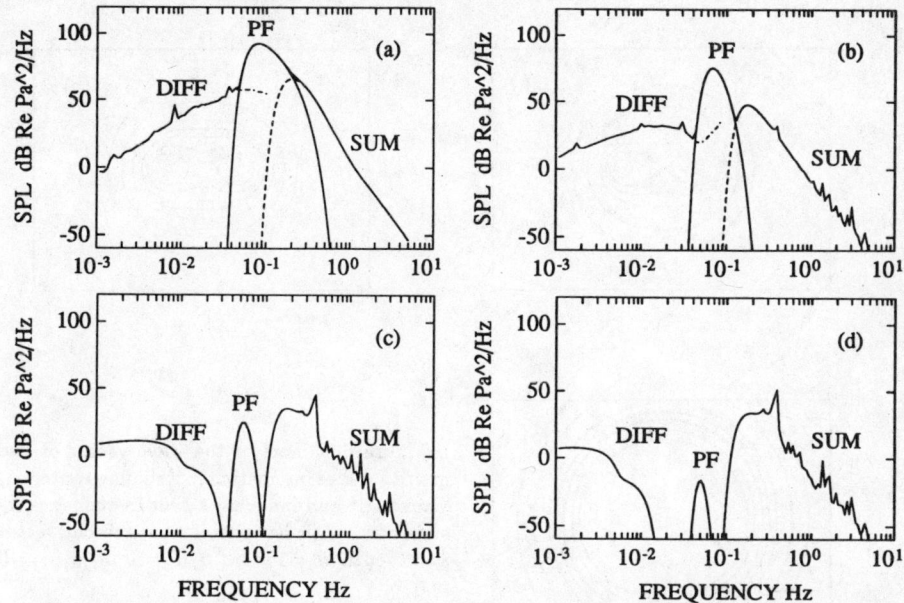

Figure 6.8: The spectral level of the total pressure field at different depths in a gravitationally deep ocean (H_1=1000 m) overlying a multilayered seabed (MDL1), under a prevailing wind of 15 ms^{-1}, and a PM sea : **a** 15 m; **b** 100 m; **c** 500 m; **d** 1000 m.

relation governing an ocean-wave train , (Eq.(5.10))

$$\omega^2 = g\gamma_+ \frac{1 + \epsilon R_b e^{(\gamma_- - \gamma_+)H_1}}{1 + R_b e^{(\gamma_- - \gamma_+)H_1}} \tag{6.29}$$

where

$$\gamma_\pm = a \mp \sqrt{a^2 - \gamma_1^2}, \qquad a \equiv \frac{g}{2\alpha_1^2}, \qquad \epsilon = \frac{\gamma_-}{\gamma_+}, \qquad \gamma_1^2 = \frac{\omega^2}{\alpha_1^2} - k^2, \tag{6.30}$$

H_1 being the water depth and $R_b(\omega, k)$ the reflection coefficient from the seafloor. When the approximations $R_b = 1$ and $a \ll \gamma_1$ apply, Eq.(6.29) reduces to

$$\omega^2 = gk \tanh(kH_1) \tag{6.31}$$

Geometric description - integration limits

Calculations similar to those leading to Fig. 6.3 can be carried out using the general dispersion relation, Eq.(6.31). Projections for the sum-frequency ($0 \leq k \leq \omega^2/g$) and difference-frequency components ($\omega^2/g \leq k$) in three different water depths are shown in Fig. 6.9 and Fig. 6.10

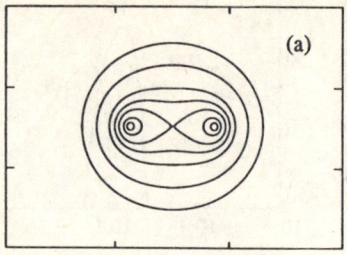
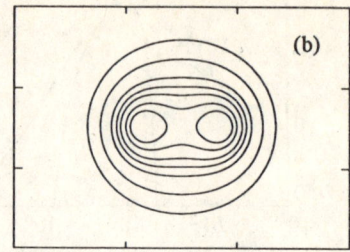
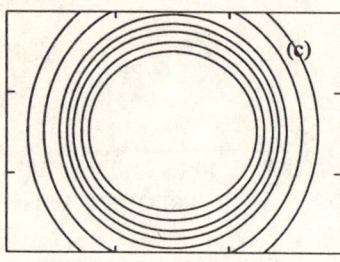

Figure 6.9: Loci of the wave vector of two gravity waves interacting in a shallow-water environment and inducing a sum-frequency pressure wave with horizontal wavenumber vector $\vec{k}$: **a** depth H_1=100 m; **b** H_1=50 m; **c** H_1=10 m.

respectively, for the same values of η used in Fig.6.3. The effect of water depth on the form of the projection is shown most clearly in Fig. 6.9. As an example consider the projection for which $\eta = 0.707$. In Fig. 6.3 and Fig. 6.9a this curve (the 3rd from the inside) has the form of the figure-eight. In Figs. 6.9b and c this has changed to a single loop, the dimension of which increases markedly as the water depth decreases. This behaviour, which is repeated for all values of η, simply reflects the change of the dispersion relation with water depth. In shallow water it has a cone-like form ($\sigma = \sqrt{gH_1}q$) rather than the horn-like form ($\sigma = \sqrt{gq}$) of the deep-water situation. The increase in the size of the projections results from the decrease in wave velocity with water depth. Figure 6.10 shows that the influence of water depth is similar in the case of the difference-frequency component, but because of the lower frequencies involved the changes are not so marked.

Generalised source spectrum

We now turn to the calculation of the source spectrum in shallow water, using the spectral-density function given by Eq.(6.5). The term

6.3 Calculation of the Source Spectrum

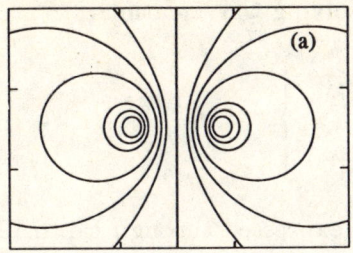

Figure 6.10: Loci of the wave vector of two gravity waves interacting in a shallow-water environment and inducing a difference-frequency pressure wave with horizontal wavenumber vector $\vec{k}$: **a** depth $H_1=100$ m; **b** $H_1=50$ m; **c** $H_1=20$ m.

$\{\vec{q}_1, \vec{q}_2\}$ is again given by Eq.(5.45) but the derivatives $\partial\sigma_i/\partial q_i$ now have to be calculated numerically using the general dispersion relation, $\sigma_i^2 = gq_i \tanh(q_i h_1)$. Again, in contrast to the deep-water case, the term $\partial(\sigma_1 + \sigma_2)/\partial q_1$ is calculated as follows. Since $\sigma_2 = \sigma_1 \chi$

$$\frac{\partial q_1}{\partial(\sigma_1 + \sigma_2)} = \frac{1}{(1+\chi)\frac{\partial\sigma_1}{\partial q_1} + \frac{\partial\chi}{\partial q_1}\sigma_1} \qquad (6.32)$$

where the derivative $\partial\chi/\partial q_1$ has again to be derived from the interaction relations of Eq.(6.6). By defining $\lambda^2(\chi) = q_2/q_1$, Eqs.(6.8) and (6.11) lead to

$$\lambda^4(\chi) = 1 - \frac{2k}{q_1}\left(\hat{t} - \frac{k}{2q_1}\right) \qquad (6.33)$$

and

$$\frac{\partial\chi}{\partial q_1} = \frac{1}{2\lambda^3\lambda'}\left[\frac{k}{q_1^2}\left(\hat{t} - \frac{k}{q_1}\right)\right] \qquad (6.34)$$

where $\lambda' = d\lambda/d\chi$. Values of λ and λ' as functions of χ can be established numerically for a given frequency, by solving the equations

$$\left.\begin{array}{l} q_1 = q_1\left(\frac{\omega}{1+\chi}\right) \\ q_2 = q_2\left(\frac{\omega\chi}{1+\chi}\right) \end{array}\right\} \quad (6.35)$$

Finally the term $|\omega^2 - g\gamma_+|^2$ can be expressed through Eq.(6.19) as

$$|\omega^2 - g\gamma_+|^2 = \omega^4 V(k,\omega) \quad (6.36)$$

while the angle ψ can be readily determined from the geometry of Fig. 6.1,

$$\cos\psi = \frac{q_1}{q_2}\left(1 - \frac{\omega^2 \hat{m}\hat{t}}{2gq_1}\right) \quad (6.37)$$

The relationship between θ_2 and θ_1 remains the same as for the deep-ocean case (see Eq.(6.14).

Examples of shallow-water spectra, based on Eq.(6.5) and calculated using these procedures, are presented in Chap. 7. In the interests of simplicity, however, calculations are restricted to the sum-frequency interactions only.

6.4 Second-order Source Spectra (Frequency)

As indicated above, we defer to Chap.7 a detailed discussion of the pressure field based on the source spectra described in this chapter. It will be helpful to that analysis, however, to present at this juncture selected pressure (frequency) spectra based on simplified versions of the frequency-wavenumber source spectra given above. We restrict consideration to the second-order field, consider only the sum-frequency interactions, but incorporate both the homogeneous and inhomogeneous components. An ocean of infinite depth (bottomless) is assumed.

Figures 6.11a and b present, respectively, such pressure spectra for PM and JONSWAP seas under a wind speed of 15 ms^{-1}. Spectra are shown for observation depths 10, 50, 100, 500, 1000 m and at infinity (the classical spectrum). The 10 m spectrum is in effect the source spectrum for that seastate. The other spectra show the decrease of the inhomogeneous component with depth. Figures 6.11c and d, on the other hand, present the (source) frequency spectrum at 10 m depth as a function of wind

6.4 Source Spectra (Frequency)

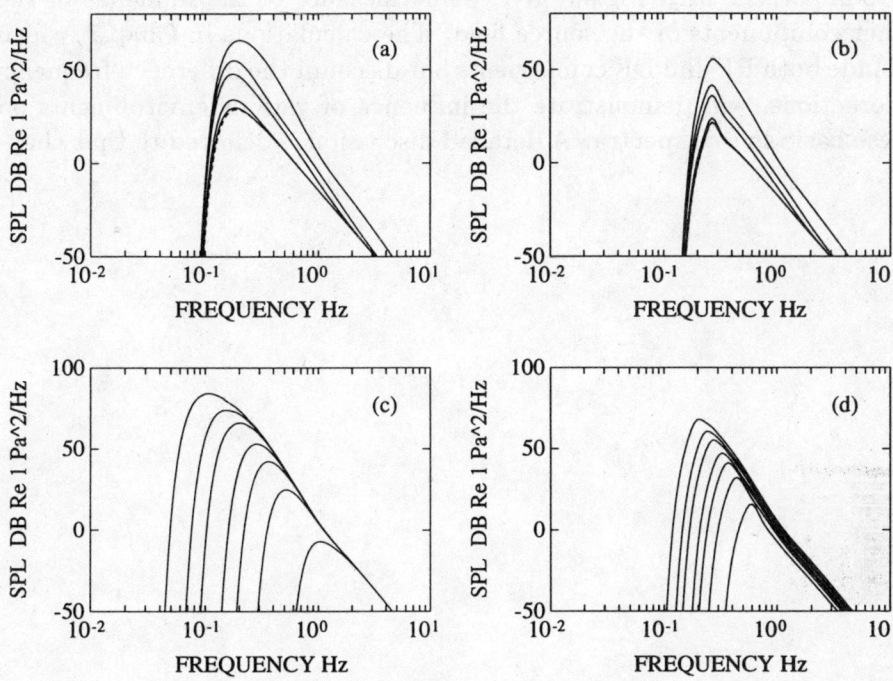

Figure 6.11: Source spectra (frequency) of the second-order pressure field (sum-frequency component); **a** and **b** the sum of the homogeneous and inhomogeneous fields at observation depths z=10, 50, 100, 500, 1000 m and at infinity, for PM and JONSWAP seas respectively and a wind speed of 15 ms^{-1}; **c** and **d** the source spectra at 10 m observation depth, for PM and JONSWAP seas and wind speeds 2.5, 5, 7.5, 10, 15, 20, and 30 ms^{-1}.

speed for the two standard sea states. Wind speeds range from 2.5 to 30 ms^{-1}.

Comparison with Fig.6.8 gives some measure of the influence of the other components of the source field. The calculations in Chap.7, which include both PF and DF components but discount the difference-frequency interactions, will demonstrate the influence of various environments on these basic source spectra. A detailed discussion is deferred to that chapter.

Chapter 7

Properties of the Wave-Induced Pressure Field

In the previous chapters the earlier theory describing wave interactions as an acoustic source has been extended in a full-wave analysis, by removing the deep-water assumption of earlier treatments, including the difference-frequency interactions, and describing the source spectrum over the whole wavenumber plane . In this chapter we extend the analysis further and consider the spectral properties of the pressure field . Coherence functions for the first- and second-order fields are developed first. As a prelude to calculations we next consider the properties of the ocean-wave field and the geoacoustic environment. Theoretical pressure spectra are then established as a function of wind speed and observation depth in a number of typical environments . The calculations are based on the generalised source spectrum but, except where stated otherwise, are restricted to the sum-frequency interactions only - see Eq.(6.18). The spectra are compared with one another and with experimental spectra recorded under comparable conditions. Consideration of the coherence properties of the pressure field completes the discussion. An equivalent analysis of the seismic response is presented in Chap. 8.

7.1 Coherence Function

7.1.1 First-Order Field

The coherence function of the first-order pressure field takes the form[57]

$$C_p^{(1)}(\vec{R}_1, \vec{R}_2, \omega) = \frac{d\omega}{(2\pi)^2} \langle p_\omega^{(1)}(\vec{R}_1) \, p_\omega^{*(1)}(\vec{R}_2) \rangle \tag{7.1}$$

Assuming an active region of infinite size, a range-independent environment and using the pressure field based on Eqs.(5.25) and (5.26), the coherence function of the first-order pressure field can be established as (see Eq.(N.11) of Appendix N)

$$C_p^{(1)}(\vec{R}_1, \vec{R}_2, \omega) = \rho_1^2 g^2 T_p^{(1)}(k, z_1) T_p^{(1)*}(k, z_2) F_a(\omega) \int_0^{2\pi} H(\theta_k) e^{ik\rho \cos(\theta_k - \theta_\rho)} d\theta_k \tag{7.2}$$

where

$$\rho e^{i\theta_\rho} \equiv \vec{r}_1 - \vec{r}_2, \vec{R}_1 = (\vec{r}_1, z_1), \vec{R}_2 = (\vec{r}_2, z_2),$$

and

$$T_p^{(1)}(k, z) = e^{\gamma_+ z} \frac{1 + R_b e^{(\gamma_- - \gamma_+)(z + H_1)}}{1 + R_b e^{(\gamma_- - \gamma_+) H_1}} \tag{7.3}$$

We recall that in the above

$$\gamma_\pm = a \mp \sqrt{a^2 - \gamma_1^2}, \qquad a \equiv g/2\alpha_1^2, \qquad \gamma_1^2 = \omega^2/\alpha_1^2 - k^2,$$

R_b is the reflection coefficient for the seabed, and $F_a(\omega)$ and $H(\theta_k)$ are respectively the frequency spectral-density function and the normalised angular-distribution function of the ocean surface-wave field.

From Eq.(7.2) and Appendix N the spectral-density relation for the first-order pressure field can be established as,

$$F_p^{(1)}(\omega, z) = \rho_1^2 g^2 |T_p^{(1)}(k, z)|^2 F_a(\omega) \tag{7.4}$$

where $k = k(\omega)$. When sea water is regarded as incompressible and the ocean infinitely deep (ie., $\alpha_1 \to \infty$, $H_1 \to \infty$, $R_b = 0$), Eq.(7.4) simplifies to

$$F_p^{(1)}(\omega, z) = \rho_1^2 g^2 F_a(\omega) e^{-2\frac{\omega^2}{g}|z|} \tag{7.5}$$

which we recognise as the well known, deep-water pressure spectrum associated with an ocean-wave field, $F_a(\omega)$.

7.1.2 Second-Order Field

General form

The coherence function of the second-order field takes the form (see Appendix N)

$$C_p^{(2)}(\vec{R}_1, \vec{R}_2, \omega) = \frac{d\omega}{(2\pi)^2} \langle p_{1\omega}^{(2)}(\vec{R}_1) p_{1\omega}^{(2)*}(\vec{R}_2) \rangle$$

$$= \frac{d\omega \omega^2 \rho_1^2}{(2\pi)^6} \int \int \langle S_b^{(2)}(\vec{k}, \omega) S_b^{(2)*}(\vec{k}', \omega) \rangle \cdot$$
$$\cdot T_p^{(2)}(k, z_1) T_p^{(2)*}(k', z_2) e^{i(\vec{k}\vec{r}_1 - \vec{k}'\vec{r}_2)} d\vec{k}\, d\vec{k}' \quad (7.6)$$

By invoking Eq.(6.1), Eq.(7.6) becomes (see Appendix N)

$$C_p^{(2)}(\vec{R}_1, \vec{R}_2, \omega) = d\omega d\vec{k} \frac{\omega^2 \rho_1^2}{(2\pi)^6} \int \langle |S_b^{(2)}(\vec{k}, \omega)|^2 \rangle T_p^{(2)}(k, z_1) T_p^{(2)*}(k, z_2) e^{i\vec{k}\vec{\rho}} d\vec{k} \quad (7.7)$$

or alternatively (N.22)

$$C_p^{(2)}(\vec{R}_1, \vec{R}_2, \omega) = \omega^2 \rho_1^2 \int \langle |B^{(2)}(\vec{k}, \omega)|^2 \rangle T_p^{(2)}(k, z_1) T_p^{(2)*}(k, z_2) e^{i\vec{k}\vec{\rho}} d\vec{k} \quad (7.8)$$

The spectral-density relation for the second-order pressure field then becomes (N.23)

$$F_p^{(2)}(\omega, z) = \omega^2 \rho_1^2 \int \langle |B^{(2)}(\vec{k}, \omega)|^2 \rangle |T_p^{(2)}(k, z)|^2 d\vec{k} \quad (7.9)$$

where the most general form of $\langle |B^{(2)}(\vec{k}, \omega)|^2 \rangle$ is defined by Eq.(6.5), or Eq.(6.18) in the case of multiple roots, and

$$T_p^{(2)}(k, z) = e^{\gamma_+ z} \frac{1 + R_b e^{(\gamma_- - \gamma_+)(z + H_1)}}{1 + \frac{\epsilon}{b} R_b e^{(\gamma_- - \gamma_+) H_1}}$$

in which

$$b = \frac{(1 - \omega^2/g\gamma_+)}{1 - \omega^2/g\gamma_-}, \quad \text{and} \quad \epsilon = \frac{\gamma_-}{\gamma_+}$$

Deep-ocean form

As we demonstrated in Sect. 6.3.1, in a deep-ocean the expression for $\langle|B^{(2)}(\vec{k},\omega)|^2\rangle$ takes the simpler form given by Eq.(6.20). In addition the relevant dispersion relation is the simple form, $\sigma_i^2 = gq_i$, and the derivatives involved can be calculated straightforwardly. If the standing-wave approximations are also assumed, (i.e., $\vec{k} = 0$, $\hat{m} = 1$, $\chi = 1$ and $\theta_{12} = \pi$), Eqs.(6.18) and (6.20) both simplify further to

$$\langle|B^{(2)}(0,\omega)|^2\rangle = \frac{g^2}{2\omega} F_a^2(\omega/2) \int_0^{2\pi} H(\theta_1) H(\theta_1 + \pi) d\theta_1 \qquad (7.10)$$

and Eq.(7.9) then becomes

$$\begin{aligned} F_p^{(2)}(\omega, z) &= \frac{\rho_1^2 g^2 \omega}{2} \int_0^{2\pi} H(\theta_1) H(\theta_1 + \pi) d\theta_1 \int_0^{2\pi} \int_0^{\omega/\alpha_1} k\, dk\, d\theta_k \\ &= \frac{\pi \rho_1^2 g^2 \omega^3}{2\alpha_1^2} F_a^2(\omega/2) I_\omega(u_{10}) \end{aligned} \qquad (7.11)$$

where

$$I_\omega(u_{10}) = \int_0^{2\pi} H(\theta_1) H(\theta_1 + \pi) d\theta_1 \qquad (7.12)$$

and is a function of the wind speed, u_{10} (see Sect. 7.2.1). We again recognise Eq.(7.11) as the classical form of the pressure spectrum given by Eq.(4.107) in Chap. 4 and discussed in Sect. 6.3.1.

7.2 Ocean-Wave Spectrum and Seafloor Structure

7.2.1 Ocean-Wave Spectrum

When turbulent air flow imparts a state of irregular motion to the ocean surface, nonlinear interactions between wave components of different frequency and direction cause wave energy to be transferred from one frequency band to another [56]. If the direction of the wind is steady and its duration is sufficiently long, the wave-energy spectrum reaches an equilibrium form which is dependent on wind speed, water depth and fetch. In a deep ocean of unlimited fetch, this equilibrium state is typically described by the Pierson-Moskowitz (PM) spectrum [65]

$$F_a(\omega) = \frac{\alpha g^2}{\omega^5} e^{-\beta(\frac{\omega_0}{\omega})^4} \qquad (7.13)$$

where $\alpha = 8.10 \times 10^{-3}$, $\beta = 0.74$, g is the gravitational acceleration, $\omega_0 = g/u_{19.5}$ and $u_{19.5}$ is the wind speed at the height 19.5 m. However, in

7.2 Wave Spectrum and Seafloor Structure

the calculation of the spectrum it is more convenient to express Eq.(7.13) as a function of frequency, f, instead of angular frequency, ω,

$$F_a(f) = \frac{\alpha g^2}{(2\pi)^4} \frac{1}{f^5} e^{-\frac{5}{4}(f_m/f)^4} \qquad (7.14)$$

where f_m, the frequency of the spectral peak, is

$$f_m = \left(\frac{4}{5}\beta\right)^{\frac{1}{4}} \frac{g}{2\pi u_{19.5}} \qquad (7.15)$$

This relation is found to describe the wave spectrum very well when the fetch exceeds 500 km[56]. In many applications the wind speed at 19.5 m height, $u_{19.5}$, is transformed to that at 10 m, u_{10}, through the relation [66]

$$u_{10} = \frac{u(z)}{1 + \frac{\sqrt{c_{10}}}{\kappa} \ln\left(\frac{z}{10}\right)} = 0.939 u_{19.5} \qquad (7.16)$$

where use is made of the average value $c_{10} = 1.5 \times 10^{-3}$ and $\kappa = 0.4$ (the von Kármán's constant).

When the fetch is limited, the equilibrium wave spectrum appears to be fetch dependent under reasonably steady wind conditions. The Joint North Sea Wave Project (JONSWAP)[67], carried out in 1970, studied this spectral dependence and established the JONSWAP function describing the fetch-limited wave spectrum as,

$$F_a(f) = \frac{\alpha g^2}{(2\pi)^4} \frac{1}{f^5} e^{[-\frac{5}{4}(\frac{f_m}{f})^4]} \gamma^q \qquad (7.17)$$

where

$$q = \exp[-(f/f_m - 1)^2/(2\sigma^2)] \qquad (7.18)$$

and

$$\sigma = \begin{cases} \sigma_a & \text{for } f \leq f_m \\ \sigma_b & \text{for } f > f_m \end{cases} \qquad (7.19)$$

Ewans and Kibblewhite[66],[15] carried out a similar study off the west coast of the North Island of New Zealand and in a markedly different geographical environment confirmed the essential form of the JONSWAP function.

Comparison with Eq.(7.14) shows that, in the fetch-limited case, the wave spectrum is enhanced around the peak frequency, f_m, by a factor γ^q. The values of the parameters f_m, σ and α are now fetch and

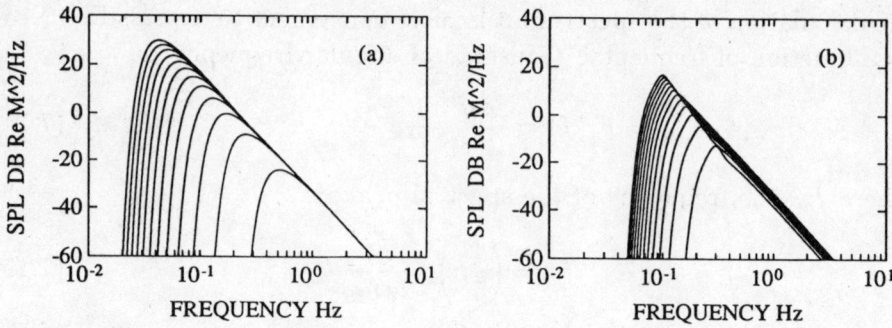

Figure 7.1: Ocean surface-wave spectra: a Pierson-Moskowitz(PM) form, b JONSWAP(J) form of the wave spectrum, wind speed 2.5-30 ms^{-1} in steps of 2.5 ms^{-1}.

wind-speed dependent. Based on the JONSWAP results and those of the New Zealand study, we elect to use the following definitions of these parameters[66] in the calculations to be described,

$$\alpha = 0.07 \left(\frac{u_{10}^2}{g\chi}\right)^{0.27}$$

$$f_m = 2.59 \frac{g^{0.27}}{u_{10}^{0.44} \chi^{0.28}}$$

$$\gamma = 2.9$$

$$\sigma = \begin{cases} 0.10 & \text{for } f \leq f_m \\ 0.13 & \text{for } f > f_m \end{cases} \qquad (7.20)$$

We note also that, according to the New Zealand studies, the Phillip's 'equilibrium range' is approached only when the wind speed exceeds 15 ms^{-1}. At lower wind speeds the wave spectral levels are lower than those of the Pierson-Moskowitz spectral form, as can be seen in Fig. 7.1.

Figures 7.1 **a** and **b** present families of PM and JONSWAP spectra for wind speed, u_{10}, ranging from 2.5 to 30 ms^{-1} in steps of 2.5 ms^{-1}. The fetch, χ, is taken to be 200 km. We see that in the high frequency range the JONSWAP spectral levels are comparable to those predicted by the PM form. At lower frequencies, however, the PM spectra have much higher levels and extend down to lower frequencies. We use both forms in the calculations to be described later.

7.2.2 Geoacoustic Environment

We now turn to the question of modelling the geoacoustic conditions of the sea floor. The first fact to recognise is that, as the frequency band of concern ranges from a few hundredths of a hertz to a few hertz (or from a few hundred meters to a few kilometers in terms of wavelength), the whole oceanic crust, from the top sediments down to the mantle, must be considered in any study of the wave-induced pressure and seismic fields. The second is that, because the geostructure of the upper oceanic crust varies significantly from place to place, it is impossible to define a single universal model to handle all situations of interest. At the same time it is impractical to present wave-induced spectra for a wide range of different models. Some compromise is required.

In a recent review Spudich and Orcutt[69] concluded that the oceanic crust can best be modelled by an upper high-gradient layer ($1 - 2s^{-1}$), followed by a layer with a much smaller velocity gradient ($\sim 0.1s^{-1}$), both overlying a thick high-velocity, homogeneous crust. Recent deep-sea drilling studies[70] have confirmed the main features of the model proposed in [69] and provided other velocity profiles to serve as a reasonable approximation of the deep crust.

Information on the acoustic properties of the upper-sedimentary layers is available from a series of parallel studies by Hamilton[71], Vidmar et al[72], Stoll[46], Trevorrow and Yamamoto[43], Dorman[37] and others. All have emphasised the complexity and variability of the top sedimentary layer, especially in coastal regions. Other studies have shown that the effects of porosity can be as important as the main elastic properties[48],[49], and a gas hydrate zone is also known to introduce significant acoustic effects, since it is characterised by very low velocities[73].

To ensure the calculations made later are based on models reasonably representative of the geoacoustic environment, the following structures have been selected for investigation.

Model MDL1

For deep oceans Collins' model [70] is chosen in the form of the 11 layer structure shown in Fig. 7.2a. We identify this model as MDL1. It has the parameters listed in Table 7.1. It is characterised by a low-speed layer of 260 m thickness, a rapid but linear increase of speed in the upper crust (down to 1420 m below the sea floor) and a low-speed layer in the lower

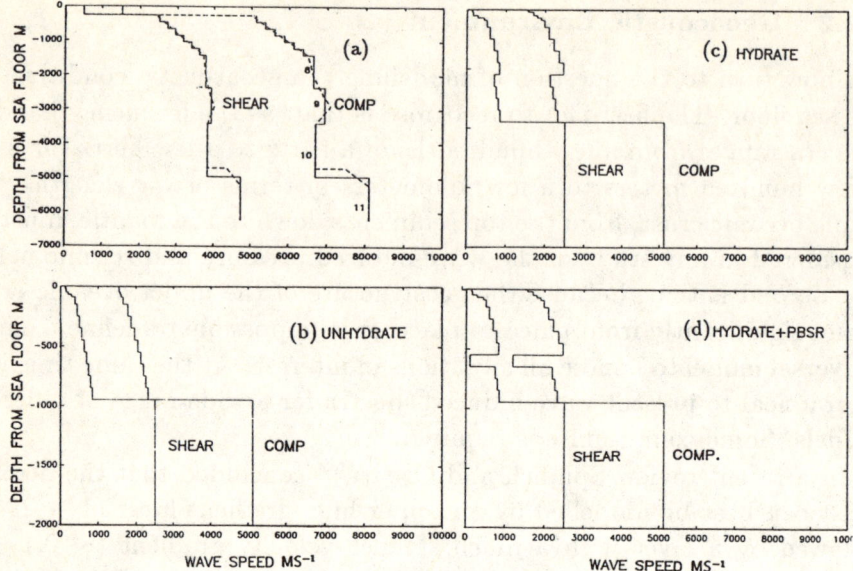

Figure 7.2: Compressional and shear-wave velocity profiles of the geoacoustic models: a MDL1 and MDL2 (in MDL2 the top layer is porous); b MDL3(1); c MDL3(2); d MDL3(3).

crust, just above the mantle (layer 10 in Fig. 7.2a corresponding to $i=11$ in Table 7.1). The authors of [69] and [70] report these features are typical of the oceanic crust structure. The linear increase of velocity in the upper crust has here been approximated in six stepped layers. Since [69] and [70] do not give a clear description of the density, we have adopted the values of 1.7 kg/m^3 for the top sediment, 2.0 for the upper-crust and 2.5 for the lower crust, following Hamilton[71] and Vidmar[72]. For the mantle basement we follow Ewing[55] and assume a density of 3.0 kg/m^3. The Q-values of the compressional and shear wave are, for convenience, assumed to be 450 and 225 in all layers.

Model MDL2

In the second model, MDL2, also represented in Fig. 7.2a, a porous sediment of 260 meters thickness is introduced as the top layer, the remainder of the structure being as for MDL1. The parameter values of the porous sediment, taken from Stoll[46], are listed in Table 7.2.

7.2 Wave Spectrum and Seafloor Structure

Table 7.1: Parameters of MDL1.

i	$H_i - H_1$ (m)	$H_{i-1} - H_i$ (m)	ρ_i (kg/m^3)	$\alpha_i(v_p)$ (m/s)	$\beta_i(v_s)$ (m/s)	$Q_{\alpha i}$	$Q_{\beta i}$
2	-260.0	260.0	1.7	1520	500.0	450	225
3	-454.2	194.2	2.0	5125	2509.2		
4	-648.3	194.1	2.0	5375	2727.5		
5	-842.5	194.2	2.0	5625	2945.8		
6	-1036.7	194.2	2.0	5875	3164.2		
7	-1230.8	194.1	2.0	6125	3382.5		
8	-1425.0	194.2	2.0	6375	3600.8		
9	-2375.0	950.0	2.5	6650	3780.0		
10	-3425.0	1050.0	2.5	6950	3935.0		
11	-4975.0	1550.0	2.5	6700	3830.0		
12	-∞	∞	3.0	8100	4740.0		

Table 7.2: Parameters of the unconsolidated porous sediment layer.

	sand	Soft Sediment
K_r	3.6×10^{10}	3.6×10^{10}
K_f	2.0×10^9	2.0×10^9
ρ_s	2656	2656
ρ_f	1000	1000
μ	2.61×10^7	2.21×10^7
K_b	4.36×10^7	3.69×10^7
$\triangle_s$	0.15	0.5
$\triangle_c$	0.15	0.5
β	0.47	0.76
K_p	5×10^{-11} ($5 \times 10^{-13} \rightarrow 10^{-10}$)	1.6×10^{-15}

Table 7.3: Parameters of MDL3(1)

		(a) Unhydrate				
i	$H_i - H_1$	$\rho_i \times 1000$	α_i	β_i	$Q_{\alpha i}$	$Q_{\beta i}$
2	-20	1.20	1535	130	73.75	49.61
3	-50	1.40	1545	155	109.15	54.58
4	-70	1.55	1575	210	118.64	68.22
5	-90	1.80	1600	300	136.44	69.97
6	-150	1.68	1649	390	143.62	77.97
7	-250	1.81	1780	448	151.60	80.26
8	-350	1.93	1864	507	170.55	85.28
9	-450	2.03	1960	570	170.55	88.06
10	-550	2.11	2048	636	170.55	90.96
11	-650	2.15	2128	702	170.55	90.96
12	-750	2.17	2202	755	181.92	94.10
13	-850	2.19	2269	810	181.92	101.07
14	-950	2.22	2330	855	194.91	109.15
15	$-\infty$	2.50	5125	2509.2	450.00	225.00

Model MDL3

The third model, MDL3, is based on three structures originally used in a study of the effects of gas hydrate layers[73]. In the first one, MDL3(1), the hydrate layer is absent. The compressional- and shear-wave velocity profiles for this model are shown in Fig. 7.2b. In the second, MDL3(2), the hydrate layer is introduced above the basement, as in Fig. 7.2c: the third, MDL3(3), which is characterised by one layer of very low rigidity (referred to as a prominent Bottom-Simulating Reflector in [73]) is shown in Fig. 7.2d.

The parameters for each of the hydrate models are given in Tables 7.3, 7.4 and 7.5 respectively. In all three cases the sediment is assumed to overlie a homogeneous basement. The Q values listed in Tables 7.3 - 7.5 have been converted from the attenuation data, η dB/wavelength, given in Table 2 of [73], using the relation

$$Q = 8.686\pi/\eta = 27.288/\eta.$$

It is to be noted that, for convenience, we have halved the number of layers by combining adjacent layers in the original profiles of [73].

7.2 Wave Spectrum and Seafloor Structure

Table 7.4: Parameters of MDL3(2)

			(b) Hydrate			
i	$H_i - H_1$	$\rho_i \times 1000$	α_i	β_i	$Q_{\alpha i}$	$Q_{\beta i}$
2	-20	1.20	1560	130	73.75	49.61
3	-50	1.40	1700	155	109.15	54.58
4	-70	1.65	1800	275	143.62	73.75
5	-90	1.85	1900	435	181.92	97.46
6	-150	1.98	1985	510	209.91	101.07
7	-250	2.10	2100	595	248.07	113.70
8	-350	2.20	2225	695	248.07	124.04
9	-450	2.20	2360	785	272.88	129.94
10	-550	2.30	2475	840	272.88	136.44
11	-650	2.15	2128	702	170.55	90.96
12	-750	2.17	2202	755	181.92	94.10
13	-850	2.19	2269	810	181.92	101.07
14	-950	2.22	2330	855	194.91	109.15
15	$-\infty$	2.50	5125	2509.2	450.00	225.00

Table 7.5: Parameters of MDL3(3)

			(c) Hydrate + PBSR			
i	$H_i - H_1$	$\rho_i \times 1000$	α_i	β_i	$Q_{\alpha i}$	$Q_{\beta i}$
2	-20	1.20	1560	130	73.75	49.61
3	-50	1.40	1700	155	109.15	54.58
4	-70	1.65	1800	275	143.62	73.75
5	-90	1.85	1900	435	181.92	97.48
6	-150	1.98	1985	510	209.91	101.07
7	-250	2.10	2100	595	248.07	113.70
8	-350	2.20	2225	695	248.07	124.04
9	-450	2.20	2360	785	272.88	129.94
10	-550	2.30	2475	840	272.88	136.44
11	-650	1.80	1200	100	59.32	27.29
12	-750	2.17	2202	755	181.92	94.10
13	-850	2.19	2269	810	181.92	101.07
14	-950	2.22	2330	855	194.91	109.15
15	$-\infty$	2.50	5125	2509.2	450.00	225.00

7.2.3 Reflection Coefficient

As any standard reference will testify [89][63] the propagation of plane waves and their interaction with boundaries is fundamental to all theoretical studies of wave physics. As we have seen the concepts involved have been critical to the plane-wave synthesis of the acoustic field generated by the wave-wave interaction process. The plane-wave reflection coefficient in turn proves to be a useful complementary concept in studying the subsequent interaction of the wave-induced pressure field with a multilayered seabed. Because of the particular properties of the source, however, the role of the inhomogeneous component is of greater relevance than normal and the behaviour of the reflection coefficient as a function of both frequency and wavenumber needs to be understood, to properly evaluate the reflection process. The analysis below is based largely on such studies we have reported earlier [48][49].

The amplitude of the reflection coefficient, $|R_b|$, for model MDL1 is shown as a function of the frequency, f, and relative wavenumber, $u = k\alpha_1/\omega$, in Figs. 7.3a and 7.3b. Figure 7.3b provides more detail for $|R_b| \leq 10$ by deleting the top part of the figure. Figure 7.4 shows the plot of $|R_b|$ for MDL2. Figures 7.5a, b and c show the same plot for the three versions of MDL3. Several points of interest are to be noted.

In Figs. 7.3a and b a series of strong peaks, indicating values of $|R_b|$ up to 150, occur in the frequency band 0.1 to 1 Hz, between $0 < log_{10}u < 0.5$ ($1 < u < 3$). As first described by Frisk[39], the high peaks appearing in the inhomogeneous region often relate to the excitation of seismic surface-wave modes in the crustal structure. At high frequencies on the other hand, the top sediment layer becomes dominant acoustically and the effects of all sublayers are suppressed. The scarp apparent at $u = 3$ ($log_{10}u = 0.5$), with a level around 10 essentially independent of frequency, corresponds to the excitation of the Sholte wave with a speed close to the shear-wave speed, 500 ms^{-1}, of the top sediment layer[55, 74]. When the top sediment (elastic) layer is replaced by an unconsolidated porous layer (sand), the level of the peaks drops significantly due largely to the intrinsic attenuation of the material – Fig. 7.4. At high frequencies the scarp is now displaced to the region $u > 10$ or ($log_{10}u > 1$). Its marked decrease in level with increasing frequency results from the low rigidity and high, frequency-dependent attenuation of the material. A more detailed description of these characteristics is available in [48] and [49].

7.2 Wave Spectrum and Seafloor Structure

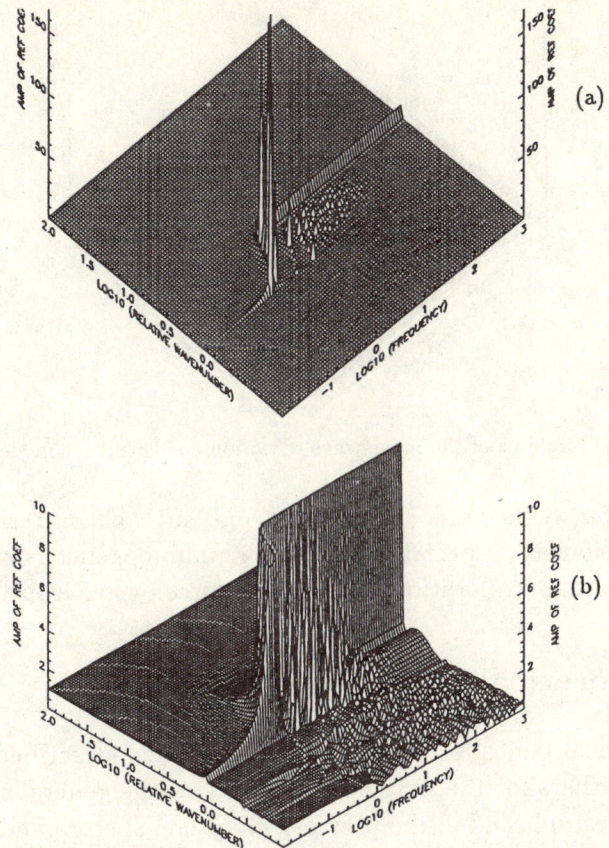

Figure 7.3: Modulus of the plane-wave reflection coefficient for model MDL1 as a function of $\log_{10}$(frequency) and $\log_{10}$(relative wavenumber): **a** emphasising the highest peaks; **b** showing the structure in more detail.

Figures 7.5a to c, relating to the hydrate structure, display similar characteristics at high and low frequencies around $u = 1$. Differences are apparent, however, in both the transition zone ($1 \leq u \leq 10$) and in the homogeneous region. To display these features more clearly the reflection loss (dB) is plotted as a function of frequency and incident angle in Figs. 7.6a to c. The same information is plotted in terms of the reflection coefficient in Figs. 7.7a to c. In the frequency band 0.1 to 1 Hz the reflection loss between incident angles 40° to 90° is clearly greater when a hydrate structure is present (compare Figs. 7.6b and c

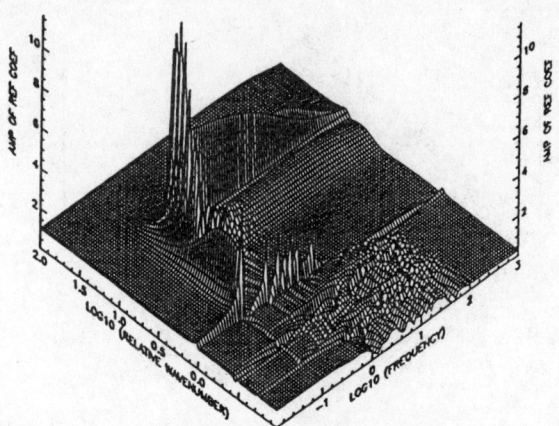

Figure 7.4: Modulus of the plane-wave reflection coefficient for model MDL2.

with Fig. 7.6a). As Frisk has pointed out such characteristics of the reflection coefficient, especially those in the inhomogeneous region, could prove useful in the exploration of such structures by geoacoustic methods.

7.3 Theoretical Pressure Spectra

Using numerical codes for the reflection coefficient described above, together with others for the source field, based on the generalised theoretical formulae established in the previous chapters, spectra and correlation functions of the wave-induced pressure field (sum-frequency component only) can be calculated for environments of any specified water depth and seabed structure . As examples, we now examine the properties of the noise field for several of the environmental models described earlier. Pressure and coherency spectra are considered in this chapter. Seismic spectra are examined in Chap. 8.

7.3.1 Dependence on Observation Depth

Deep-water environments

Figure 7.8 presents a set of spectra of the total pressure field generated by a well developed (PM) sea in a deep-ocean environment under a wind of 15 ms^{-1}. The water column is assumed to be 4000 m deep. The

7.3 Theoretical Pressure Spectra

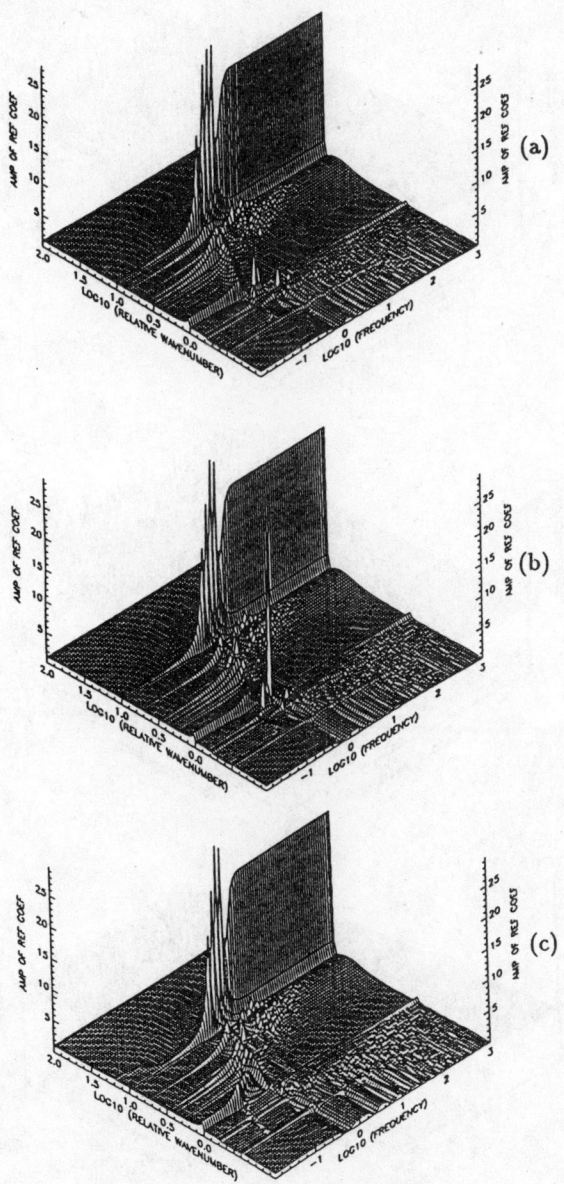

Figure 7.5: Modulus of the plane-wave reflection coefficient for model MDL3: **a** with no hydrate layer; **b** with a hydrate layer; and **c** with a hydrate layer in association with a thin gas layer.

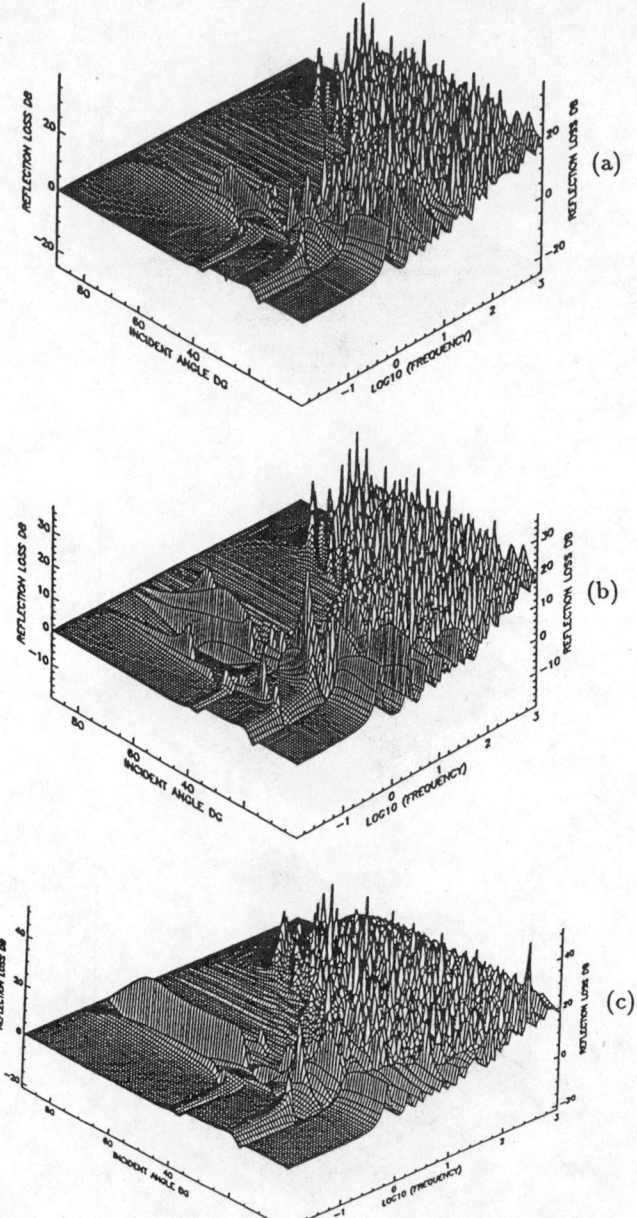

Figure 7.6: Reflection loss for model MDL3 as a function of frequency and incident angle: **a** with no hydrate layer; **b** with a hydrate layer; **c** with a hydrate layer in association with a thin gas layer.

7.3 Theoretical Pressure Spectra

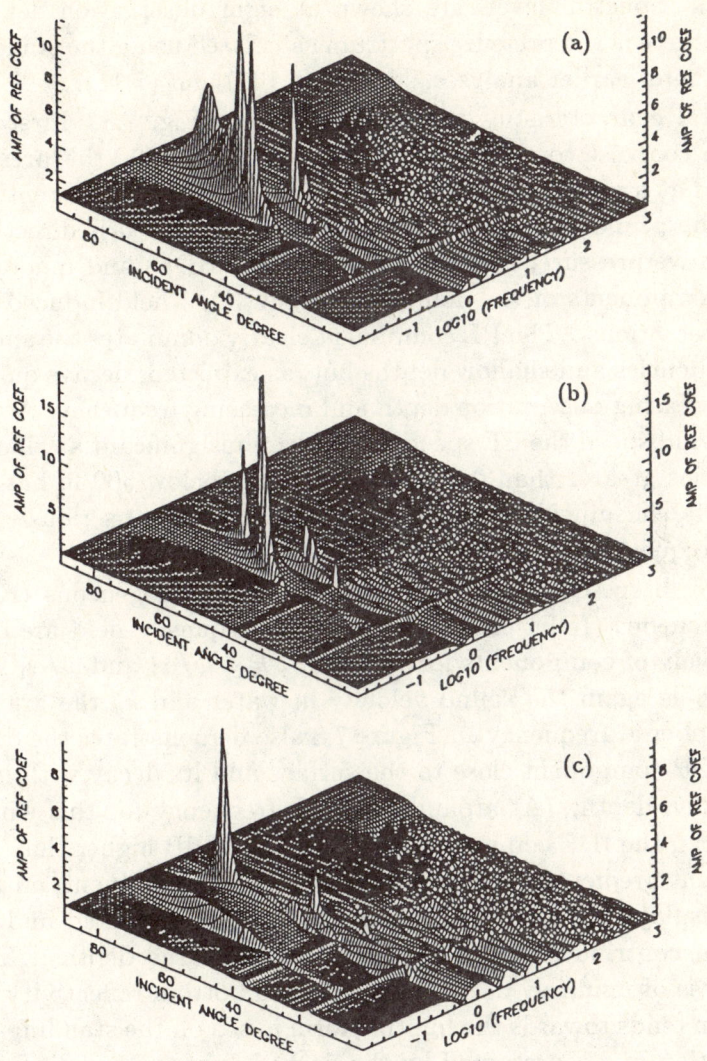

Figure 7.7: Modulus of the plane-wave reflection coefficient for model MDL3 as a function of $\log_{10}$(frequency) and incident angle: **a** with no hydrate layer; **b** with a hydrate layer; **c** with a hydrate layer in association with a thin gas layer.

underlying multilayered seabed has the structure described in Fig. 7.2a (MDL1). Spectral levels are shown at eight observation depths. The dashed curve is the pressure spectrum calculated using the standing-wave expression of earlier analyses, see Eqs.(4.107) and (7.11).

Several characteristics can be identified in Fig. 7.8. First we recall that, in contrast to the source spectra of Fig. 6.11, the pressure field observed at a given depth, z, is here the sum of three different contributions: the primary-frequency component (PF), which is directly related to the wave pressure, and the homogeneous (HDF) and inhomogeneous (IDF) components of the double-frequency (DF) field induced by wave-wave interactions. The PF component clearly dominates the spectrum at low frequencies and shallow depths but, as expected, decays quickly with both increasing observation depth and increasing frequency. At the nominated wind speed the PF spectral levels are insignificant at all depths for frequencies greater than 0.2-0.3 Hz, but only below 500 m has the peak level '*' fallen sufficiently that the DF spectrum is not seriously distorted at low frequencies.

As we discussed in previous sections, the homogeneous (HDF) and inhomogeneous (IDF) parts of the double-frequency field are composed respectively of components for which $0 \leq k \leq \omega/\alpha_1$ and $\omega/\alpha_1 < k \leq k_g$, where α_1 is again the sound velocity in water and k_g the gravitational wavenumber at frequency ω. Figure 7.8 also demonstrates the dominance of the IDF component close to the surface and its decay with increasing observation depth. (At around the peak frequency for this wind speed, say 0.2 Hz, the IDF component can be 30 to 40 dB higher than the HDF component predicted by Eq.(7.11).) The HDF component, on the other hand, displays the negligible depth dependence expected and becomes the main contributor to the pressure field at large depths. In the extreme case of infinitely deep water and zero bottom reflectivity the total spectrum tends towards the limiting form based on the standing-wave approximation and represented by the dashed curve. We note finally that the mean level of the deepest (4000 m) spectrum is approximately 3 dB higher than its classical equivalent. This increase is the result of the finite reflectivity associated with the seabed model MDL1.

The fine structure apparent in the deepest spectra is introduced by normal mode modulation of the pressure field. To demonstrate this we replot the data of Fig. 7.8 in the form of contour plots in Fig. 7.9. We first note that the peaky, depth-dependent modulation of the spectral level is

7.3 Theoretical Pressure Spectra

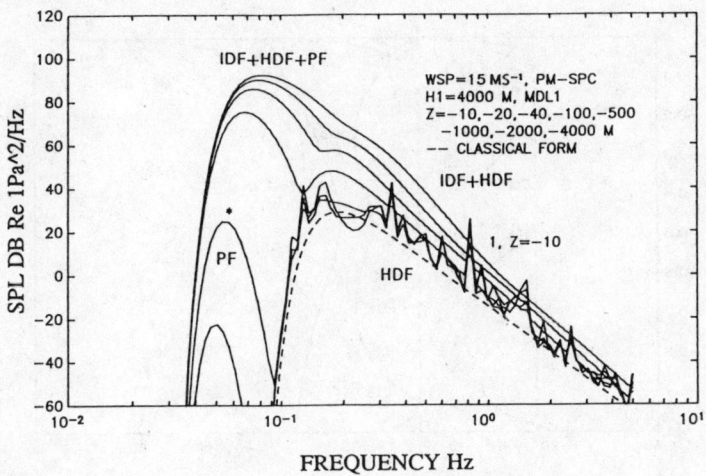

Figure 7.8: Wave-induced pressure spectra at 10 m, 20 m, 40 m, 100 m, 500 m, 1000 m, and 4000 m below the sea surface, for an ocean 4000 m deep, model MDL1, a PM sea and a wind speed of 15 ms^{-1}.

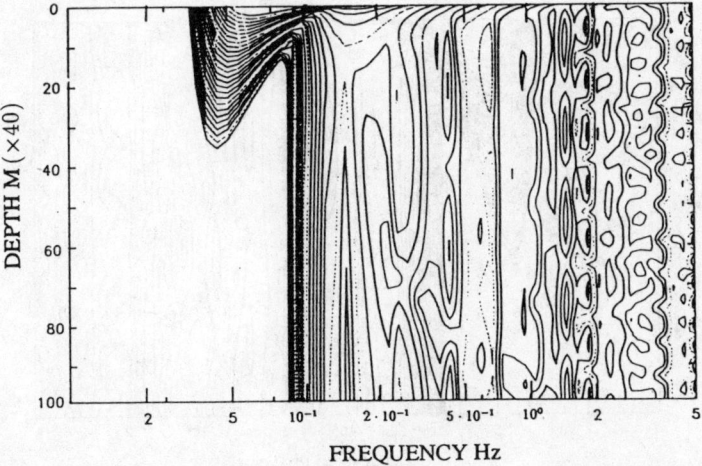

Figure 7.9: A contour plot of the spectral level of the depth-dependent, wave-induced pressure field, for an ocean 4000 m deep, model MDL1, a PM sea and a wind speed of 15 ms^{-1}.

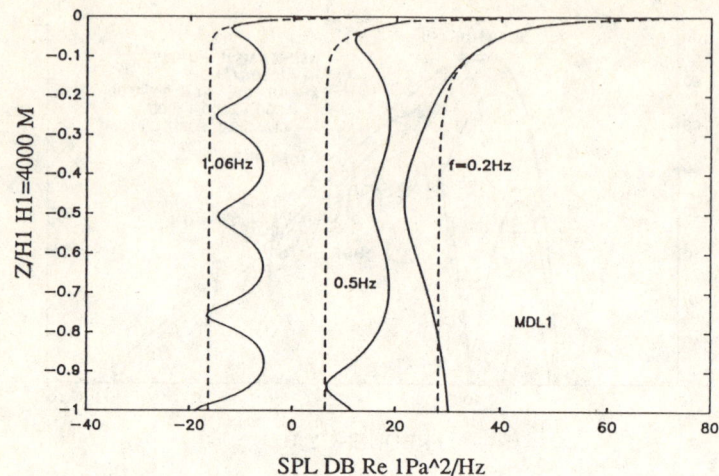

Figure 7.10: Depth-dependence of the wave-induced pressure spectral level at 0.2, 0.5 and 1.06 Hz, for the environment and conditions of Fig. 7.8.

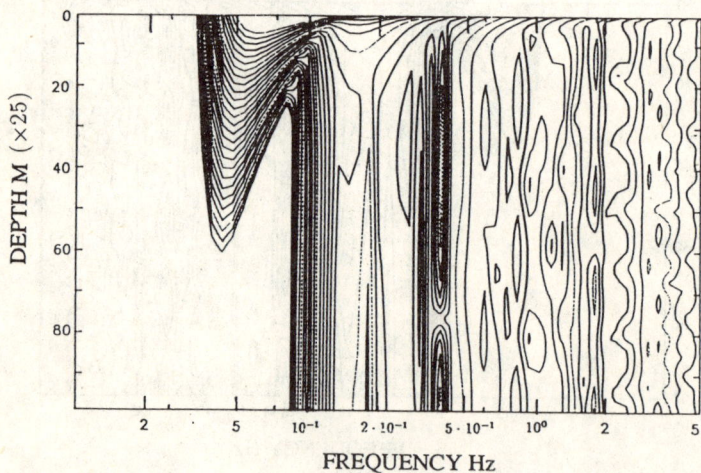

Figure 7.11: A contour plot of the spectral level of the depth-dependent, wave-induced pressure field, for an ocean 2500 m deep, model MDL3(3), a PM sea and a wind speed of 15 ms^{-1}.

7.3 Theoretical Pressure Spectra

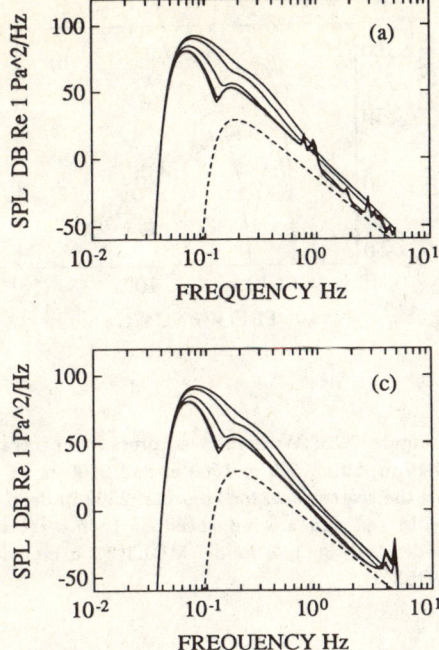

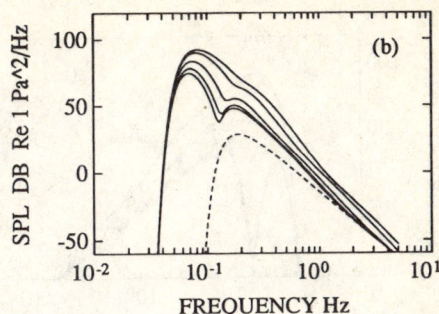

Figure 7.12: Wave-induced pressure spectra at 10 m, 20 m, 40 m, 80 m, and 100 m below the sea surface, for an ocean 100 m deep, a PM sea and a wind speed of 15 ms^{-1}: **a** model MDL(1); **b** model MDL0, $R_b = 0$; **c** model MDL0, $R_b = 1$.

only apparent at frequencies in excess of cut-off for this water-depth and that the number of peaks (loops) within the water layer increases with frequency. To show the depth dependence arising from mode effects more clearly, spectral profiles at frequencies 0.2, 0.5 and 1.06 Hz are plotted in Fig. 7.10. (The dotted curves again relate to the classical spectrum for a bottomless-ocean.) The fine structure in the spectra of Fig. 7.8 and the dominant loops in Fig. 7.9 occur at the same frequencies. (The slight frequency shift is caused by a scaling of the axes. In Fig 7.9 the frequency extends only to 5 Hz). The contour levels range from -60 to +70 dB in steps of 5 dB.

By way of confirmation of these effects, the total pressure field (under a 15 ms^{-1} sea) for the substructure MDL3(3) is presented in contour form in Fig. 7.11, using the same range and steps as in Fig. 7.9. The water-depth in this case is 2500 m. Comparison of the two figures also shows effects which are related to the hydrate layer, but the mode-related loop structure clearly reflects the change in water depth.

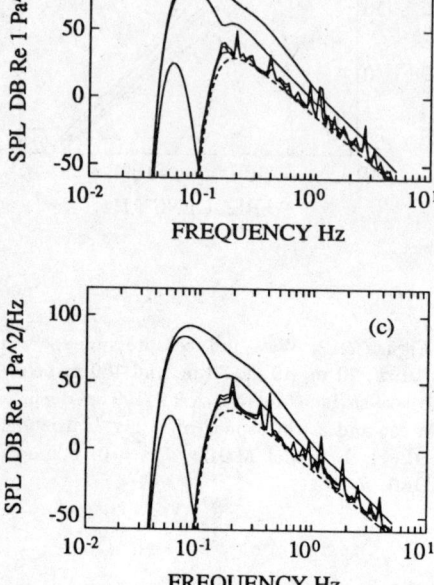
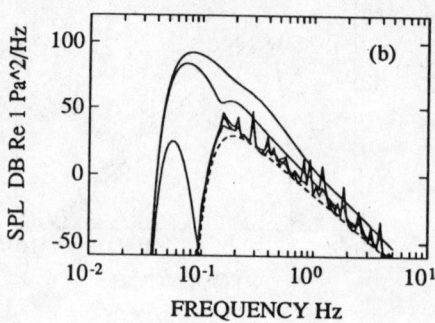

Figure 7.13: Wave-induced pressure spectra at 10 m, 50 m, 500 m 1500 m and 2500 m below the sea surface, for an ocean 2500 m deep, a PM sea and a wind speed of 15 ms^{-1}: a model MDL3(1); b model MDL3(2); c model MDL3(3).

Shallow-water environments

It is now instructive to consider the behaviour of the pressure field in a shallow-water environment. We again use model MDL1 and a PM sea under a 15 ms^{-1} wind, but reduce the water depth to 100 m. The resulting depth-dependent spectra are presented in Fig. 7.12a. Comparison with the spectrum for the same observation depth in Fig. 7.8 shows little change in spectral level but, as expected, the mode related fine structure in Fig. 7.12 only appears at much higher frequencies.

Spectra for the same water depth and sea state, but with $R_b = 0$ and $R_b = 1$, are presented in Fig. 7.12b and c respectively. Comparison of Figs. 7.12a,b and c shows several features of interest. First, as expected, the spectral levels based on model MDL1 lie between the extremes represented by Fig. 7.12b and c - also see below. Secondly, the absence of fine structure in Fig. 7.12b and its appearance only at high frequencies in Fig. 7.12c is consistent with the values of the reflection coefficient in the two cases. Thirdly, the expected doubling of the pressure field with the change in R_b from 0 to 1 is manifest in the 3 dB increase in the deeper spectral levels in Fig. 7.12c.

7.3 Theoretical Pressure Spectra

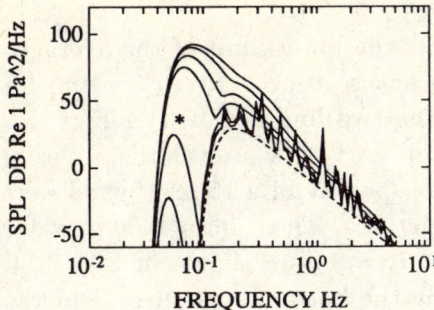

Figure 7.14: Wave-induced pressure spectra at 10 m, 20 m, 40 m, 100 m, 500 m, 1000 m, 2000 m and 4000 m below the sea surface, for an ocean 4000 m deep, a seabed with $R_b = 1$, a PM sea and a wind speed of 15 ms^{-1}.

7.3.2 Influence of the Seabed Structure

Hydrate structures

The three hydrate models allow us to examine the influence of the seabed structure on the pressure field in more detail. A PM sea under a 15 ms^{-1} wind and a water depth of 2500 m are assumed. Figure 7.13a presents the depth-dependent spectra for model MDL3(1), the structure in which the hydrate layer is absent - see Fig. 7.2**b**. Equivalent spectra for structures MDL3(2) and MDL3(3) - see Figs. 7.2**c** and 7.2**d** - are given in Figs. 7.13**b** and **c**. (In all figures the dotted curve again represents the classical deep-water spectrum.) It is apparent that the presence or absence of the hydrate layer has little effect on the pressure spectra in a deep-water environment. It is appropriate to recall, however, the marked influence such a feature had on the reflection coefficient - see Sect. 7.2.3 This influence was also reported in [70] - see for example Fig. 7.5 of that reference. In that case the behaviour of the reflection coefficient at 10 Hz around grazing angles of 60 to 70 degrees was found to be notably different when a hydrate layer was present. This observation is consistent with the results of Figs. 7.5, 7.6 and 7.7 of this contribution. These figures show further, however, that even more marked effects occur in other parts of the wavenumber-frequency presentation of the reflection coefficient. The possible exploitation of these features in seabed exploration is the subject of ongoing studies.

Range of spectral variation

Since the dashed spectrum represents the lower limit of the averaged DF pressure field at large depths, the spectral levels at $z \leq -10$m for a given sea state must all lie within the two limits shown in Figs. 7.8 and 7.12 (the dashed curve and the top curve), whatever the environment. To confirm this point the pressure spectra for a 15 ms^{-1} wind were calculated for a number of different structures. Those for a 4000 m water layer and an absolutely hard bottom ($R_b = 1$) are shown in Fig. 7.14. Those for a 100 m water layer overlying the bottom structures depicted by models MDL2, MDL3(1) and MDL3(3), are shown in Figs. 7.15a, b and c respectively. As expected all spectra lie between the two extremes.

7.3.3 Influence of the Ocean-Wave Spectrum

To demonstrate the influence of fetch, depth-dependent pressure spectra for the environment of Fig. 7.15a, but using the JONSWAP wave spectrum (fetch of 200 km) instead of the PM spectrum, are shown in Fig. 7.16. As expected, the levels at high frequencies are similar for the two sea states, but the peaks associated with the PF and DF components are lower and narrower in the case of the JONSWAP sea and spectral levels fall off more quickly below the peak. Since the fetch is generally limited in shallow waters for off-shore winds, pressure spectra are also calculated for the three typical structures, MDL1, MDL2 and MDL3(3) (all with a 50 m water layer), to show the general features of the spectra in such circumstances. The results are shown in Fig. 7.17. Comparison with Figs. 7.12a, 7.15a and 7.15c clearly demonstrates the influence of the form of the ocean-wave field and the need to select that appropriate to a given situation.

7.3 Theoretical Pressure Spectra

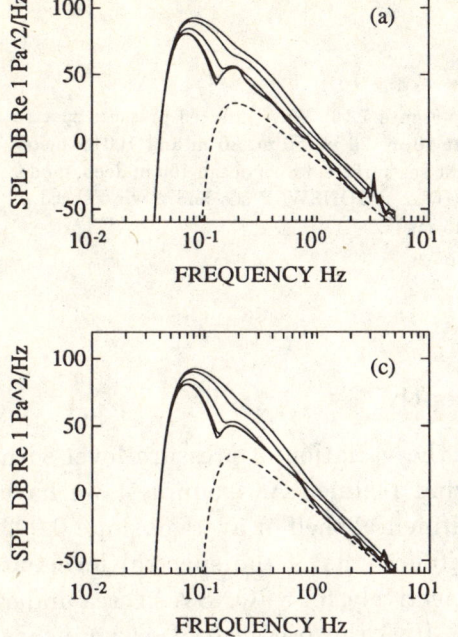

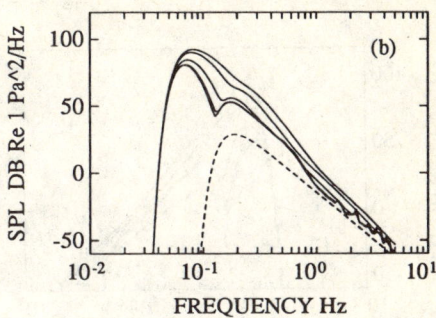

Figure 7.15: Wave-induced pressure spectra at 10 m, 20 m, 50 m, 80 m and 100 m below the sea surface, for an ocean 100 m deep, a PM sea and a wind speed of 15 ms^{-1}: **a** model MDL2; **b** model MDL3(1); **c** model MDL3(3).

7.3.4 Dependence on Wind Speed

Figures 7.18**a** and 7.18**b** demonstrate the wind dependence of the pressure spectra at an observation depth of 10 m for wind speeds ranging from 2.5 to 30 ms^{-1}. Model MDL1, with water depth 100 m, is used in both calculations. The PM spectrum is used in Fig. 7.18a and the JONSWAP version in Fig. 7.18b. Figures 7.19**a**, **b**, and 7.20**a**, **b** provide the same comparisons for water depths of 50 m and 30 m. These plots show that at low wind speeds the spectral levels of both the PF and DF peaks are lower in the case of the developed sea but the frequency of the spectral peaks is higher. At high wind speeds, on the other hand, the spectral levels are higher and the frequency of the dominant spectral peak lower, than in the case of the JONSWAP sea. These features are clearly related to the character of the ocean-wave spectra shown in Figs. 7.1**a** and **b**.

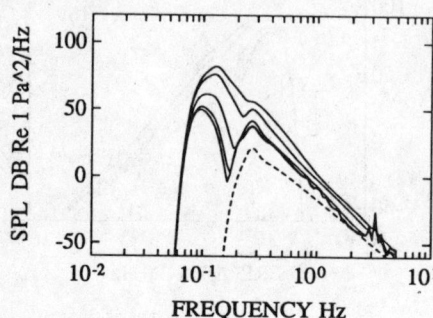

Figure 7.16: Wave-induced pressure spectra at 10 m, 20 m, 50 m, 80 m and 100 m below the sea surface, for an ocean 100 m deep, model MDL2, a JONSWAP sea and a wind speed of 15 ms^{-1}.

7.3.5 Dependence on Water Depth

It is of practical interest to examine the variation of pressure level with water depth under a steady-state wind regime. As examples, we have calculated the spectral level on a continental shelf of average slope 0.002 [76]. Figures 7.21a, b, c present contour plots of the spectral levels at frequencies 0.21, 0.50 and 1.05 Hz respectively, for a JONSWAP sea under a steady off-shore wind of 15 ms^{-1}. It is the presence of a significant PF component at 0.2 Hz (see Fig. 7.20b) that makes the contours in Fig. 7.21a so different from those at the two higher frequencies. We also note that in Figs. 7.21a and c the level (-18.7 dB), represented by the symbol '*', is that predicted by the deep-water relation, Eq.(7.11). As the levels at 0.5 Hz all exceed 3 dB, the symbol '*' does not appear in Fig. 7.21b. The increase in level apparent in this figure towards the edge of the shelf is due to bottom reflection and mode effects. In contrast the drop in level to below -18.7 dB beyond 100 km in Fig. 7.21c is the result of interference between the incident and reflected components of the pressure field. All figures demonstrate clearly that, at frequencies below 1 Hz, predictions based on Eq.(7.11) will be unreliable in continental shelf regions.

7.3 Theoretical Pressure Spectra

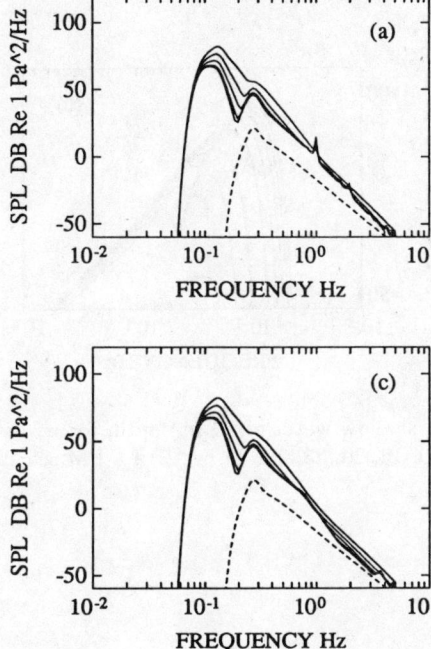

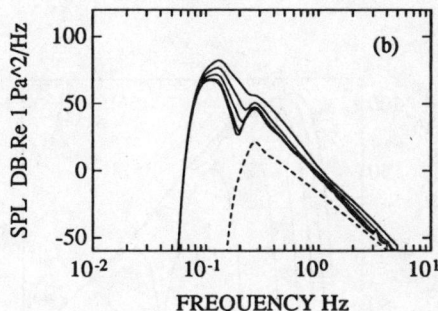

Figure 7.17 Wave-induced pressure spectra at 10 m, 20 m, 30 m, 40 m and 50 m below the sea surface, for an ocean 50 m deep, a JONSWAP sea and a wind speed of 15 ms^{-1}: a model MDL1; b model MDL2; c model MDL3(3).

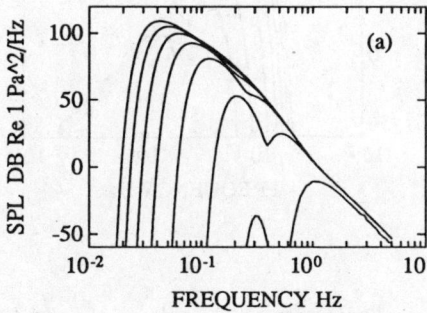

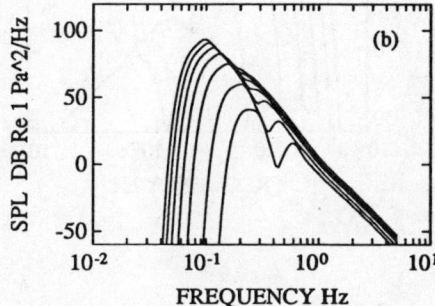

Figure 7.18: Wave-induced pressure spectra in shallow water of 100 m depth, for model MDL1, and wind speeds ranging from 2.5, 5, 10, 15, 20, 25 and 30 ms^{-1}: a a PM sea; b a JONSWAP sea.

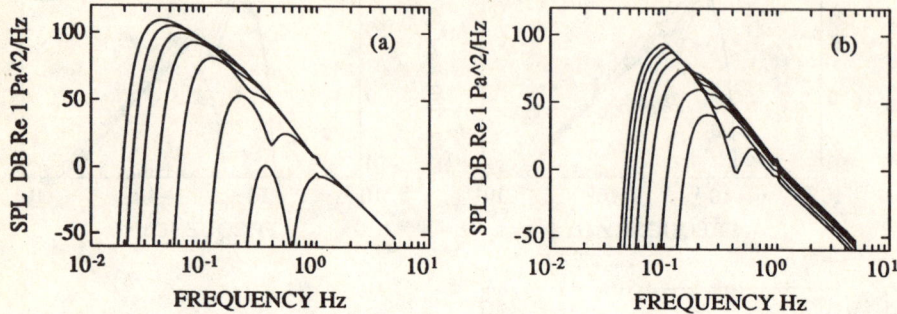

Figure 7.19: Wave-induced pressure spectra in shallow water of 50 m depth, for model MDL1, and wind speeds ranging from 2.5, 5, 10, 15, 20, 25 and 30 ms^{-1}: a a PM sea; b a JONSWAP sea.

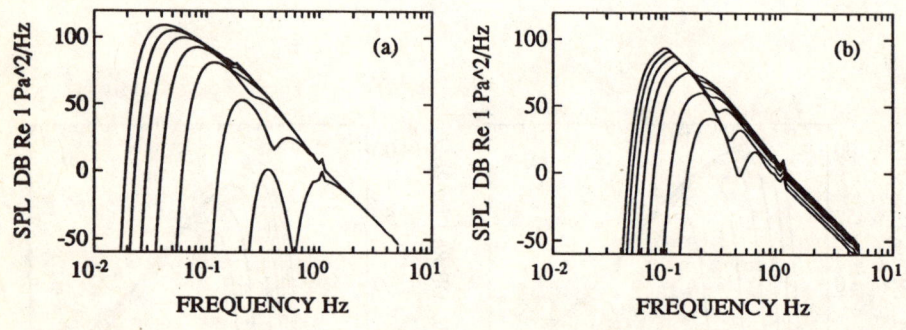

Figure 7.20: Wave-induced pressure spectra in shallow water of 30 m depth, for model MDL1, wind speeds ranging from 2.5, 5, 10, 15, 20, 25 and 30 ms^{-1}: a a PM sea; b a JONSWAP sea.

7.3 Theoretical Pressure Spectra

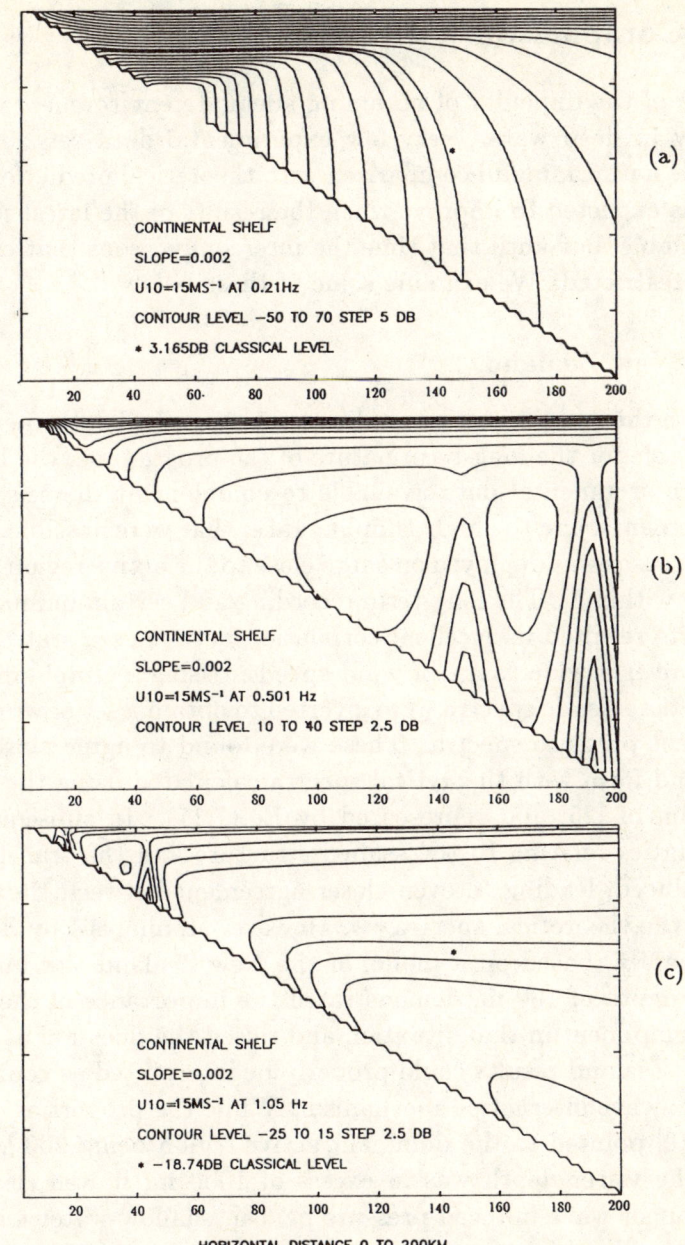

Figure 7.21: Contours of the wave-induced pressure level over a continental shelf, extending 200 km from the shore with a slope of 0.002, for a JONSWAP sea and a wind speed of 15 ms^{-1}: **a** 0.2 Hz; **b** 0.5 Hz; **c** 1.05 Hz.

7.4 Comparison with Field Data

Because of the difficulty of obtaining adequate environmental data, particularly in deep water, very few experimental data sets are currently available for meaningful comparison with theoretical predictions. The situation is expected to improve when the results of the latest programmes are available, but until that time the intercomparisons that can be made remain restricted. We examine some of these below.

7.4.1 New Zealand

Although the comparison was indirect this New Zealand experiment[19][20] was notable for the long-term nature of the programme, the high quality of the environmental data available to complement the seismo-acoustic measurements, and the high sampling rates that were possible because the seismic sensor was deployed on shore close to the active region rather than offshore within it. The long-term recording and certain unique properties of the site resulted in excellent correlation between sea-state and seismic spectra over a wide range of wind speeds. Using a simple model of the seabed, the seismic spectra were inverted to obtain a set of wind/seastate-dependent pressure spectra. These were found to agree closely in level, shape and form with theoretical spectra calculated using the deep-water formalism of [25] and represented by Eq.(7.11). In subsequent papers uncertainties relating to the seabed structure and the transfer function were reduced, leading to even closer agreement between the experimental and the theoretical spectra [35]. However an analysis by Schmidt and Kuperman[34], based on a model of the New Zealand site but assuming a water depth of 100 m, demonstrated the importance of the inhomogeneous component in that situation and raised the question as to whether the New Zealand results could properly be interpreted as confirmation of the wave-wave interaction mechanism. While the properties of the spectra in [19] pointed to the dominant active region being 200 km offshore, where the water depth was in excess of 1000 m, it was clear that the behaviour of wave-induced pressure field in shallow-water environments needed clarification.

This clarification was provided in a series of studies. In [38] a generalised treatment of the wave-wave interaction process was developed,

7.4 Comparison with Field Data

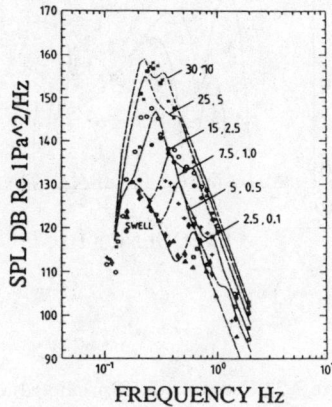

Figure 7.22 Comparison of the experimental spectra derived from the seismic data with the oretical pressure spectra in which the effect of the swell and the interaction of a local residual sea is included.

which demonstrated clearly the relative importance of the homogeneous and inhomogeneous components of the pressure field as a function of observation depth. This was followed by an investigation which demonstrated that, in the particular circumstances of the New Zealand experiment, the seismic response recorded onshore was due primarily to the homogeneous component[41]. Whatever the depth of water in the active region, therefore, the inverted pressure spectra reported by Kibblewhite and Ewans[19] were actually equivalent to those that would be recorded in deep water. The close agreement between the experimental wind-dependent-spectra and their theoretical equivalent based on the deep-water relation, Eq.(7.11), demonstrated in [19] and [38] and represented in Fig. (7.22), was thus reaffirmed as positive confirmation of the general validity of the original deep-water analysis given by Longuet-Higgins and others.

7.4.2 Eleuthra

In a review of very low-frequency noise, Nichols[22] included results from hydrophones installed at various depth on the bottom off the island of Eleuthra. In Fig. 7.23 the pressure spectra (1/3 octave) at three depths are compared with theoretical predictions based on Eq.(7.9) of this contribution.

In Fig. 7.23a the circles represent the experimental data under sea

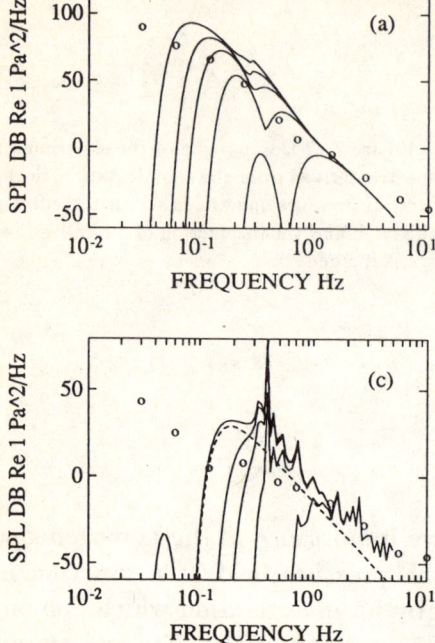

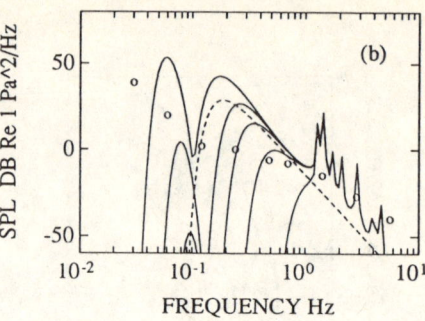

Figure 7.23 Comparison of predicted and experimental noise spectra measured in water off Eleuthra: **a** water depth 13 m; **b** water depth 300 m; **c** water depth 1200 m.

state 3 (wind speed 5-7ms^{-1}), and the curves the predicted spectra for wind speeds 2.5, 5.0, 7.5, 10 and 15 ms^{-1}. Because the wind was from the NE (onshore) the PM spectrum has been used in the calculations, and because no geoacoustic parameters were reported, a hard bottom ($R_b = 1$) has been assumed (as observed earlier the predicted levels are likely to be a few dB lower in a more realistic environment). It can be seen that above 0.1 Hz the measured data agree reasonably well with the theoretical spectra for wind speeds 5 and 7.5 ms^{-1}, but two anomalous data points at lower frequencies mar the overall agreement. At the time these data were reported only the third-octave level close to 0.2 Hz could be related satisfactorily to theoretical predictions (based on the Wenz spectrum) and this was identified with the first-order pressure field for the prevailing seastate. The 5 ms^{-1} spectrum in Fig. 7.23a supports this interpretation. Of equal significance, however, is the agreement at higher frequencies based on the predictions for the second-order field. Only the two low-frequency points now remain anomalous and it is recognised that these levels may be due to another source mechanism[40] or to measurement error.

Figures 7.23b and c compare theoretical spectra ($R_b = 1$) with experimental data recorded at 300 m and 1200 m . The comparison here is clearly less rewarding but a lack of reliable wind data makes it difficult to draw a firm conclusion. The sharp peaks in the theoretical spectra are due to normal modes in the water layer - see Sect. 7.3.1.

7.4.3 Pacific Ocean

While numerous reports[21],[32],[33],[37],[40],[77],[78]–[80] confirm the general character of the ULF seismo-acoustic spectrum in deep water, the difficulties of providing wind and wave data longterm have not allowed a full analysis of the effects observed. The definitive experiment is therefore still to be reported and any quantitative comparison of theory and experiment based on available data must necessarily be restricted. Nevertheless we here attempt such a comparison with data typical of deep-water sites.

In Fig. 7.24a we compare theoretical spectra with data reported by Sutton and Barstow[33]. In compiling the analytical spectra we have used the model of the seabed they described as an approximation of the recording site. Theoretical spectra are shown for a PM sea and wind speeds 15, 20 and 25 ms^{-1}. The difference-frequency component is included by basing the calculation on Eq.(6.27) - see Sect. 6.3.

It can be seen that: (i) good agreement exists above 0.08 Hz if wind speed at the site was around 20 ms^{-1}; (ii) the experimental values within the noise notch (0.02 - 0.1 Hz) are higher than predicted by some 10 - 20 dB; (iii) a rise in spectral levels below 0.02 Hz occurs in both the theoretical and experimental spectra, but the spectral mismatch suggests that the difference-frequency component is not the most significant source in this band; (iv) the PF component does not influence the levels at this depth(the PF component shown in the "noise notch" is associated with the maximum wind speed).

While the theoretical predictions in Fig. 7.24a do not replicate the experimental data at very low frequencies, they do suggest an interpretation of the data of [33] different from that given by Sutton and Barstow. For instance, we believe the spectra with a peak around 0.1 Hz are a manifestation of the DF pressure field produced by local winds at the edge of the storm centre (i.e. at the OBS site) and not by a swell field which has propagated from the path of the hurricane. The wind speeds at the storm centre would be expected to produce wave spectral peaks (PM sea) at frequencies around 0.02 to 0.03 Hz rather than the 0.06

to 0.07 Hz (see Fig. 7.1) responsible for the main peak in the pressure spectra. The frequencies observed are more consistent with local winds of 20 ms^{-1}.

The low coherence between the pressure and seabed displacement reported in the noise notch, suggests the possibility of noise contamination. The 10-20 dB mismatch in the theoretical and experimental levels could be further confirmation of this possibility.

Finally, while the mismatch below 0.02 Hz suggests that difference-frequency interactions are not the most significant source in this band, the wind- and depth-dependence of the difference-frequency pressure field is consistent with the variability in this band reported by Sutton and Barstow. This prompted us to examine other influences which might enhance the spectral levels of this component. The limit on wind speed ($\sim$ 20 ms^{-1}), imposed by the frequency of the main peak in the experimental spectra, precludes simply increasing the wind speed to achieve higher levels in the theoretical spectra . The presence of a very-long period component in the local wave field was recognised as providing another interaction mechanism by which the difference-frequency field could be enhanced. The possibility that such swell components could be incoming from the hurricane-generated seas led us to examine the influence of long-period swell on the analytical spectra. (Our experience with Southern Ocean swells[91] provided us with a reasonable approximation of the spectral form of this component and the seastates reported in [33] allowed a reasonable assessment of its peak spectral level.) We found that the spectral levels below 0.02 Hz were increased significantly, but not uniformly across the band. On the basis of these limited calculations we conclude that, while the difference-frequency component can contribute to the variability in the infragravity band (see Fig. 8.5), it is not the prime source. It has been proposed that long gravity-water waves are responsible for the pressure effects observed and predicted that the levels in this band should be fairly constant with time[40],[80]. Although the significant temporal variations reported by Sutton and Barstow (and also in Sect.8.6.1) are contrary to this generalisation, the sources detailed in [40] and [80] still seem to be the most likely candidates.

Figure 7.24b compares theoretical spectra (dotted curves) for wind speeds 15, 20 and 25 ms^{-1} with data reported by Webb and Cox[32]. By way of contrast with Fig. 7.24a the analytical spectra are based on an ocean of infinite depth and restricted to the sum-frequency component

7.4 Comparison with Field Data

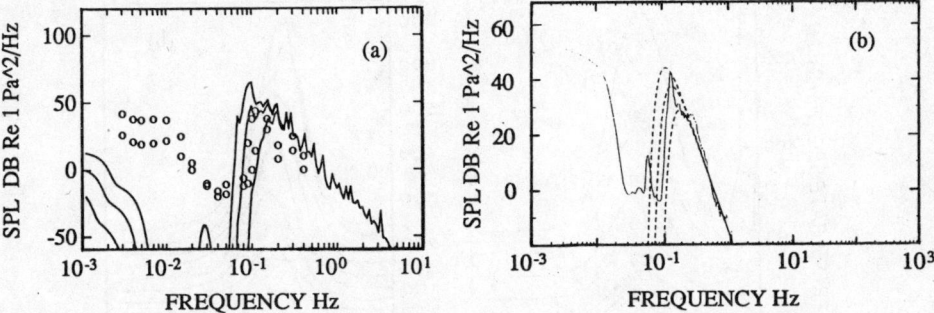

Figure 7.24: Comparison of predicted and measured noise spectra at deep-water sites: a 4000 m in depth; b over 3000 m in depth on the Pacific Rise.

only. Agreement is closest for the 20ms^{-1} spectrum. This wind speed is compatible with the conditions reported, but a lack of precise environmental data reduces the significance of the comparison. As with Fig. 7.24a the rise in level below the noise notch is due to another source.

7.4.4 Woronara Dam - Australia

In a test of his theory of wave-generated noise Cato[52] conducted an experiment in a water supply reservoir south of Sydney. The site provided an opportunity for greater experimental control than is possible at sea but other complications were involved. Noise and wave-height spectra and wind speed were measured simultaneously. The distinctive 2 to 1 ratio of the frequency of the peak in the noise spectra to that of the wave spectra was clearly evident and the general nature of the results confirmed the role of wave interactions as a significant noise source in the ocean. Predictions of the far-field noise spectrum to be expected in a deep bottomless ocean on the basis of the Woronara results are shown to be broadly consistent with other measurements and predictions, but detailed comparisons are not practical.

7.4.5 Upper Levels of the Ocean

A comparison of theoretical spectra with spectra measured in the upper levels of a deep ocean was reported in [36]. Pressure-rate spectra recorded at various depths by Cox and Jacobs[77],[78] were compared with synthetic spectra calculated using the generalised expression for the second-order field - see Fig. 7.25. It will be recalled that this expression, Eq.(6.18), incorporates the inhomogeneous component in the second-order pressure

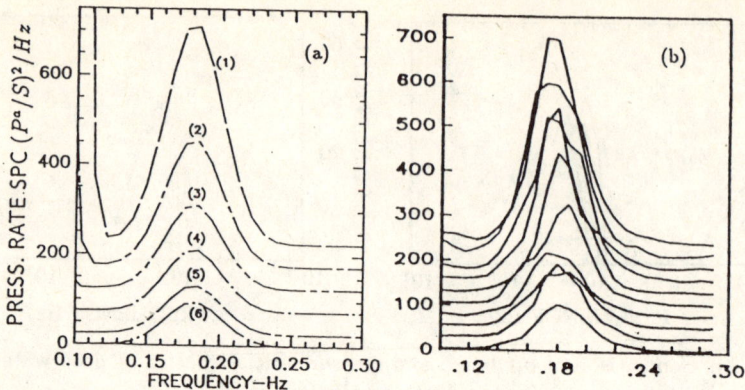

Figure 7.25: A comparison of synthetic pressure rate (dP/dt) spectra with measured data(right) for depths 110, 150, 190, 230, 270, 290 m below the surface. The spectral curves are each spaced upward by 50 $(Pa/s)^2$/Hz. The experimental spectra, reproduced from Fig. 3 of Ref.[78], are measured at depths 110-290 m and the curves are again spaced (by 25$(Pa/s)^2$/Hz) from each other for clarity. (This figure first appeared in Natural Physical Sources of Underwater Sound - Sea Surface Sound(2), Ed. B.R. Kerman, Kluwer Academic Publishers (1993) - see Ref.[36]. Reprinted by permission of Kluwer Academic Publishers).

field but not the first-order field. As the instrument used by Cox and Jacobs does not record the linear effects of the overhead waves (see Sect. 4.1.2 of [77]) the use of Eq.(6.18) is appropriate. A pleasing level of agreement between observation and prediction is apparent.

Overall these results give one confidence in the validity of the extended formalism, but additional comparisons with the results of other experimental programmes are awaited with interest.

7.5 The Equivalent Source Level

A brief discussion of the "equivalent source level" of the pressure field generated by surface-wave motion is appropriate at this point. From the above analysis we recall that the pressure field so generated is composed of three main components, the primary frequency component (PF) and the inhomogeneous and homogeneous components (IDF and HDF) of the double-frequency (DF) field. Of these the first two are evanescent, decaying exponentially with increasing depth. The third does not decay and

7.5 The Equivalent Source Level

propagates to depth. We have seen also that in shallow waters and in the upper part of deep oceans the PF and IDF components dominate, being up to 40 dB (or more) higher than the HDF component. At large depths on the other hand the reverse is true. It is also notable that the noise field is quite different from that associated with a uniform distribution of omni-directional point sources. The IDF component has a wind-dependent directivity, a finite upper limit to the horizontal wavenumber (in the case of the sum-frequency interactions - see Sect. 6.3.1) and a specific amplitude function different from that of a point source. Further, the horizontal wavenumber of the PF component relates to frequency through a depth-dependent dispersion relation, and as the next section demonstrates the second-order pressure field has a finite spatial correlation distance in the horizontal plane, again in contrast to that of a distribution of ordinary monopoles or dipoles.

All these properties have a marked bearing on the concept of an equivalent source strength for the wave-induced pressure field and its use as a source in seabed investigations. In acoustics it is customary to define an "equivalent source level" as the sound level measured in the far field extrapolated back to the reference point (unit distance) on the basis of some spreading law. The equivalent source level so defined can then be used to predict the sound field at a defined distance by incorporating appropriate propagation losses. In the case of the wave-induced pressure field a source level defined in terms of the HDF component is useful for deep-observation depths, where the IDF component is negligible, but for reasons given below the concept is fraught with difficulty in other situations.

Source spectra, based on the HDF component alone, are shown in Figs. 7.26a and b for wind speeds ranging from 2.5 to 30 ms^{-1}, and for both the PM and JONSWAP forms of the ocean-wave spectrum. Comparison with figures presented previously, especially Figs. 7.12 to 7.15, shows that the pressure levels measured at any finite depth from the surface, or on the bottom of a shallow sea, are higher than those of the equivalent HDF source defined above by an amount corresponding to the sum of the PF and IDF components, viz:

$$F_{pi}(\omega, z) = \int \int_{k=\omega/\alpha_1}^{k_{\max}} S_p^{(1)}(\omega, \vec{k}) T_p^{(1)}(\omega, k, z) \, d\vec{k} +$$

$$\int \int_{k=\omega/\alpha_1}^{k_{\max}} S_p^{(2)}(\omega, \vec{k}) T_p^{(2)}(\omega, k, z) \, d\vec{k} \quad (7.21)$$

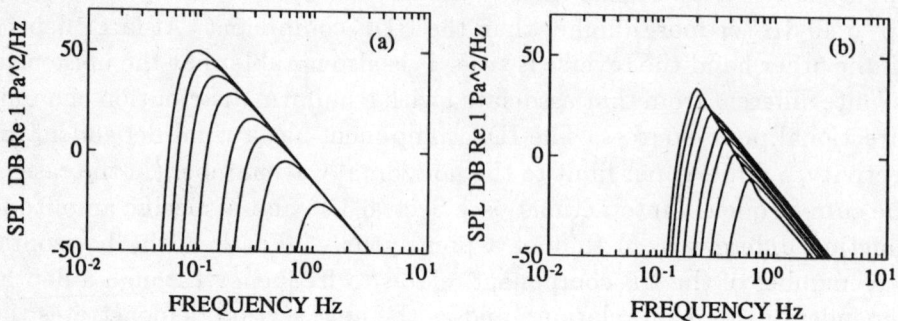

Figure 7.26: The classical deep-water form of the pressure spectra generated by nonlinear wave interactions under wind speeds of 2.5, 5, 7.5, 10, 15, 20, 25, and 30 ms^{-1}, calculated using **a** the PM and **b** the JONSWAP form of the ocean-wave spectrum.

where $S_p^{(i)}(\omega, \vec{k})$ is the source term and $T_p^{(i)}(\omega, k, z)$ describes the medium response. Variations in the parameters of the geoacoustic model will thus clearly influence the frequency distribution of the source energy. Even though the average levels of the noise field will always remain within the bounds of the two extreme cases, defined by $R_b = 0$ and $R_b = 1$ - see Fig. 7.12- it is of importance to determine how best to deal with the inhomogeneous component in particular, since many attempts have been made in recent years to estimate the source pressure level through a measurement of the associated seismic spectra.

In the literature two ways of recording seismic activity are described. One involves an ocean-bottom seismometer (OBS) installed on (or below) the ocean bottom. The other uses a land-based system (LBS) with the seismometers placed onshore near the coast. In the OBS case the seismic spectrum can (depending on observation depth) clearly include PF, IDF and HDF components corresponding to those in the pressure field. It is therefore necessary in this case to subtract the PF and IDF components (and any propagation effects) from the total spectrum to establish an equivalent source level based on the HDF component alone.

Because of the propagation processes involved in transmitting energy from off shore to the land-based sensor, matters are even more complicated for the LBS installation. It was shown in [41] that this process is critically dependent upon the seabed model. We recall in particular that:

7.5 The Equivalent Source Level

(i) When the bottom is an elastic half-space the energy propagating to the land does so mainly through the Rayleigh wave for which the horizontal wavenumber, k_r, is normally less than ω/α_1. Further, if the water in the active region is reasonably deep, the energy recorded by a LBS (or an OBS) will be mainly due to the HDF component, as in this case the IDF component is small and the seabed has a narrow-band seismic response in the homogeneous region.

(ii) When the bottom is an unconsolidated (liquid) half-space, on the other hand, an OBS in shallow-water will experience a strong IDF component, because the seabed response is in this case effectively constant throughout the entire k-range, $0 \leq k < \infty$. (As we noted earlier, in waters 50 to 100 m deep the contribution from the IDF component can be 30 to 40 dB higher than that of the HDF component.) In the case of an OBS sensor this component must be subtracted to establish the true level of the HDF field. For the LBS the situation is again different. Since there is no surface-wave mode to act as a carrier, little IDF energy propagates outside the active region. The seismic signals received outside the active region are largely due to diffraction of the HDF field.

(iii) In more realistic environments incorporating low-velocity layers, the seismic response at the seabed is characterised by a multiband structure extending beyond $k = \omega/\alpha_1$. In this case part of the IDF energy will be recorded by the LBS by virtue of the propagation of low-velocity surface-wave modes . The response of the seabed to the IDF energy is therefore model dependent at both the OBS and LBS sites. If the upper layer of the seabed is an unconsolidated sediment, the IDF component will be especially pronounced at the OBS site, because the layer "softens" the whole structure and increases the seabed displacement produced by the incident pressure field. The influence of the IDF component at the LBS site, on the other hand, will be much lower. In this case the only energy received is propagated by waves travelling on the subinterfaces and the exponential decay of the IDF component reduces the level of excitation at the effective interfaces. The calculation in [41] shows that, in the case of an upper unconsolidated layer of 100 m thickness, the seismic response at the LBS to a given offshore pressure field can be 30 to 40 dB lower than that at an OBS inside the active region.

It is apparent therefore that any inversion to establish the conventional deep-water source pressure spectrum from seismic records must be made on the basis of 'site' considerations. An awareness of the differences between the response at LBS and OBS sites can satisfactorily explain apparent anomalies in wave-interaction source levels, which have been reported in a number of studies in the past.

The situation is even more critical when the ocean-wave field is used as a low-frequency pressure source in measurements of seabed parameters. Even if the pressure field is measured directly, an unambiguous application of inversion procedures requires a knowledge of its composition. Given the complexities involved, it seems more appropriate to abandon the concept of the standard deep-water source spectrum of the wave-induced pressure field and use a new definition based on the pressure field at the average plane, $z = 0$. Pressure levels and spectral format for any wind speed and seastate can then be predicted at any observation depth, by using the appropriate formulation and including the propagation effects appropriate to the environment under consideration.

7.6 Normalised Coherency Spectra

7.6.1 The Formalism

The normalised coherency spectrum of two random processes is defined as[57]

$$N_{12}(\omega) = \frac{\text{Cov}\{dZ_1(\omega), dZ_2(\omega)\}}{[\text{Var}\{dZ_1(\omega)\}\text{Var}\{dZ_2(\omega)\}]^{1/2}}$$

where Cov and Var stand for the covariance and variance respectively and $dZ_1(\omega)$ and $dZ_2(\omega)$ are the Fourier-Stieltjes transforms of the random processes

$$X_{1,2}(t) = \int_{-\infty}^{\infty} e^{-i\omega t} dZ_{1,2}(\omega)$$

Regarding $X_1(t)$ and $X_2(t)$ as the pressure field at two different locations in space, $X(t, \vec{R}_1)$ and $X(t, \vec{R}_2)$, we can write

$$N(\vec{R}_1, \vec{R}_2, \omega) = \frac{\text{Cov}\{dZ(\omega, \vec{R}_1), dZ(\omega, \vec{R}_2)\}}{[\text{Var}\{dZ(\omega, \vec{R}_1)\}\text{Var}\{dZ(\omega, \vec{R}_2)\}]^{1/2}}$$

$$= \frac{\langle [p_\omega(\vec{R}_1) - \langle p_\omega(\vec{R}_1)\rangle] [p_\omega(\vec{R}_2) - \langle p_\omega(\vec{R}_2)\rangle]^*\rangle}{[\langle |p_\omega(\vec{R}_1) - \langle p_\omega(\vec{R}_1)\rangle|^2\rangle]^{1/2} [\langle |p_\omega(\vec{R}_2) - \langle p_\omega(\vec{R}_2)\rangle|^2\rangle]^{1/2}}$$

7.6 Normalised Coherency Spectra

Since the pressure field is assumed to be a zero-mean process, the coherence spectrum becomes

$$N_p(\vec{R}_1, \vec{R}_2, \omega) = \frac{\langle p_\omega(\vec{R}_1) p_\omega^*(\vec{R}_2) \rangle}{[\langle |p_\omega(\vec{R}_1)|^2 \rangle]^{1/2} [\langle |p_\omega(\vec{R}_2)|^2 \rangle]^{1/2}} \quad (7.22)$$

or

$$N_p(\vec{R}_1, \vec{R}_2, \omega) = \frac{C_p(\vec{R}_1, \vec{R}_2, \omega)}{\sqrt{C_p(\vec{R}_1, \vec{R}_1, \omega)} \sqrt{C_p(\vec{R}_2, \vec{R}_2, \omega)}} \quad (7.23)$$

(For convenience, in the following text we suppress the symbol ω in the expressions $N_p(\vec{R}_1, \vec{R}_2, \omega)$ and $C_p(\vec{R}_1, \vec{R}_2, \omega)$ unless its retention is necessary for clarity.) If we regard the first- and second-order fields as being independent processes we can write the terms in Eq.(7.23) as

$$C_p(\vec{R}_1, \vec{R}_2) = C_p^{(1)}(\vec{R}_1, \vec{R}_2) + C_p^{(2)}(\vec{R}_1, \vec{R}_2) \quad (7.24)$$

where

$$C_p^{(1)}(\vec{R}_1, \vec{R}_2) = \langle p_\omega^{(1)}(\vec{R}_1) p_\omega^{(1)*}(\vec{R}_2) \rangle \frac{d\omega}{(2\pi)^2} \quad (7.25)$$

and

$$C_p^{(2)}(\vec{R}_1, \vec{R}_2) = \langle p_\omega^{(2)}(\vec{R}_1) p_\omega^{(2)*}(\vec{R}_2) \rangle \frac{d\omega}{(2\pi)^2} \quad (7.26)$$

First-order field:

As is shown in Appendix N, the covariance function of the first-order pressure field is (see Eq.(N.11))

$$C_p^{(1)}(\vec{R}_1, \vec{R}_2, f) = \rho_1^2 g^2 T_p^{(1)}(k, z_1) T_p^{(1)*}(k, z_2) F_a(f) \int_0^{2\pi} H(\theta_k) e^{ik\rho \cos(\theta_k - \theta_\rho)} d\theta_k \quad (7.27)$$

where $k = k(\omega)$ and use has been made of the definitions

$$\vec{R}_1 = (\vec{r}_1, z_1, \vec{z}_0) \quad \vec{R}_2 = (\vec{r}_2, z_2, \vec{z}_0),$$

with $\vec{z}_0$ the unit vector in the direction of the z axis, and, as defined in Sect 7.1,

$$\rho = |\vec{r}_2 - \vec{r}_1| \quad \theta_\rho = \arg(\vec{r}_1 - \vec{r}_2)$$

and,

$$T_p^{(1)}(k, z) = e^{\gamma_+ z} \frac{1 + R_b e^{(\gamma_- - \gamma_+)(z + H_1)}}{1 + R_b e^{(\gamma_- - \gamma_+) H_1}} \quad (7.28)$$

Second-order field:

The same function for the second-order field (the sum-frequency component) is given by Eq.(N.21) in Appendix N. For convenience in later calculations this is expressed here in the alternative form (see Eq.(N36)):

$$C_p^{(2)}(\vec{R}_1, \vec{R}_2, f) = \frac{\rho_1^2 g^2 \omega}{2} F_a^2(f/2) I_s \int_0^{k_g} \int_0^{2\pi} \left[\int_0^{2\pi} W(\vec{k}, \theta_1) d\theta_1 \right] \cdot$$

$$\cdot T_p^{(2)}(k, z_1) T_p^{(2)*}(k, z_2) e^{i\rho \cos(\theta_k - \theta_\rho)} k \, d\theta_k \, dk \quad (7.29)$$

where

$$W(\vec{k}, \theta_1) = \frac{1}{\pi \omega g^2 I_s V(k,\omega)} |\{\vec{q}_1, \vec{q}_2\}|^2 \frac{\sigma_1^2 \sigma_2^2}{q_2} \left(\frac{\partial \sigma_1}{\partial q_1}\right) \left(\frac{\partial \sigma_2}{\partial q_2}\right) \cdot$$

$$\cdot \frac{\partial q_1}{\partial (\sigma_1 + \sigma_2)} \frac{F_a(f_1) F_a(f_2)}{F_a^2(f/2)} H(\theta_1) H(\theta_2) \quad (7.30)$$

$$V(k, \omega) = |1 - \frac{g\gamma_+}{\omega^2}|^2$$

and the other definitions apply. I_s is again the term accounting for the angular distribution of the energy of the ocean-wave field, and is defined as (see Sect. 7.2.1 and Appendix N):

$$I_s = \int_0^{2\pi} H(\theta_1) H(\theta_1 + \pi) d\theta_1$$

and the transfer function $T_p^{(2)}(k, z)$ takes the form

$$T_p^{(2)}(k, z) = e^{\gamma_+ z} \frac{1 + R_b e^{(\gamma_- - \gamma_+)(z+H_1)}}{1 + \frac{\epsilon}{b} e^{(\gamma_- - \gamma_+) H_1}} \quad (7.31)$$

with

$$\frac{\epsilon}{b} \equiv \frac{1 - \omega^2/g\gamma_-}{1 - \omega^2/g\gamma_+}$$

In practice two kinds of coherency spectra are of most interest, the vertical and horizontal (longitudinal). They are defined as

$$N_p(\vec{r}, \Delta z) = \frac{C_p(\vec{r}, \Delta z)}{\sqrt{C_p(\vec{r}, z_1)} \sqrt{C_p(\vec{r}, z_2)}} \quad (7.32)$$

and

$$N_p(\rho, \theta_\rho, z) = \frac{C_p(\rho, \theta_\rho, z)}{\sqrt{C_p(\vec{r}_1, \theta_\rho, z)} \sqrt{C_p(\vec{r}_2, \theta_\rho, z)}} \quad (7.33)$$

where θ_ρ is the angle relative to the wind direction. Both are considered below. For convenience we again ignore ω in these expressions.

7.6.2 Vertical Coherence Function

First-order

Without loss of generality we need only consider the case $\rho = 0$ when calculating the vertical coherence from Eqs.(7.27) and (7.29). There are two extreme cases, when either the first-order field or the second-order field dominates. In the first case we have (Eq.(7.32))

$$N_p^{(1)}(\vec{r}, \Delta z) = \frac{C_p^{(1)}(\vec{r}, z_1, z_2)}{\sqrt{C_p^{(1)}(\vec{r}, z_1, z_1)}\sqrt{C_p^{(1)}(\vec{r}, z_2, z_2)}}$$

$$= \sqrt{\frac{T_p^{(1)}(k, z_1)T_p^{(1)*}(k, z_2)}{T_p^{(1)*}(k, z_1)T_p^{(1)}(k, z_2)}}$$

$$= e^{i[\phi(z_2) - \phi(z_1)]} \quad (7.34)$$

where $\phi(z_1)$ and $\phi(z_2)$ are the phase angles of $T_p^{(1)}(k, z)$ at z_1 and z_2. Since for the first-order field $\gamma_\pm \simeq \mp k$ and $R_b \simeq 1$, $\phi(z_1)$ and $\phi(z_2)$ are very close to zero if attenuation is negligible, and

$$N_p^{(1)}(\vec{r}, \Delta z) \cong 1 \quad (7.35)$$

Second-order

When the second-order field dominates (as mentioned above we restrict our discussion to the sum-frequency component)

$$N_p^{(2)}(\vec{r}, \Delta z) = \frac{\int_0^{k_e} \left[\int_0^{2\pi}\int_0^{2\pi} W(\vec{k}, \theta_1)d\theta_1 d\theta_k\right] T_p^{(2)}(k, z_1)T_p^{(2)*}(k, z_2)k dk}{[F(\vec{r}, z_1)]^{1/2}[F(\vec{r}, z_2)]^{1/2}} \quad (7.36)$$

where

$$F(\vec{r}, z_i) = \int_0^{k_e} \left[\int_0^{2\pi}\int_0^{2\pi} W(\vec{k}, \theta_1)d\theta_1 d\theta_k\right] |T_p^{(2)}(k, z_i)|^2 k dk$$

Examples

Figure 7.27 presents, as examples, the vertical coherence function at frequencies 0.25, 0.5 and 1.0 Hz, for a water layer of 100 m depth, the seabed model MDL1 and a steady wind of 15 ms^{-1}. In these plots the upper-left corner represents $(z_1 = 0, z_2 = 0)$ and the dotted line $z_1 = z_2$. It is seen that in the top layers the pressure field is highly coherent, as is to be expected, since the PF component is dominant. With increase of frequency the highly coherent zone shrinks. We note further that in the top layers the coherence function is not symmetrical. This is due to the exponential decay of the PF and IDF components. Symmetry improves with increasing frequency and observation depth, as the HDF component becomes the dominant feature of the pressure field.

Finally we note that for the standing-wave approximation, (in which only the homogeneous component is considered, $k = 0$ and $R_b = 0$), the vertical coherence function of the second-order field degenerates to

$$N_p(\vec{r}, \Delta z) = e^{-i\frac{\omega}{\alpha_1}(z_1 - z_2)} \tag{7.37}$$

and demonstrates a sinusoidal variation of coherence in the vertical direction.

7.6.3 Horizontal Coherence Functions

From Eqs.(7.27) and (7.29) the two horizontal coherence functions are given by

$$C_p^{(1)}(\rho, \theta_\rho, z) = \rho_1^2 g^2 |T_p^{(1)}(k, z)|^2 F_a(f) \int_0^{2\pi} H(\theta_k) e^{ik\rho \cos(\theta_k - \theta_\rho)} d\theta_k \tag{7.38}$$

and

$$C_p^{(2)}(\rho, \theta_\rho, z) = \frac{\rho_1^2 g^2 \omega}{2} F_a^2(f/2) I_s \int_0^{k_e} \int_0^{2\pi} \left[\int_0^{2\pi} W(\vec{k}, \theta_1) d\theta_1 \right] |T_p^{(2)}(k, z)|^2 \cdot e^{ik\rho \cos(\theta_k - \theta_\rho)} k \, d\theta_k \, dk \tag{7.39}$$

In the two extreme cases where either the first- or the second-order field dominates

$$N_p^{(1)}(\rho, \theta_\rho, z) = \int_0^{2\pi} H(\theta_k) e^{ik\rho \cos(\theta_k - \theta_\rho)} d\theta_k \tag{7.40}$$

and

$$N_p^{(2)}(\rho, \theta_\rho, z) = \frac{\int_0^{k_e} \int_0^{2\pi} \left[\int_0^{2\pi} W(\vec{k}, \theta_1) d\theta_1 \right] |T_p^{(2)}(k, z)|^2 e^{ik\rho \cos(\theta_k - \theta_\rho)} k \, d\theta_k \, dk}{\int_0^{k_e} \int_0^{2\pi} \left[\int_0^{2\pi} W(\vec{k}, \theta_1) d\theta_1 \right] |T_p^{(2)}(k, z)|^2 k \, d\theta_k \, dk} \tag{7.41}$$

7.6 Normalised Coherency Spectra

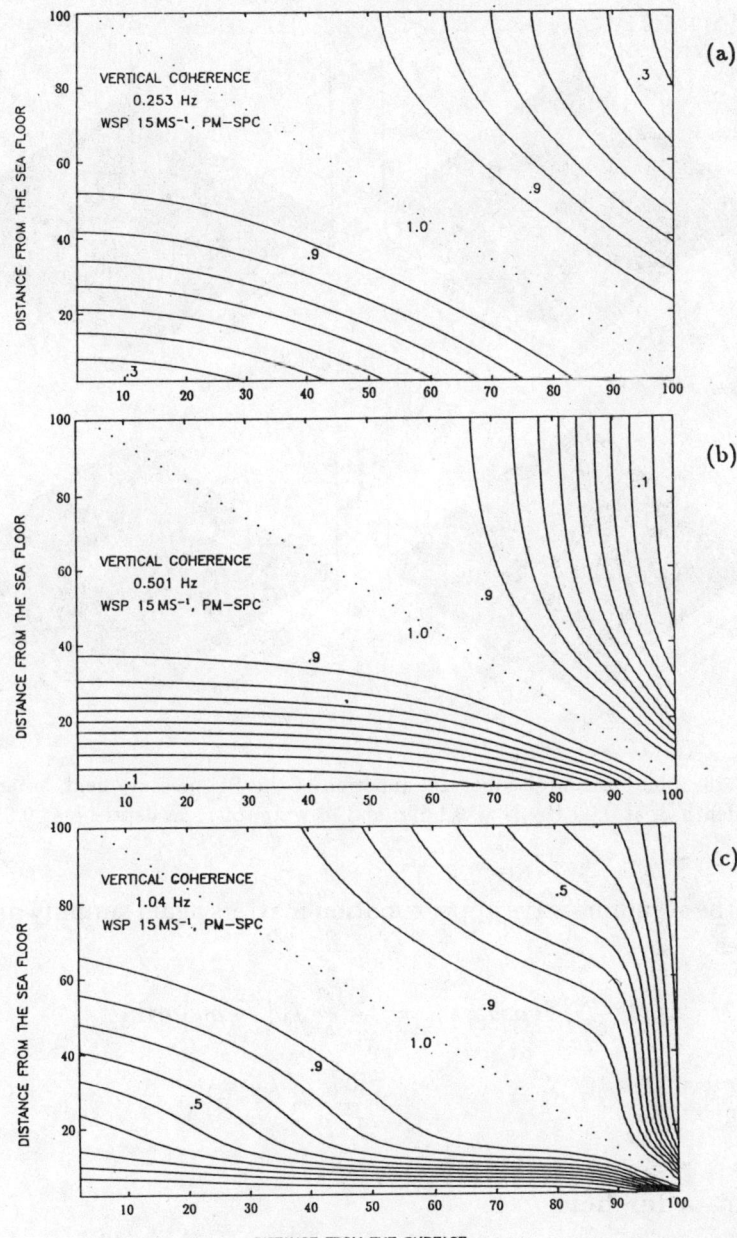

Figure 7.27: The vertical coherence function of the wave-induced pressure field (wind speed 15 ms^{-1}) in an ocean environment 100 m in depth: **a** at 0.25 Hz; **b** at 0.5 Hz; **c** at 1.0 Hz.

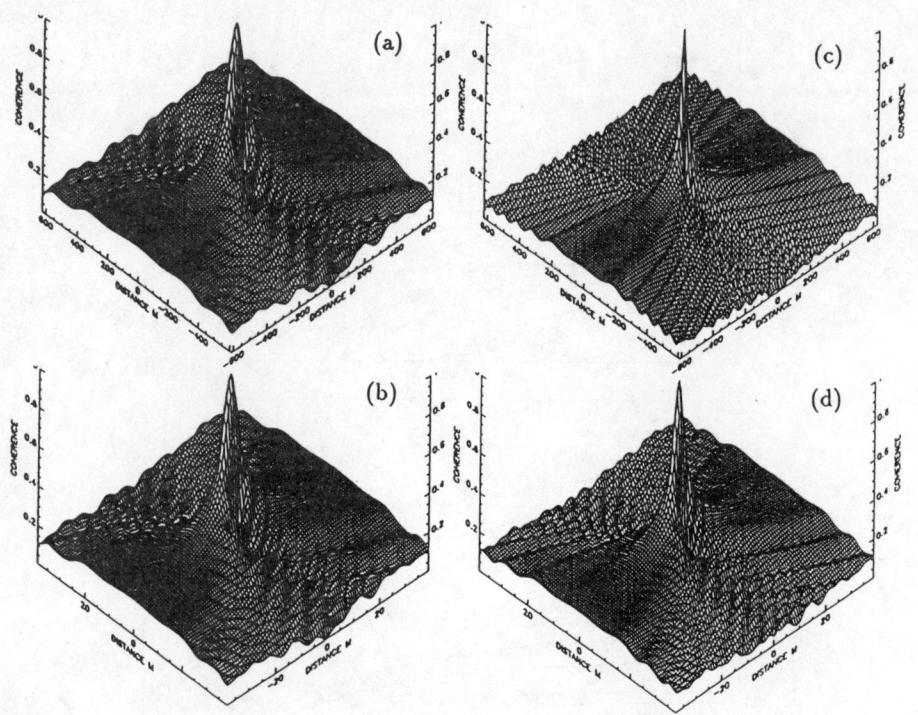

Figure 7.28: The horizontal coherence function of the PF pressure field: in an ocean of infinite depth, **a** at 0.05 Hz; **b** at 0.1 Hz; and in water of 15 m depth, **c** at 0.05 Hz; **d** at 0.1 Hz.

Under the standing-wave approximation matters again simplify and Eq.(7.41) becomes

$$N_p^{(2)}(\rho, \theta_\rho, z) = \frac{1}{\pi} \int_0^1 J_0\left(\frac{\omega}{\alpha_1}\rho u\right) u \, du$$

$$= \frac{\alpha_1}{\pi \omega \rho} J_1\left(\frac{\omega}{\alpha_1}\rho\right) \quad (7.42)$$

First-order field

According to Eq.(7.40) the coherence function of the first-order pressure field has the form of the zero-order Bessel function, $J_0(k\rho)$, where k is the horizontal wavenumber of the surface gravitational wave, the radius of coherence being the wavelength of the surface wave. Typical features

7.6 Normalised Coherency Spectra

of this function are demonstrated in Figs 7.28a to d. The calculations are made for a PM sea, wind speed 15 ms^{-1}, frequencies 0.05, 0.1 Hz, and with a value of 2 assumed for the index s of the directivity function $cos^{2s}\theta$. Figures 7.28a and b describe the deep-ocean case and Figs. 7.28c and d that for a water depth of 15 m. As expected from Eq.(7.40), the coherence curve is periodic in the wind-wave direction and smooth normal to it. The coherence distance is also seen to be much greater in deep water than in shallow. This is due to the decrease in wave speed with decreasing water depth. In both cases the coherence radius is reduced, as expected, by an increase in frequency.

Second-order field

Figures 7.29a to c show the coherence function of the second-order pressure field at depths 0, 100 and 1000 m below the surface, when the ocean depth is infinite and the effects of the seabed are ignored. The frequency of 0.2 Hz used corresponds closely to the spectral peak for a PM sea and wind speed of 15 ms^{-1}. The difference in the scale of the horizontal coordinates in the three figures, emphasises how much smaller the radius of coherence is close to the surface. This is simply because the DF pressure field is dominated by the IDF component in the upper levels of the water column. Further, as the horizontal wavenumber of the inhomogeneous component extends to 0.96 times the gravity wavenumber, the radius of coherence close to the surface is roughly twice the value of the dominant wave length, i.e. $g/\pi f^2 \simeq 78$m. With increasing observation depth the IDF component decays, the HDF component becomes more important and the radius of coherence increases. At depths where the HDF component finally dominates, the radius of coherence approaches the order of the acoustic wavelength, which is about 7.5 km at the frequency involved.

It is also apparent from Figs. 7.29a and b that the radius of coherence is not totally symmetrical about the origin, but forms an ellipse with the long axis in the direction of $\theta = 0$. This behaviour reflects the directionality of the IDF component in the prevailing wind direction. Only at large depths, where the HDF component is dominant, does the function tend to the omnidirectional form shown in Fig. 7.29c. This situation is represented more clearly in Figs. 7.30a to c where the spectra are plotted in the horizontal wavenumber plane. A definite directivity is

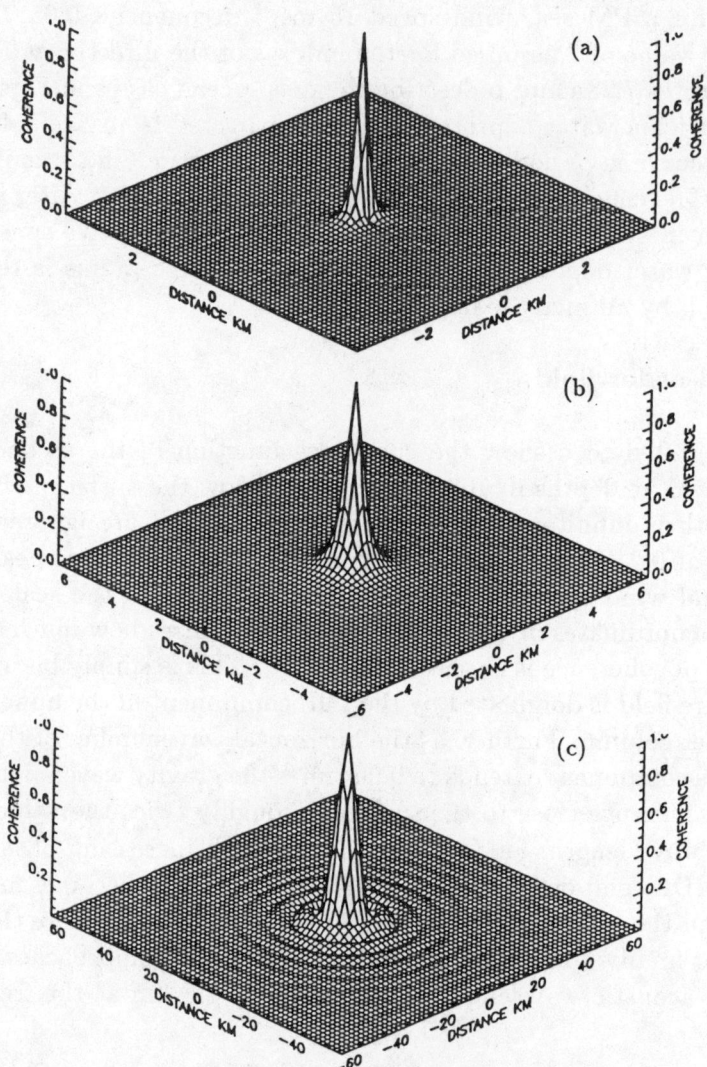

Figure 7.29: The horizontal coherence function of the DF pressure field, for a PM sea, wind speed 15 ms^{-1}, in an ocean of infinite depth: a at 0 m; b at 100 m; and c at 1000 m below the surface.

7.6 Normalised Coherency Spectra

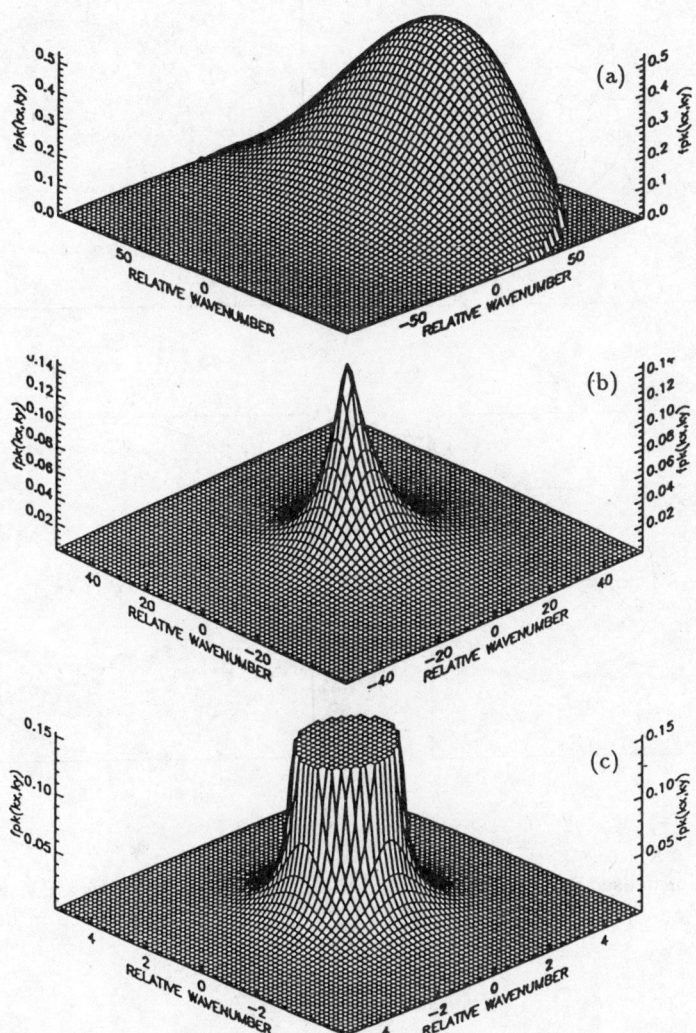

Figure 7.30: Distribution of the DF pressure spectral levels in the horizontal wavenumber plane, for a PM sea, wind speed 15 ms^{-1}, in an ocean of infinite depth: **a** at 0 m; **b** at 100 m; and **c** at 1000 m below the surface.

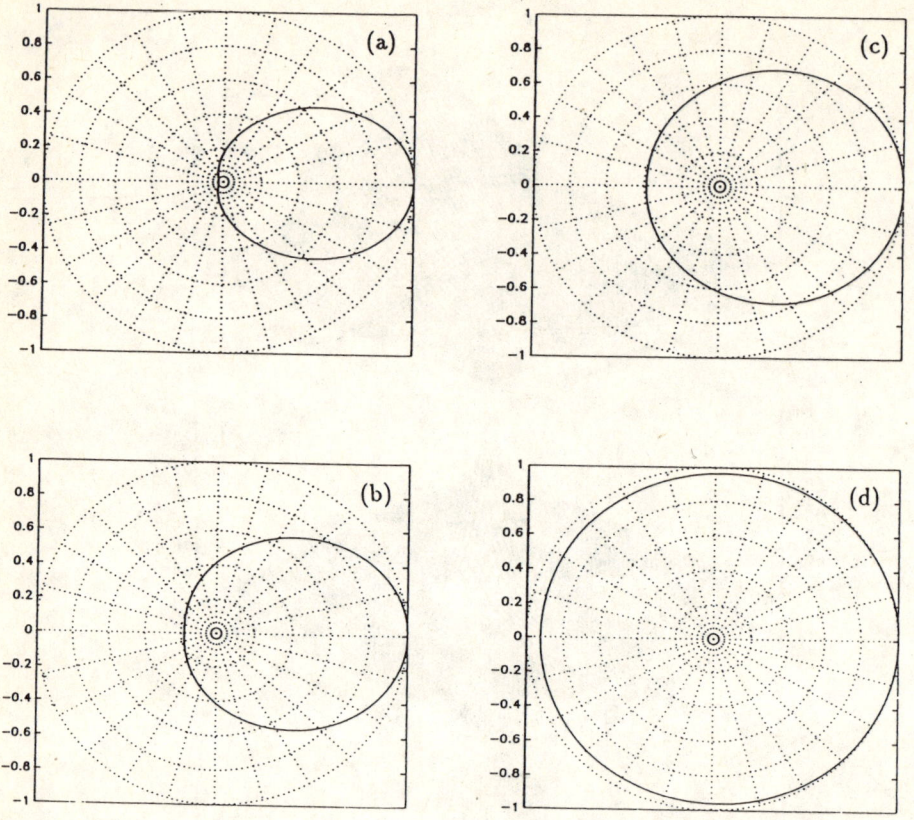

Figure 7.31: Normalised angular distribution of the DF pressure field, for a PM sea, wind speed 15 ms^{-1}, $f = 0.2$ Hz: **a** at 0 m; **b** at 50 m; **c** at 100 m; **d** at 1000 m below the surface.

7.6 Normalised Coherency Spectra

the result of the rapid decay of this component. To show the variation of the directivity of the DF field with observation depth more clearly, Figs. 7.31a to d present normalized horizontal wavenumber spectra at depths $z=0$, 50, 100 and 1000 m for a PM sea, a wind speed of 15 ms^{-1} and frequency 0.2 Hz. The wind direction is set at zero degrees.

The analysis of the coherence properties of the wave-induced noise field has important implications for inversion investigations and any study concerned with the propagation of the related noise field. The observations reported in [19] demonstrated, for instance, that in shallow water seismic noise can be excited by the direct coupling of sea-surface noise into the seabed. The excitation involves both the homogeneous and inhomogeneous components of the pressure field and in shallow water environments the latter can be particularly effective in transferring energy into the waveguide. In deep water, on the other hand, where the homogeneous component alone is effective, the excitation of the high-wavenumber (low-velocity) seismic waves observed at the seafloor is not so well understood. Studies such as those by Schreiner and Dorman [37] and Webb[80] have sought to demonstrate alternative mechanisms by which wave-induced energy can be transferred into the sedimentary layers. One of these involves scattering at basin boundaries[37]. The coherence properties of the ambient noises have implications for the design of arrays deployed for such studies, for inversion studies and for any application where wavenumber processing can be used to suppress short-wavelength noise while leaving desired longer-wavelength signals unimpaired. The analysis in this section and in Chap. 8 should be helpful to such investigations.

Chapter 8

The Seismic Response

8.1 Introduction

In this section we examine the seismic response to the wave-induced pressure field. Theoretical studies to establish the pressure and seismic fields arising from wave action have usually considered the seismoacoustic response either inside the active region or so far from it that only a few interface-wave modes are involved in the propagation of energy. In the first of these scenarios the active region is considered infinite in size. In the second the observation point is taken to be at a distance large compared with the dimensions of an active region of finite size. In [41] we presented an analysis of the response at intermediate ranges. Such an analysis is of practical importance in many situations, particularly in respect of the onshore measurements of noise generated on the continental shelf. We use this as the basis of the discussion to follow. As in [41] we make the simplifying assumptions that the properties of the sea-surface and the geoacoustic model are range-independent, but extend discussion to include all components of the seismic response. We recognise the greater value of an analysis based on a range-dependent model and are currently attempting to provide this extension. However this book is concerned primarily with the source and its immediate effects. We believe the simpler treatment of propagation should still prove informative for many applications.

Given a range-independent environment there are still two ways to model the problem. The first assumes a water layer overlying a sequence of parallel strata, with the external pressure field acting at the sea-surface. In the second the seafloor is considered as being in direct contact with the atmosphere and the wave-induced pressure field is adjusted to incorporate the effects of propagation from the ocean surface to the seafloor. The first model was used by Longuet-Higgins [25]. It is more suitable for the prediction of seismic levels at the seafloor. The second, adopted by Hasselmann[26], is more suitable for the prediction of microseisms outside the active region, for instance on shore. The second approach is more convenient for our present purposes, with the source region defined as the area of the seafloor immediately below the active region at the sea-surface. The analysis then involves the derivation of Green's functions to relate the movement of the seafloor to a point-pressure source acting on it and the use of these functions to establish expressions for the three components of the seismic response, both inside an active region of infinite size and as a function of distance from one of finite size. As extensions of [41] we introduce the general dispersion relation into wave-interaction theory, consider both the first- and second-order fields, and include porosity as a parameter of the seabed sediments.

8.2 The Green's Functions

We introduce the Green's functions, $G_v(\vec{r}, \vec{r}_0)$ and $G_h(\vec{r}, \vec{r}_0)$, linking the vertical and horizontal displacements, u_v and u_h, of the sea floor at a point $\vec{r}$ relative to a monotonic point pressure source at $\vec{r}_0$, through (see Appendix P)

$$u_{v,h}(\vec{r}, \omega) = \int_S p_b(\vec{r}_0, \omega) G_{v,h}(\vec{r}, \vec{r}_0) \, d\vec{r}_0 \qquad (8.1)$$

where S is the "source" region on the sea floor. The required Green's functions can be derived by expanding both the unit point-pressure source and the displacement field in the sedimentary medium (assumed porous viscoelastic) in terms of plane waves and applying the boundary-continuity conditions to determine the amplitudes of these waves. Owing to the continuity of the vertical component of the particle velocity at the water-sediment and sediment-sediment interfaces, the vertical displacement for any multilayered geoacoustic structure can be accounted for by simply introducing the plane-wave reflection coefficient (pressure) from the seafloor. To account for the effects of the structure on the horizontal displacement of the seafloor, however, we must at least know the seismic field in the first layer, i.e. we also require the transmission coefficients of the compressional and shear waves (and that of the slow wave if a porous-elastic sediment is involved) from water to the top layer and

8.2 The Green's Function

the corresponding reflection coefficients from the lower boundary of that layer.

In the following discussion the reflection coefficient of the seafloor is, as before, denoted by R_b, the transmission coefficients of the fast- and slow-compressional waves and the shear wave of the solid frame by T_1, T_2 and T_s respectively, and the corresponding reflection coefficients from the subinterface as R_{p1}, R_{p2} and R_s. Further, from the expression for the displacement potential on the sea floor, generated by the point-pressure source, $p_{00}\delta(\vec{r}-\vec{r}_0)$, acting at point $\vec{r}_0$ on it

$$\Phi_1(\vec{r}-\vec{r}_0,\omega) = \frac{A_0}{2\pi}\int_{-\infty}^{\infty} e^{i\vec{k}(\vec{r}-\vec{r}_0)}d\vec{k} \tag{8.2}$$

we can express the displacement produced by the three waves in the frame of the sedimentary layer as (see Appendices O and P for the definition of the coefficients of the plane-wave components in the layers)

$$\Phi_{2s1}(\vec{r},z) = \frac{A_0}{2\pi}\int_{-\infty}^{\infty}\frac{T_1}{(1+R_b)}\left[e^{-i\gamma_{21}(z+H_2)} + \epsilon_{p1}e^{i\gamma_{21}(z+H_2)}\right]e^{i\vec{k}(\vec{r}-\vec{r}_0)}d\vec{k}$$

$$\Phi_{2s2}(\vec{r},z) = \frac{A_0}{2\pi}\int_{-\infty}^{\infty}\frac{T_2}{(1+R_b)}\left[e^{-i\gamma_{22}(z+H_2)} + \epsilon_{p2}e^{i\gamma_{22}(z+H_2)}\right]e^{i\vec{k}(\vec{r}-\vec{r}_0)}d\vec{k}$$

$$\Psi_{2s}(\vec{r},z) = \frac{A_0}{2\pi}\int_{-\infty}^{\infty}\frac{T_s}{(1+R_b)}\left[e^{-i\gamma_{23}(z+H_2)} + \epsilon_{p1}e^{i\gamma_{23}(z+H_2)}\right]e^{i\vec{k}(\vec{r}-\vec{r}_0)}d\vec{k}$$

$$\tag{8.3}$$

where

$$A_0 = \frac{p_{00}}{(2\pi)\omega^2\rho_1}$$

$\gamma_{21} = \sqrt{l_{c12}^2 - k^2}, \quad \gamma_{22} = \sqrt{l_{c22}^2 - k^2}, \quad \gamma_{23} = \sqrt{l_{s2}^2 - k^2}, \quad k^2 = k_x^2 + k_y^2,$

and l_{c12}, l_{c22} and l_{s2} are respectively the wavenumbers of the fast-and slow-compressional waves and the shear wave (see Appendix F). Introducing the displacement vector,

$$\vec{u}_0 = \nabla(\Phi_{2s1} + \Phi_{2s2}) + \nabla \times \vec{\Psi}_{2s} \tag{8.4}$$

where $\vec{\Psi}_{2s} = (\vec{k}_0 \times \vec{z}_0)\Psi_{2s}$ with $\vec{k}_0$ and $\vec{z}_0$ unit vectors in the direction of the horizontal wavenumber vector and the z-axis, we finally establish the required Green's functions as (Appendix P, Eq.(P.26))

$$G_{v,h}(\vec{r},\vec{r}_0) = \frac{-i\omega}{2\pi\rho_1\alpha_1^3}\int_0^{\infty}\left(\frac{1-R_b}{1+R_b}\right)R_{v,h}J_{0,1}(u\xi)\sqrt{1-u^2}\,u\,du \tag{8.5}$$

where R_v, J_0, and R_h, J_1 relate to G_v and G_h respectively,

$$\xi = \frac{\omega}{\alpha_1}|\vec{r} - \vec{r}_0|$$

$$R_v = \frac{\gamma_{21}A_{21}(1 - \bar{\epsilon}_{p1}) + \gamma_{22}A_{22}(1 - \bar{\epsilon}_{p2}) + kA_{23}(1 + \bar{\epsilon}_s)}{\gamma_{21}A_{21}(1 - \bar{\epsilon}_{p1})(1 - \delta_{12}) + \gamma_{22}A_{22}(1 - \bar{\epsilon}_{p2})(1 - \delta_{22}) + kA_{23}(1 + \bar{\epsilon}_s)(1 - \delta_{32})}$$

(8.6)

and

$$R_h = -i \cdot$$
$$\cdot \frac{kA_{21}(1 + \bar{\epsilon}_{p1}) + kA_{22}(1 + \bar{\epsilon}_{p2}) - \gamma_{23}A_{23}(1 - \bar{\epsilon}_s)}{\gamma_{21}A_{21}(1 - \bar{\epsilon}_{p1})(1 - \delta_{12}) + \gamma_{22}A_{22}(1 - \bar{\epsilon}_{p2})(1 - \delta_{22}) + kA_{23}(1 + \bar{\epsilon}_s)(1 - \delta_{32})}$$

(8.7)

In the above

$$\bar{\epsilon}_{p1} \equiv \epsilon_{p1}e^{i\gamma_{21}h_2}, \qquad \bar{\epsilon}_{p2} \equiv \epsilon_{p2}e^{i\gamma_{22}h_2}, \qquad \bar{\epsilon}_s \equiv \epsilon_s e^{i\gamma_{23}h_2},$$

$h_2 = H_2 - H_1$ is the thickness of the sublayer, and the parameters δ_{12}, δ_{22} and δ_{32} are the amplitude ratios of the three Biot waves in the pore fluid and the solid frame, - see Chap. 3.

8.3 The Coherence Function of the Seismic Field

Referring to Sect.7.6 and Eq.(8.1), the coherence function of the seismic field at the seafloor can be expressed as

$$C_{v,h}^{(i)}(\vec{r}_1, \vec{r}_2, \omega) = \langle u_{v,h}^{(i)}(\vec{r}_1, \omega) u_{v,h}^{(i)*}(\vec{r}_2, \omega) \rangle \frac{d\omega}{(2\pi)^2}$$

$$= \int \int_S \langle p_b^{(i)}(\vec{r}_{01}, \omega) p_b^{(i)*}(\vec{r}_{02}, \omega) \rangle \frac{d\omega}{(2\pi)^2} G_{v,h}(\vec{r}_1, \vec{r}_{01}) G_{v,h}^*(\vec{r}_2, \vec{r}_{02}) \, d\vec{r}_{01} \, d\vec{r}_{02}$$

(8.8)

Defining

$$p_b^{(i)}(\vec{r}_0, \omega) = \frac{1}{(2\pi)^2} \int p_b^{(i)}(\vec{k}, \omega) e^{i\vec{k}\vec{r}_0} d\vec{k}$$

8.3 Coherence Function

and

$$C_p^{(i)}(\vec{r}_{01}, \vec{r}_{02}, \omega) = \langle p_b^{(i)}(\vec{r}_{01}, \omega) p_b^{(i)*}(\vec{r}_{02}, \omega)\rangle \frac{d\omega}{(2\pi)^2}$$

$$= \frac{d\vec{k}\, d\omega}{(2\pi)^6} \int \langle |p_b^{(i)}(\vec{k}, \omega)|^2 \rangle e^{i\vec{k}(\vec{r}_{01}-\vec{r}_{02})} d\vec{k}$$

$$= \int f_p^{(i)}(\vec{k}, \omega, -H_1) e^{i\vec{k}(\vec{r}_{01}-\vec{r}_{02})} d\vec{k} \quad (8.9)$$

where

$$f_p^{(i)}(\vec{k}, \omega, -H_1) = \frac{d\vec{k}\, d\omega}{(2\pi)^6} \langle |p_b^{(i)}(\vec{k}, \omega)|^2 \rangle$$

Eq.(8.8) can be expressed as

$$C_{v,h}^{(i)}(\vec{r}_1, \vec{r}_2, \omega) =$$
$$\int f_p^{(i)}(\vec{k}, \omega, -H_1) \left[\int_S \int_S G_{v,h}(\vec{r}_1, \vec{r}_{01}) G_{v,h}^*(\vec{r}_2, \vec{r}_{02}) e^{i\vec{k}(\vec{r}_{01}-\vec{r}_{02})} d\vec{r}_{01}\, d\vec{r}_{02} \right] d\vec{k}$$

(8.10)

or alternatively as

$$C_{v,h}^{(i)}(\vec{r}_1, \vec{r}_2, \omega) = \int f_p^{(i)}(\vec{k}, \omega, -H_1) \Gamma_{v,h}(\vec{k}, \omega, \vec{r}_1) \Gamma_{v,h}^*(\vec{k}, \omega, \vec{r}_2)\, d\vec{k} \quad (8.11)$$

where

$$\Gamma_{v,h}(\vec{k}, \omega, \vec{r}) = \int_S G_{v,h}(\vec{r}, \vec{r}_0) e^{i\vec{k}\vec{r}_0} d\vec{r}_0 \quad (8.12)$$

Further, the pressure spectrum on the seafloor can be expressed in terms of the equivalent source-pressure spectra of the first- and second-order fields at the sea surface and a transfer function, ie.,

$$f_p^{(i)}(\vec{k}, \omega, -H_1) = f_{p0}^{(i)}(\vec{k}, \omega) |T_p^{(i)}(\vec{k}, -H_1)|^2 \quad (8.13)$$

where, referring to Eqs.(N.6), (N.8), (N.9) and (N.21) of Appendix N,

$$f_{p0}^{(1)}(\vec{k}, \omega) = \rho_1^2 g^2 F_a(\omega) H(\theta_k) \frac{d\omega}{k\, dk} \quad (8.14)$$

$$f_{p0}^{(2)}(\vec{k}, \omega) = \rho_1^2 \omega^2 \langle |B^{(2)}(\vec{k}, \omega)|^2 \rangle \quad (8.15)$$

$$|T_p^{(1)}(\vec{k}, -H_1)|^2 = \left| \frac{1 + R_b}{1 + R_b e^{(\gamma_- - \gamma_+)H_1}} e^{-\gamma_+ H_1} \right|^2 \quad (8.16)$$

$$|T_p^{(2)}(\vec{k}, -H_1)|^2 = \left|\frac{1+R_b}{1+(\epsilon/b)R_b e^{(\gamma_- - \gamma_+)H_1}} e^{-\gamma_+ H_1}\right|^2 \qquad (8.17)$$

The coherence function then becomes

$$C_{v,h}^{(i)}(\vec{r}_1, \vec{r}_2, \omega) = \int f_{p0}^{(i)}(\vec{k}, \omega)|T_p^{(i)}(\vec{k}, -H_1)|^2 \Gamma_{v,h}(\vec{k}, \omega, \vec{r}_1)\Gamma_{v,h}^*(\vec{k}, \omega, \vec{r}_2)\,d\vec{k} \qquad (8.18)$$

and the power-density spectrum

$$F_{v,h}^{(i)}(\vec{r}, \omega) = \int f_{p0}^{(i)}(\vec{k}, \omega)|T_p^{(i)}(\vec{k}, -H_1)|^2 |\Gamma_{v,h}(\vec{k}, \omega, \vec{r})|^2\,d\vec{k} \qquad (8.19)$$

Equations (8.18) and (8.19) can be used to predict numerically the spectral characteristics of the seismic field in any given wave guide. In the following examples two important extreme cases are considered. They provide a useful 'check' on the numerical calculations and a test of the reliability of the procedures prior to considering more general situations.

8.3.1 Source Region of Infinite Size

We start with the simplest case in which the active wave region is infinite in size. In this situation the function $\Gamma_v(\vec{k}, \omega, \vec{r})$ in Eq.(8.12) takes the form

$$\begin{aligned}\Gamma_v(\vec{k}, \omega, \vec{r}_1) &= \int_S G_v(\vec{r}_1, \vec{r}_{01}) e^{i\vec{k}\vec{r}_{01}} d\vec{r}_{01} \\ &= \int_0^\infty f_v(u) u \left\{ \int J_0\left(u\frac{\omega}{\alpha_1}|\vec{r}_1 - \vec{r}_{01}|\right) e^{i\vec{k}\vec{r}_{01}} d\vec{r}_{01} \right\} du \end{aligned}$$

From Eq.(P.26) of Appendix P (or Eq.(8.5))

$$f_{v,h}(u) = \frac{-i\omega}{2\pi\rho_1\alpha_1^3}\left(\frac{1-R_b}{1+R_b}\right) R_{v,h}(u)\sqrt{1-u^2} \qquad (8.20)$$

so that

$$\Gamma_v(\vec{k}, \omega, \vec{r}_1)\Gamma_v^*(\vec{k}, \omega, \vec{r}_1) = e^{i\vec{k}(\vec{r}_1-\vec{r}_2)} \int_0^\infty \int_0^\infty f_v(u') f_v^*(u'') u' u'' \cdot \\ \cdot \left\{ \int J_0(k'\rho') e^{-i\vec{k}\vec{\rho}'} d\vec{\rho}' \int J_0(k''\rho'') e^{i\vec{k}\vec{\rho}''} d\vec{\rho}'' \right\} du'\,du''$$

where

$$\vec{\rho}' = \vec{r}_1 - \vec{r}_{01}, \qquad \vec{\rho}'' = \vec{r}_2 - \vec{r}_{02}, \qquad k' = \frac{\omega}{\alpha_1}u', \qquad k'' = \frac{\omega}{\alpha_1}u''$$

8.3 Coherence Function

The inner integration becomes

$$I' = \int J_0(k'\rho')e^{-i\vec{k}\vec{\rho}'}d\vec{\rho}' = 2\pi \int_0^\infty J_0(k'\rho')J_0(k\rho')\rho'\, d\rho'$$

$$= \frac{2\pi}{k}\delta(k'-k) = \frac{2\pi\alpha_1^2}{\omega^2}\frac{1}{u}\delta(u'-u)$$

and similarly

$$I_2'' = \frac{2\pi\alpha_1^2}{\omega^2}\frac{1}{u}\delta(u''-u)$$

This leads finally to

$$\Gamma_v(\vec{k},\omega,\vec{r}_1)\Gamma_v^*(\vec{k},\omega,\vec{r}_2) = \frac{1}{\rho_1^2\alpha_1^2\omega^2}\left|\frac{1-R_b}{1+R_b}\right|^2|R_v(u)|^2|1-u^2|e^{i\vec{k}(\vec{r}_1-\vec{r}_2)} \quad (8.21)$$

Similarly, from Eq.(P.13) of Appendix P, the horizontal components can be established as

$$\Gamma_x(\vec{k},\omega,\vec{r}_1) = \int G_x(\vec{r}_1,\vec{r}_{01})e^{i\vec{k}\vec{r}_{01}}d\vec{r}_{01}$$

$$= \frac{-1}{(2\pi)^2\omega^2\rho_1}\int \frac{k_x'}{k'}\left(\frac{1-R_b}{1+R_b}\right)\gamma_1'R_h e^{i\vec{k}'\vec{r}_1}\left\{\int e^{-i(\vec{k}'-\vec{k})\vec{r}_{01}}d\vec{r}_{01}\right\}d\vec{k}'$$

$$= \frac{-1}{\omega^2\rho_1}\frac{k_x}{k}\left(\frac{1-R_b}{1+R_b}\right)\gamma_1 R_h e^{i\vec{k}\vec{r}_1}$$

$$\Gamma_y(\vec{k},\omega,\vec{r}_1) = \frac{-1}{\omega^2\rho_1}\frac{k_y}{k}\left(\frac{1-R_b}{1+R_b}\right)\gamma_1 R_h e^{i\vec{k}\vec{r}_1}$$

It follows that

$$\Gamma_h(\vec{k},\omega,\vec{r}_1)\Gamma_h^*(\vec{k},\omega,\vec{r}_2) = \Gamma_x(\vec{k},\omega,\vec{r}_1)\Gamma_x^*(\vec{k},\omega,\vec{r}_2) + \Gamma_y(\vec{k},\omega,\vec{r}_1)\Gamma_y^*(\vec{k},\omega,\vec{r}_2)$$

$$= \frac{1}{\rho_1^2\omega^2\alpha_1^2}\left|\frac{1-R_b}{1+R_b}\right|^2|R_h(u)|^2|1-u^2|e^{i\vec{k}(\vec{r}_1-\vec{r}_2)} \quad (8.22)$$

Substituting Eqs.(8.21) and (8.22) into (8.18) and (8.19) then leads to the required coherence functions and power-density spectra.

First-order field

The coherence functions and power-density spectra for the first-order field now become

$$C_{v,h}^{(1)}(\vec{r}_1,\vec{r}_2,\omega) = F_{v,h}^{(1)}(\omega)e^{i\vec{k}(\vec{r}_1-\vec{r}_2)}\Big|_{k=k(\omega)} \quad (8.23)$$

and

$$F_{v,h}^{(1)}(\omega) = \frac{g^2}{\omega^2 \alpha_1^2} F_a(\omega) \left| \frac{1-R_b}{1+R_b e^{(\gamma_- - \gamma_+)H_1}} \right|^2 \left| e^{-\gamma_+ H_1} \right|^2 |R_{v,h}(u)|^2 |1-u^2| \quad (8.24)$$

Expressed in terms of frequency, f, instead of ω, Eq.(8.24) becomes

$$F_{v,h}^{(1)}(f) = \frac{g^2}{(2\pi f)^2 \alpha_1^2} F_a(f) \left| \frac{1-R_b}{1+R_b e^{(\gamma_- - \gamma_+)H_1}} \right|^2 \left| e^{-\gamma_+ H_1} \right|^2 |R_{v,h}(u)|^2 |1-u^2|$$

where u relates to the frequency through the appropriate dispersion relation.

Second-order field

The equivalent expressions for the second-order field are

$$C_{v,h}^{(2)}(\vec{r}_1, \vec{r}_2, \omega) = \int f_{v,h}^{(2)}(\vec{k}, \omega, -H_1) e^{i\vec{k}(\vec{r}_1 - \vec{r}_2)} d\vec{k} \quad (8.25)$$

and

$$F_{v,h}^{(2)}(\omega) = \int f_{v,h}^{(2)}(\vec{k}, \omega, -H_1) d\vec{k} \quad (8.26)$$

where

$$\begin{aligned}f_{v,h}^{(2)}(\vec{k}, \omega, -H_1) &= \frac{1}{\alpha_1^2} \langle |B^{(2)}(\vec{k}, \omega)|^2 \rangle \left| \frac{1-R_b}{1+(\epsilon/b)R_b e^{(\gamma_- - \gamma_+)H_1}} \right|^2 \left| e^{-\gamma_+ H_1} \right|^2 \cdot \\ &\quad \cdot |R_{v,h}(u)|^2 |1-u^2| \end{aligned} \quad (8.27)$$

Therefore,

$$F_{v,h}^{(2)}(\omega) = \frac{\omega^2}{\alpha_1^4} \int \left| \frac{1-R_b}{1+(\epsilon/b)R_b e^{(\gamma_- - \gamma_+)H_1}} \right|^2 \left| e^{-\gamma_+ H_1} \right|^2 |R_{v,h}(u)|^2 |1-u^2| \cdot$$
$$\cdot \left\{ \int_0^{2\pi} \langle |B^{(2)}(\vec{k}, \omega)|^2 \, d\theta_k \rangle \right\} u \, du$$

Using the substitution

$$|E_v(u, \omega, -H_1)|^2 = \left| \frac{1-R_b}{1+\frac{\epsilon}{b}R_b e^{(\gamma_- - \gamma_+)H_1}} \right|^2 \left| e^{-\gamma_+ H_1} \right|^2$$

and Eq.(N.20) of Appendix N for $\langle |B^{(2)}(\vec{k}, \omega)|^2 \rangle$, we finally establish

$$F_{v,h}^{(2)}(\omega) = \frac{\omega g^2}{2\alpha_1^4 V(k,\omega)} F_a^2(\omega/2) \cdot$$
$$\cdot \int_0^\infty |E_v(u, \omega, -H_1)|^2 |R_{v,h}|^2 \bar{I}(u, \omega) |1-u^2| u \, du \quad (8.28)$$

8.3 Coherence Function

where
$$V(k,\omega) \equiv \left|1 - \frac{g\gamma_+}{\omega}\right|^2$$

and
$$\bar{I}(u,\omega) = \int_0^{2\pi} I^{(2)}(u,\theta_k,\omega)\, d\theta_k \tag{8.29}$$

with
$$I^{(2)}(u,\theta_k,\omega) = \frac{2}{\omega g^2} \int_0^{2\pi} |\{\vec{q}_1,\vec{q}_2\}|^2 \frac{\sigma_1^2 \sigma_2^2}{q_2} \left(\frac{\partial \sigma_1}{\partial q_1}\right)\left(\frac{\partial \sigma_2}{\partial q_2}\right) \frac{\partial q_1}{\partial(\sigma_1+\sigma_2)} \cdot$$
$$\cdot \frac{F_a(\sigma_1) F_a(\sigma_2)}{F_a^2(\omega/2)} H(\theta_1) H(\theta_2)\, d\theta_1 \tag{8.30}$$

Except for the modifying factor, $V(k,\omega)$, the vertical component of Eq.(8.28) is identical to Eq.(16) of [41].

When the deep-ocean dispersion relation can be used Eq.(8.30) reduces to a form which has its equivalent in Eq.(6.20)

$$I^{(2)}(u,\theta_k,\omega) \cong \int_0^{2\pi} \frac{(1-\cos\theta_{12})^2 \chi^2}{(1+\chi)^2 \left[\chi^2 - \left(\frac{1}{2}\hat{m}\hat{t}+1\right)\chi + \left(1-\frac{1}{2}\hat{m}\hat{t}\right)\right]} \cdot$$
$$\cdot \frac{F_a(\frac{\omega}{1+\chi}) F_a(\frac{\omega\chi}{1+\chi})}{F_a^2(\omega/2)} H(\theta_1) H(\theta_2)\, d\theta_1 \tag{8.31}$$

8.3.2 Outside a Source Region of Finite Size

The other extreme case of interest occurs when observations are made at locations sufficiently far from an active region of finite size that the asymptotic form of the Bessel function can be used in evaluating the seismic response, viz.,

$$J_\nu(u\frac{\omega}{\alpha_1}|\vec{r}_1 - \vec{r}_0|) \simeq \left(\frac{2\alpha_1}{\pi\omega u|\vec{r}_1 - \vec{r}_0|}\right)^{1/2} \cos(\frac{\omega}{\alpha_1} u|\vec{r}_1 - \vec{r}_0| - \frac{\pi}{4} - \nu\frac{\pi}{2})$$

If r_1 is considerably greater than the equivalent radius of the active region, this expression can be further simplified to

$$J_0(u\frac{\omega}{\alpha_1}|\vec{r}_1 - \vec{r}_0|) \simeq \left(\frac{2\alpha_1}{\pi\omega u r_1}\right)^{1/2} \cos\left[\frac{u\omega}{\alpha_1}(r_1 - \frac{\vec{r}_1 \vec{r}_0}{r_1}) - \frac{\pi}{4}\right]$$

and

$$J_1(u\frac{\omega}{\alpha_1}|\vec{r}_1 - \vec{r}_0|) \simeq \left(\frac{2\alpha_1}{\pi\omega u r_1}\right)^{1/2} \sin\left[\frac{u\omega}{\alpha_1}(r_1 - \frac{\vec{r}_1 \vec{r}_0}{r_1}) - \frac{\pi}{4}\right]$$

The Green's functions, Eq.(8.5), can now be written as

$$G_{v,h}(\vec{r}_1, \vec{r}_{01}) = \left(\frac{\alpha_1}{2\pi\omega r_1}\right)^{1/2} e^{-i\nu\pi/2} \left\{ \int_0^\infty f_{v,h}(u') e^{i\left[\frac{\omega}{\alpha_1}u'(r_1 - \frac{1}{r_1}\vec{r}_1\vec{r}_{01}) - \frac{\pi}{4}\right]} \sqrt{u'} du' \right.$$
$$\left. + (-1)^\nu \int_0^\infty f_{v,h}(u') e^{-i\left[\frac{\omega}{\alpha_1}u'(r_1 - \frac{1}{r_1}\vec{r}_1\vec{r}_{01}) - \frac{\pi}{4}\right]} \sqrt{u'} du' \right\} \quad (8.32)$$

where $\nu = 0$ for the vertical component and $\nu = 1$ for the horizontal component. The function $\Gamma_{v,h}$ of Eq.(8.12) then takes the form

$$\Gamma_{v,h}(\vec{k}, \omega, \vec{r}_1) = \left(\frac{\alpha_1}{2\pi\omega r_1}\right)^{1/2} e^{-i\nu\pi/2} \left\{ \int_0^\infty f_{v,h}(u') e^{i(\frac{\omega}{\alpha_1}u'r_1 - \frac{\pi}{4})} I(\vec{k}, u') \sqrt{u'} du' \right.$$
$$\left. + (-1)^\nu \int_0^\infty f_{v,h}(u') e^{-i(\frac{\omega}{\alpha_1}u'r_1 - \frac{\pi}{4})} I(\vec{k}, u') \sqrt{u'} du' \right\} \quad (8.33)$$

where

$$I(\vec{k}, u') = \int\int e^{\frac{i\omega}{\alpha_1} r_{01}[u\cos(\theta_{01} - \theta_k) + u'\cos(\theta_{01} - \theta_1)]} r_{01} d\theta_{01} dr_{01}$$
$$= \int\int e^{\frac{i\omega}{\alpha_1} a r_{01}\cos(\theta_{01} - \phi)} r_{01} d\theta_{01} dr_{01} \quad (8.34)$$

or if, for convenience, the area S is assumed circular with radius r_s,

$$I(\vec{k}, u') = 2\pi \int_0^{r_s} J_0\left(\frac{\omega a}{\alpha_1} r_{01}\right) r_{01} dr_{01} = \frac{2\pi r_s \alpha_1}{\omega a} J_1\left(\frac{\omega a}{\alpha_1} r_s\right) \quad (8.35)$$

where

$$a = \left[u^2 + u'^2 + 2uu'\cos(\theta_k - \theta_1)\right]^{1/2} \quad (8.36)$$

Substituting Eq.(8.35) into Eq.(8.33) leads, after some manipulation, to

$$\Gamma_v(\vec{k}, \omega, \vec{r}_1) = (2\frac{\alpha_1}{\omega})^{3/2} \sqrt{\frac{S}{r_1}} \int_0^\infty f_v(u') \frac{1}{a} J_1\left(\frac{\omega a r_s}{\alpha_1}\right) \cdot$$
$$\cdot \cos\left(\frac{\omega}{\alpha_1} u' r_1 - \frac{\pi}{4}\right) \sqrt{u'} du' \quad (8.37)$$

and

$$\Gamma_h(\vec{k}, \omega, \vec{r}_1) = (2\frac{\alpha_1}{\omega})^{3/2} \sqrt{\frac{S}{r_1}} \int_0^\infty f_h(u') \frac{1}{a} J_1\left(\frac{\omega a r_s}{\alpha_1}\right) \cdot$$
$$\cdot \sin\left(\frac{\omega}{\alpha_1} u' r_1 - \frac{\pi}{4}\right) \sqrt{u'} du' \quad (8.38)$$

Equations (8.37) and (8.38) can also be expressed in the form

$$\Gamma_{v,h}(\vec{k}, \omega, \vec{r}_1) = 2(\frac{\alpha_1}{\omega})^{3/2} \sqrt{\frac{S}{r_1}} \left\{ (-1)^\nu \int_0^\infty f_{v,h}(u') \frac{1}{a} J_1(\xi_s a) \cos(\xi_1 u') \sqrt{u'} du' + \right.$$
$$\left. \int_0^\infty f_{v,h}(u') \frac{1}{a} J_1(\xi_s a) \sin(\xi_1 u') \sqrt{u'} du' \right\} \quad (8.39)$$

8.4 Spectral Characteristics

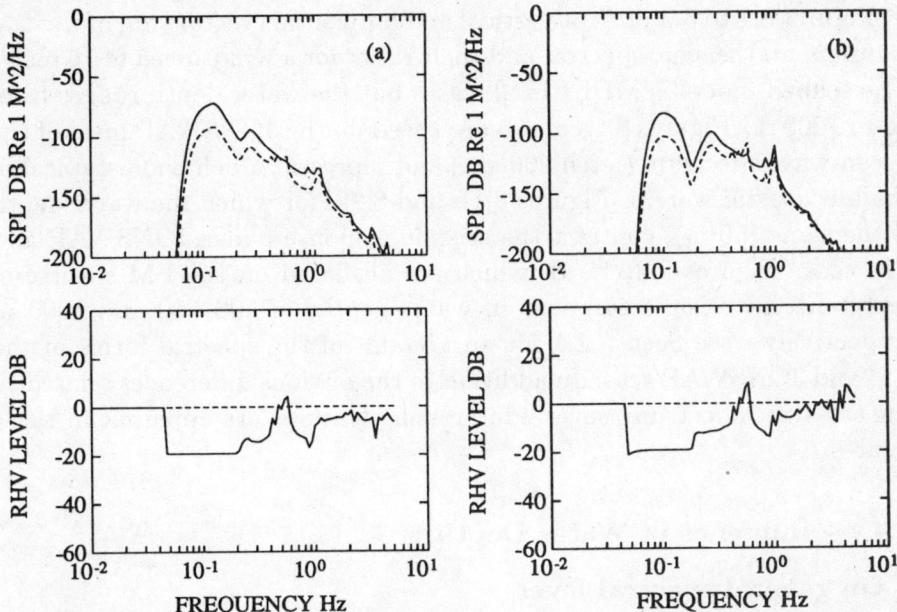

Figure 8.1: Spectral levels of the vertical (solid) and horizontal (dashed) components of the seabed displacement, and their ratio, for model MDL1, wind speed 15 ms^{-1}, and a JONSWAP sea: **a** $H_1 = 15$ m; **b** $H_1 = 50$ m.

which is more suitable for the use of cosine and sine transforms. Here

$$\xi_s = \frac{\omega}{\alpha_1} r_s, \quad \xi_1 = \frac{\omega}{\alpha_1} r_1 \tag{8.40}$$

and, as before, $\nu = 0$ for Γ_v and $\nu = 1$ for Γ_h.

It is worth noting that by expressing the Bessel functions in Eq.(8.5) in terms of the corresponding Hankel functions and applying the residue theorem in the calculation of the Green's functions, the coherence function can be established as the sum of interface wave modes, a form more suited to the study of propagation effects and the coherence properties of the seismic field at distance from the source.

8.4 Spectral Characteristics of the Seismic Field

In this section we use Eqs.(8.28) and (8.29) to calculate the wave-induced seismic spectra for an active region of infinite size and examine their dependence on the spectral form of the ocean-wave field, water depth, wind speed and seabed structure.

Figures 8.1 to 8.3 present vertical and horizontal (vector sum of the two components) seismic spectra, and their ratio, for a wind speed of 15 ms^{-1}. The seabed model is MDL1 in all cases but the water depth ranges from 15 to 1000 m. Figures 8.1a and b are based on the JONSWAP form of the ocean-wave spectrum (fetch 200 km) and represent a behaviour typical of shallow coastal waters. Figures 8.1b and 8.2a, for which the water depth is the same (50 m), contrast the seismic response under JONSWAP and PM seas. Figures 8.2b to d, which are all based on the PM spectrum, represent the seismic response in water depths of 100, 500 and 1000 m respectively - see Sect. 7.2.1 for an account of the spectral forms of the PM and JONSWAP seas. In addition to the obvious differences related to the sea-state spectrum, several interesting features are apparent in these figures.

8.4.1 Influence of Water Depth

On general spectral level

Figures 8.1a and b, based on model MDL1 and the JONSWAP sea, demonstrate that the horizontal and vertical spectral levels both drop when the water depth increases from 15 to 50 m. The decrease is about 10 dB for the PF component (the first peak) and about 10 to 15 dB for the DF component. A dependence on water depth is also apparent in the spectra of Fig. 8.2, again based on model MDL1 but a PM sea. These changes can be roughly equated to the effect of the factor $20\, log_{10}|e^{-\gamma_+ H_1}|$ in Eq.(8.27), the value $\gamma_+ \sim k$ being defined by the dispersion relation $\omega^2 = gk \cdot tanh(kH_1)$. Under the influence of this factor the PF component reduces progressively as H_1 increases, while the main peak of the DF component tends to a stable value around - 100 dB re 1 m^2/Hz at depths where the inhomogeneous component becomes negligible. (The minor peaks at higher frequencies are related to mode effects - see Sect. 7.3.)

On the spectral ratio

Of more interest is the effect of water depth on the ratio of the spectral levels of the horizontal and vertical components. As Figs. 8.1a and b show the ratio is basically negative for $H_1 = 15$ and 50 m, except for a prominent peak around 0.5 to 0.7 Hz. At even greater water depths the ratio becomes positive above 0.3 Hz and exceeds 20 dB when $H_1 =$

8.4 Spectral Characteristics

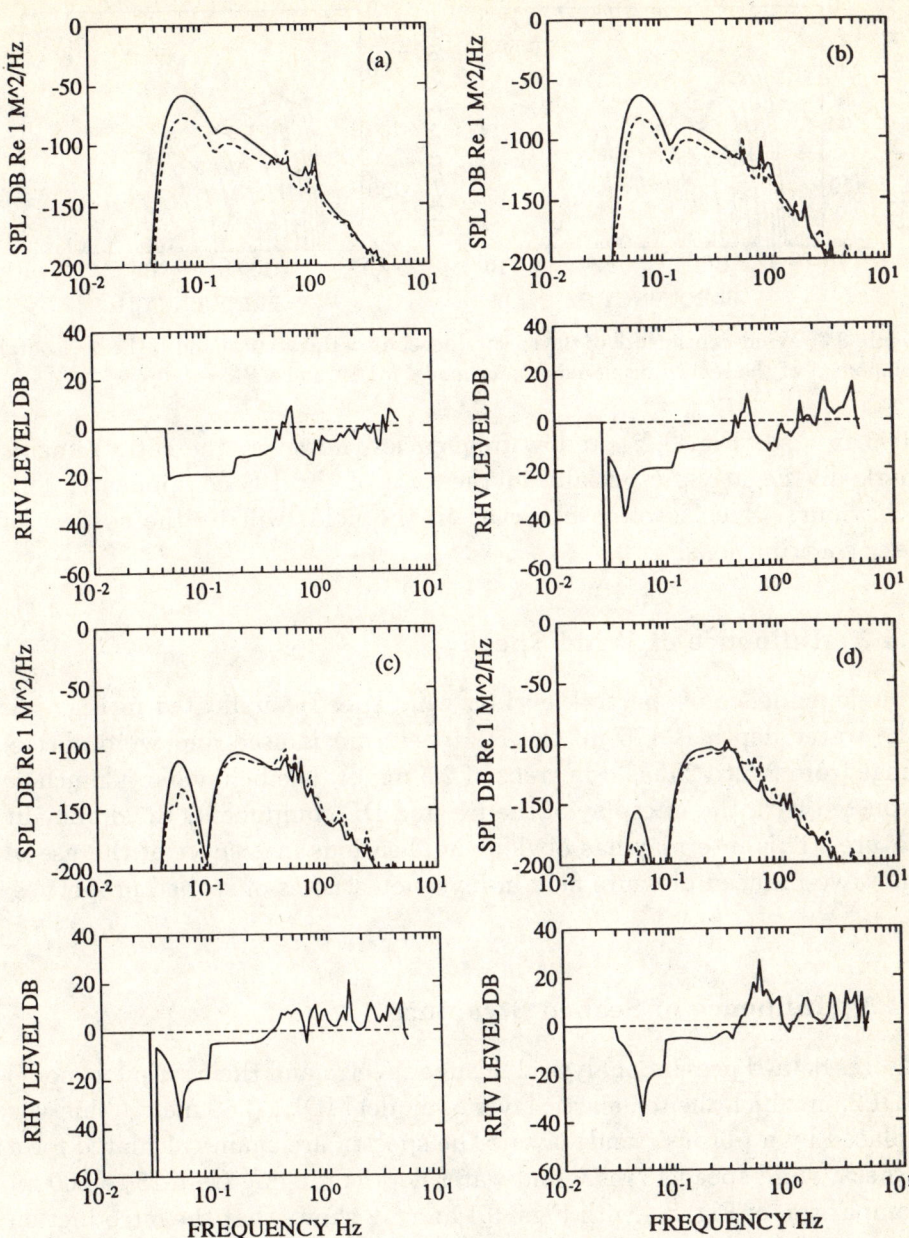

Figure 8.2: Spectral levels of the vertical (solid) and horizontal (dashed) components of the seabed displacement, and their ratio, for model MDL1, wind speed 15 ms^{-1}, and a PM sea: a $H_1 = 50$ m; b $H_1 = 100$ m; c $H_1 = 500$ m; d $H_1 = 1000$ m.

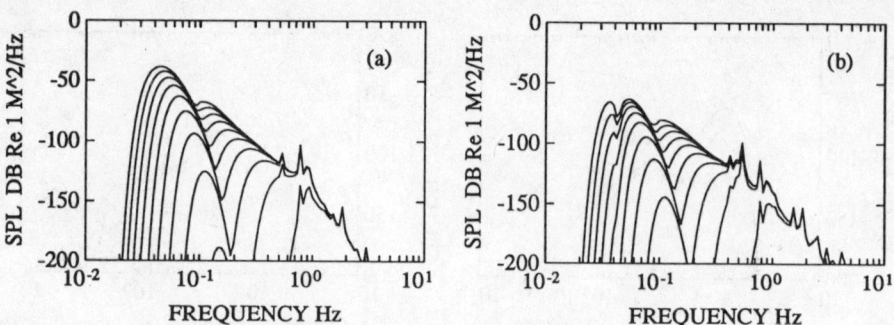

Figure 8.3: Wind-dependence of the spectral levels of a the vertical and b the horizontal component of the seabed displacement, for model MDL1 and a PM sea.

1000 m - see Fig. 8.2. At low frequencies, however, the ratio remains markedly negative, especially in the case of the PF component. This behaviour, which is often observed in the field, will be the subject of later contributions.

8.4.2 Influence of Wind Speed

The dependence of spectral level on wind speed is indicated in Fig. 8.3. The water depth is 100 m, the PM spectrum is used and wind speeds range from 2.5 to 25 ms^{-1} in steps of 2.5 ms^{-1}. The behaviour is much as expected with the interplay of the PF and DF components the dominant feature. This interplay has obvious implications in respect of the use of the wave-induced pressure field in inversion studies of seabed properties.

8.4.3 Influence of Seabed Structure

Figures 8.4a-d present analytical seismic spectra and their ratio for model MDL2, in which the top elastic layer of model MDL1 (260 metres thick) is replaced by a porous (sand) layer. The spectra are again calculated for a PM sea, wind speed 15 ms^{-1} and water depths ranging from 15 to 500 m. Comparison of Fig. 8.4 with Figs. 8.1 and 8.2 shows that the introduction of the porous layer results in a significant increase in spectral level. This increase is the result of the "softening" of the top layer discussed in Sect. 7.5 and is similar to that reported in our earlier paper[41]. In that case the seismic spectral level inside the active wave region was shown to increase by up to 40 to 50 dB when the solid basement was replaced by an unconsolidated (liquid) layer. The behaviour of the spectral ratio, however, remains similar to that observed for MDL1.

Figures 8.5a to c present spectra and their ratio for the three hydrate structures defined in Sect. 7.2.2. The wind speed is again 15 ms^{-1} and the sea PM in form, but the water depth is 1000 m in all cases. Since the water is deep the horizontal components are once again higher than the vertical (up to 20 dB), in this case at all frequencies above 0.1 Hz.

8.4.4 Influence of Layer Thickness

To examine the influence of the thickness of the upper sedimentary layer we assume, for convenience, that the seabed can be represented by an unconsolidated layer overlying a solid, homogeneous halfspace, characterised by compressional and shear-wave speeds of 4100 ms^{-1} and 2367 ms^{-1} respectively and Q-values of 450 and 225 (shear-wave). The water depth is taken to be 50 m. The thickness of the unconsolidated layer is allowed to vary from 10 - 1000 m and two versions of it are considered. In the first the layer is taken to be porous with the properties of that in model MDL2; in the other it is regarded as elastic with a density 3000 kg m^{-3} and compressional and shear-wave speeds identical to those of the porous layer. The wind speed is again 15 ms^{-1} and a PM sea is assumed.

In the absence of colour Figs. 8.6a to d attempt to present the behaviour of the ratio H/V, by contour plots of the level, $20\log_{10}(H/V)$, as a function of frequency and thickness. Figures 8.6a and b describe the behaviour in the elastic layer by presenting contour levels for $H/V \leq 1$ and $H/V > 1$ respectively. Figures 8.6c and d give the same information for the porous layer. The contour levels shown in the plots range from -20 to 0 in a and c and from 0 to 20 in b and d, the step being 5 dB in each case. These plots show clearly trends related to the mode structure of the system, but a quantitative description of the patterns shown is beyond the scope of the book. The greater complexity of the contours in the case of the elastic layer is the result of the lower attenuations involved.

8.5 Range Dependence of the Seismic Response

We now consider the dependence of the seismic response on distance from the centre of an active region of finite size. A comprehensive study of this behaviour should ideally include range-dependent environments but, as

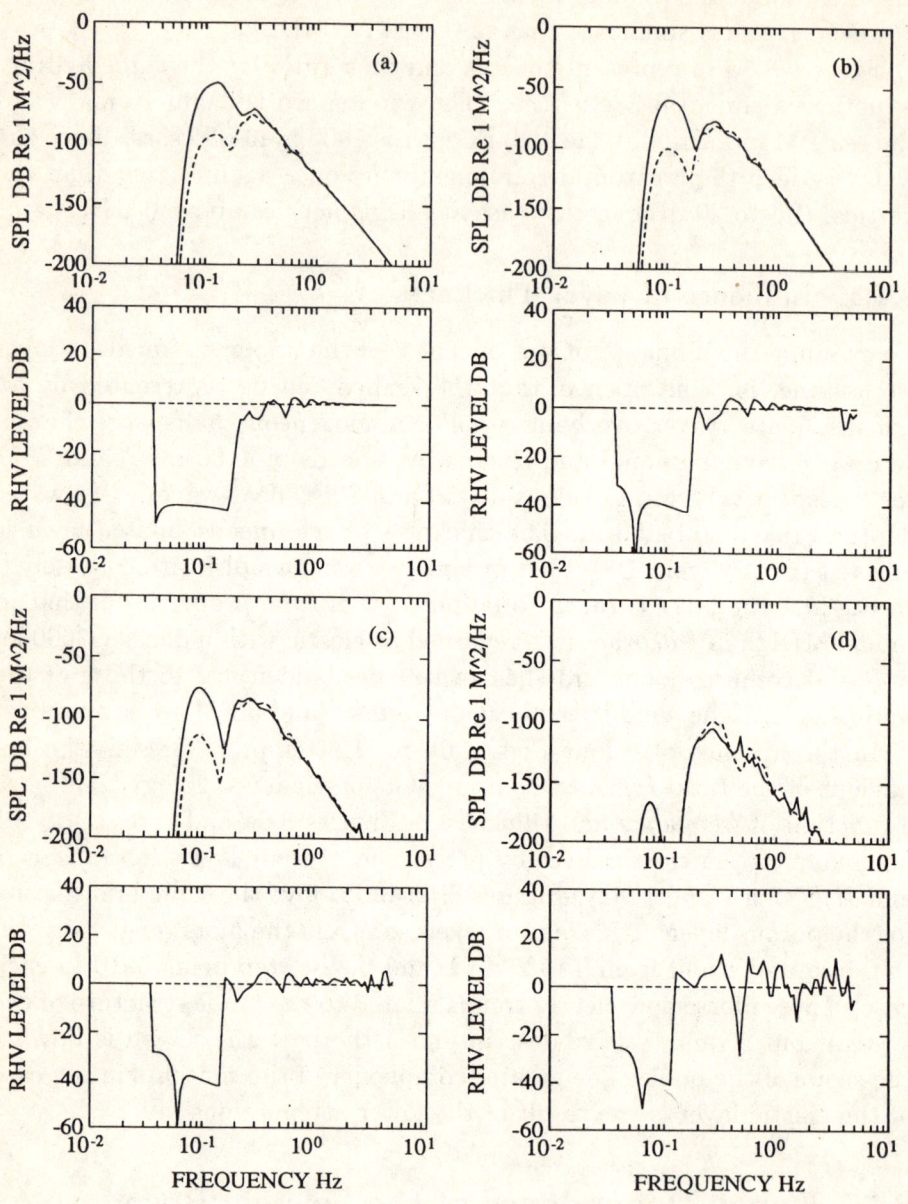

Figure 8.4: Spectral levels of the horizontal and vertical components of the seabed displacement, and their ratio, for model MDL2, wind speed 15 ms^{-1} and a PM sea: a $H_1 = 15$ m; b $H_1 = 50$ m; c $H_1 = 100$ m; d $H_1 = 500$ m.

8.5 Range Dependence

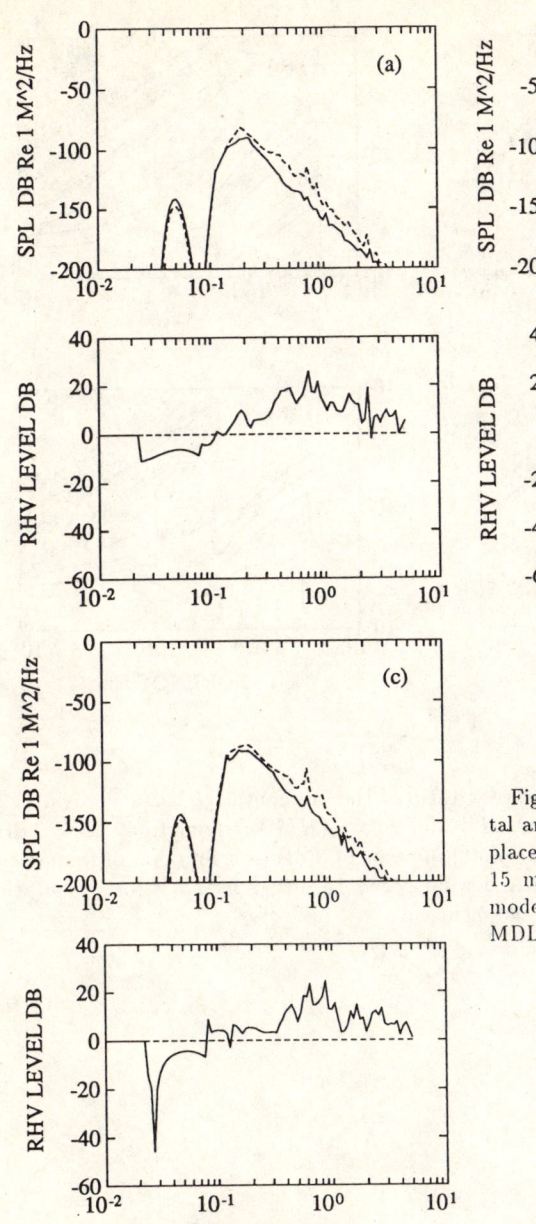

Figure 8.5 Spectral levels of the horizontal and vertical components of the seabed displacement, and their ratio, for a wind speed 15 ms^{-1} and a PM sea, $H_1 = 1000$ m: a model MDL3(1); b model MDL3(2); c model MDL3(3).

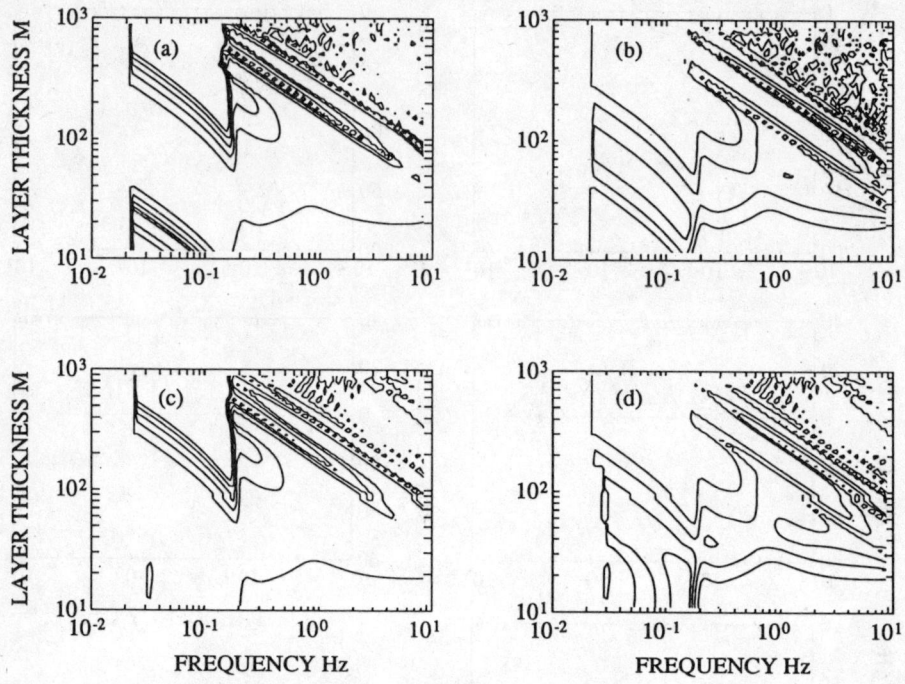

Figure 8.6: Contour plots of the amplitude ratio of the horizontal (H) and vertical (V) components of the seabed displacement, H/V, for $20\log_{10}(H/V)$ ranging from -20 to 0 dB (left plots) and from 0 to 20 dB (right plots) in steps of 5 dB as a function of frequency and layer thickness, under a PM sea and wind speed 15 ms^{-1}: **a** and **b** for the elastic sediment and **c** and **d** for the porous sediment.

8.5 Range Dependence

Table 8.1: Parameters of the geoacoustic models

n	Density ρ_n kg m^{-3}	Thickness h_n m	Comp.Vel.(m/s) α_n	Shear Vel.(m/s) MDK1	MDK2	Q Values Q_{an}	Q_{bn}
1	1000	100	1500	-	-	-	-
2	1700	400	1560	-	-	10	-
3	1900	850	2000	700	1154	500	300
4	2300	1675	3100	1000	1789	500	300
5	2500	1775	4100	1500	2367	500	300
6	2500	-	5000	2500	2885	500	300

we commented earlier, this extension is beyond the scope of this book. By reference to [41], however, we can demonstrate some interesting effects for range-independent structures, in particular the drop in the seismic response as the observation point moves beyond the border of the active region. This drop, which can be as much as 30-40 dB in unconsolidated sediments, results because in this case the IDF component is not a free mode of the system. This response behaviour is clearly of importance to the analysis and interpretation of OBS and LBS records.

In [41] these effects are demonstrated in terms of the vertical component of the seabed displacement at 0.22 Hz, a frequency close to the peak of the seismic spectrum for a JONSWAP sea under a wind speed of 30 ms^{-1}. Using Green's function analysis the seismic response was established as a function of distance from the centre of the active region, for a number of geoacoustic models . (To avoid confusion we base the discussion below on these models rather than those described in Chap. 7.) The size of the active region used in the numerical calculation, 80 x 80 km^2, was chosen as a compromise between computation cost and the dimension necessary to demonstrate clearly the range-decay characteristics of the wave-induced seismic response in the transition region.

We first examine the simplest model, MD2, characterised by an unconsolidated (liquid) bottom. The decay of spectral level (at 0.22 Hz) with distance from the centre of an active area of finite size is shown by curve (1) in Fig. 8.7a. In this figure the horizontal lines (2) and (3) indicate respectively the seismic response to the total pressure field (involving both the inhomogeneous and homogeneous components) and that to the homogeneous component alone, when the active region is infinite in size. The plot confirms that, for this model, the inhomogeneous component dominates the response inside the active region, but also demonstrates the rapid decrease in the influence of the inhomogeneous component, as the observation point is moved beyond the bounds of the active region. Outside the active region the seismic response arises primarily from the diffracted field associated with the homogeneous component of the pressure field (as no modes exist). Since the wavelength of this component ($\sim$7 km) is comparable with the dimensions of the active region, the diffracted field persists out to several hundred kilometers even though it drops a further 40 dB in this distance - see Fig. 8.7a.

Figure 8.7b presents the same information for a model involving a solid half-space — MD1(1). Since for this model the pole of the bottom-response function is located in the homogeneous region, the difference between level (2) (−94.5 dB) and level (3) (−94.6 dB) is only 0.1 dB in this case. The calculated spectral level for a region of finite size is again given by curve (1). Although the seismic response inside the active region appears to be some 10 dB lower than that for a region of infinite size, level (2), this is simply a consequence of the size assumed for the finite active region. Because the dominant contribution in this model comes from the homogeneous component, the size of the active region used in the calculation is not sufficiently large, compared with the wavelengths in the homogeneous field, to produce a closer agreement in this case. This contrasts with the case of the liquid bottom discussed above.

The symbols "++" in Fig. 8.7b indicate the range dependence of the spectral level predicted on the basis of the far-field approximation,

$$F_{mv}^{(2)} = A(f,r)|E_v(ur)|^2 e^{-2\eta r}/r$$

in which only a single Rayleigh pole is considered. Here, η is the attenuation coefficient of the Rayleigh wave, $\eta = (\omega/2c_r)Q_r^{-1}$, Q_r^{-1} is the linear combination of Q_{a2}^{-1} and Q_{b2}^{-1}, i.e., $Q_r^{-1} = BQ_{a2}^{-1} + (1-B)Q_{b2}^{-1}$, c_r is the Rayleigh-wave phase velocity and B a constant determined by the model parameters. The term $A(f,r)$, involving the wave spectrum and bottom impedance, can be readily established by using the residual theorem[55]. As expected, curve (1) tends to the far-field approximation at large distances. If, on the other hand, the bottom structure includes low-velocity layers, poles will also be present in the inhomogeneous region. Low-velocity interface modes will then be excited (when the water is sufficiently shallow) and be available to propagate energy to regions (inland or deep ocean) far from the source region. Schreiner and Dorman[37] believe this mechanism is largely responsible for the wave-induced seismic

8.5 Range Dependence

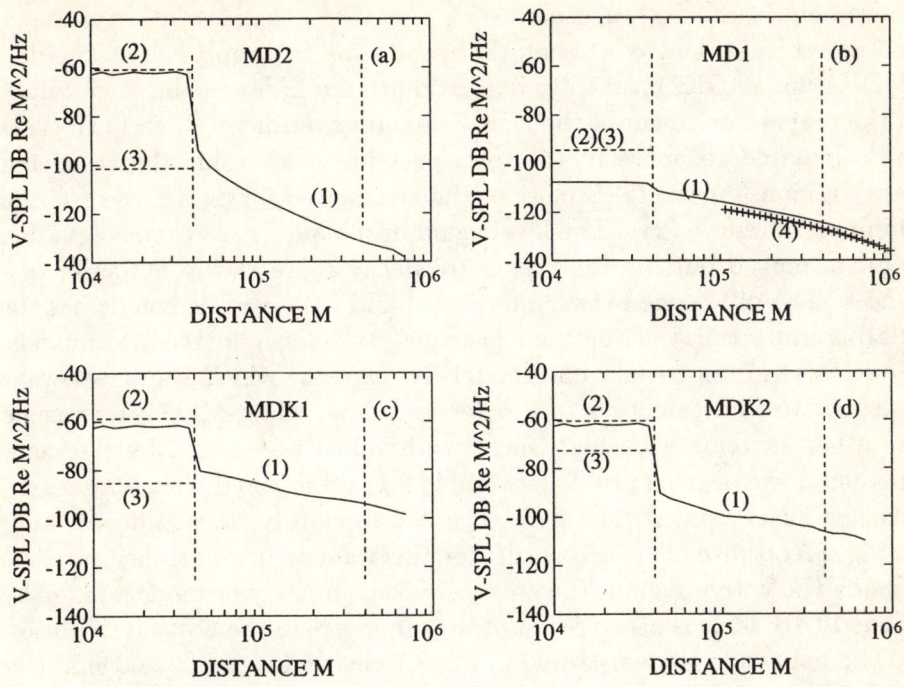

Figure 8.7: Decay of the vertical component of the seabed displacement with distance from the centre of the active region: a overlying an unconsolidated bottom; b overlying an elastic halfspace; c overlying a multilayered bottom characterized by low shear-wave speed; d overlying a multilayered bottom characterized by high shear-wave speed.

-noise field in the deep ocean.

Figures 8.7c and d present the results for two multilayered models (MDK1 and MDK2), taken to approximate the limits of the uncertainty in the real environment of the New Zealand experiment[19]. In both cases the calculated response to the total pressure field inside the restricted active region (curve (1)) is close to that estimated for an active region of infinite size (curve (2)). The level again drops quickly as the observation point is moved outside the region, to decay more slowly at larger distances. The difference between levels (2) and (3) again demonstrates the relative importance of the homogeneous component in the two models. Model MDK1 represents one model extreme, in which the shear-wave speed in the bottom is taken to be very low, while MDK2 represents the other extreme with high shear-wave velocities defined by the compressional wave-speed profile (see Table 8.1) established from actual geophysical survey data. The difference can obviously be significant, and, as Fig. 8.7c shows, is nearly 30 dB for this model. In the transition zone outside the active region, the spectral levels in the two models differ by about 10 dB. This is also a result of the difference in the bottom rigidities.

The behaviour demonstrated in Fig.8.7 clearly has significant implications for the interpretation of the seismic response to a wave-generated pressure field. For instance, Fig. 8.7 indicates that if a storm centre passes across an OBS deployed in shallow water (depths up to 500 m), the seismic response will display dramatic changes as the inhomogeneous component comes in and out of play. At those times when the OBS is inside the active region it will be influenced by both the IDF and HDF components of the incident pressure field. As the weather system moves away levels will change quickly as the influence of the inhomogeneous component dies out. Secondly, the seismic response to an offshore wavefield, recorded by an LBS onshore, will be largely due to the homogeneous component of the pressure field. We examined the implications of this behaviour for the definition of source level in Sect.7.5. As we will see in the next section, it is also relevant to the monitoring of offshore wavefields by land-based systems.

8.6 Potential Applications of the Source Field

As we have seen from the discussions throughout this book, the wave-induced pressure field represents a low-frequency acoustic source of unique properties. It is characterised by high spectral levels even at modest wind speed, it is effective at frequencies and wavenumbers that can not be achieved by any controlled source, its wind dependence ensures an operating bandwidth that is quite broad, and the pressure levels and low frequencies involved can result in the insonification of the deepest strata in the ocean crust. Furthermore its dependence on sea state, wind speed and geoacoustical environment are now sufficiently well understood for the source to have real potential as an effective tool in geophysical investigations . We discuss some of these possibilities briefly below.

8.6.1 Monitoring the Ocean-Wave Field

When microseisms first attracted attention they were seen as providing a means of monitoring the movement of weather systems at sea. More recently onshore sensors have been used to monitor ocean swell [87]. The results presented in Sects. 7.4.1 and 7.5, in particular, show that our current understanding of the processes involved now makes it possible to monitor the full spectral and directional properties of an offshore wave-field with a land-based array, to a degree of accuracy sufficient for many purposes. The savings in cost over conventional wave-recording systems can be considerable. We reported recently on such an application in a wave-energy resource assessment in the Southern Ocean [90],[91]. The wave-energy results are interesting in their own right, but it is more general features of the study, relating to the ULF noise field and its propagation, which prompts us to include this commentary on it here.

As the geophysical structure off the coast of interest is not well known, the transfer function relating the offshore wave-field to the onshore seismic response was established experimentally by a short-term calibration experiment using a waverider. A theoretical transfer function was also calculated using the limited geophysical data available. It was clear later that, if the geological structure had been known sufficiently well, the theoretical transfer function would have been adequate without recourse to such a calibration.

Good correlation between wave and seismic spectra was obtained. Most of the spectral and temporal detail in the waverider data was reflected in the seismic record, although the latter usually displayed more stationarity and smaller deviations from the mean value. This characteristic is not due to the transfer function but to the fact that, whereas the waverider measures the wavefield at a single point, the seismometer presents an average response to the whole offshore region. This averaging process can lead to significant differences in individual spectra but the longer term statistics provided by the two systems are much more consistent and quite adequate for most purposes. An extension to a simple three component seismic array was also shown to give a reliable picture of the angular distribution of the incoming wavefield.

Figure 8.8 presents, as an example of the short-term statistics obtained, a comparison of the significant wave-height derived from the waverider and from the inverted seismic spectra. The sampling period involved was 20 minutes every 4 hours. As can be seen the averaging in the seismic-based data, mentioned above, can lead to differences of 3 dB in individual spectra. The differences are, however, much smaller in the longer term statistics. Examples of monthly averages of the ocean-wave spectrum derived from the seismic spectra are shown in Fig. 8.9. These and longer term statistics[91] were more than adequate for the purposes involved.

It is of interest to note that the main peak of the acceleration spectra, shown in Fig.8.10 around 0.15 Hz, has a value of 10 to 15 dB (re 1 $\mu m^2 s^{-4} Hz^{-1}$), which is close to that observed by Schreiner and Dorman[37] at the deep-ocean floor (0.1 to 0.2 Hz, 10 dB re 1 $\mu m^2 s^{-4} Hz^{-1}$). This implies that wave-induced microseisms observed on shore and in the deep ocean may both originate in shallow areas, where the inhomogeneous component can play a dominant role in transferring energy from the sea surface to the ocean bottom.

The H/V ratio

Features of the seismic spectra, which are relevant to other aspects of the phenomena discussed in this book, are exemplified by Fig. 8.10. This figure presents the average seismic spectra (acceleration) for the month of March 1992, recorded by the three acceleration seismometers making up the three-component directional array. The spectra are in this case not restricted to the main double-frequency peak of the wave-induced response, but are extended above and below this band to demonstrate differences in the response of the two horizontal sensors (EW and NS) and the vertical sensor (V), which were a notable feature at all times.

A detailed analysis of the spectral behaviour indicated has yet to be made, but it is of interest to note the clear relationship at 0.5 to 0.7 Hz

8.5 Range Dependence

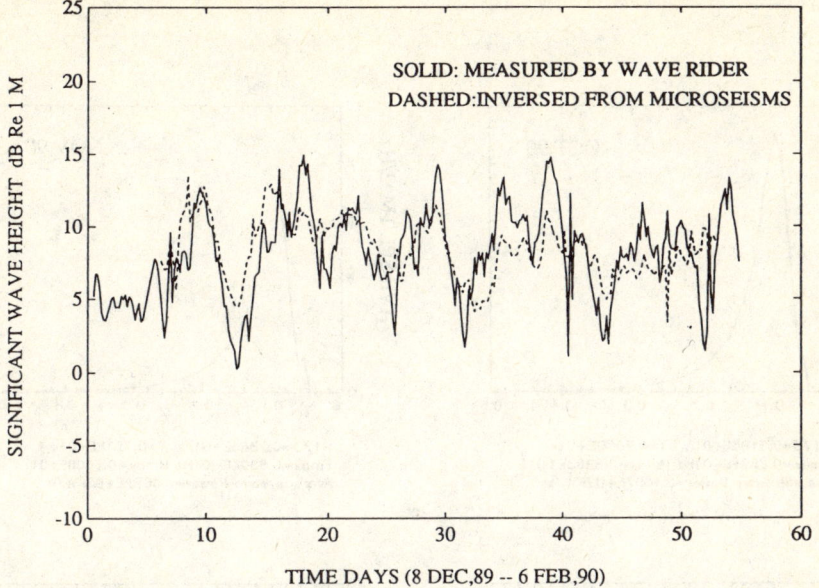

Figure 8.8: A comparison of short-term histories of the significant waveheight measured from waverider and inverted seismic spectra.

between the experimental spectra of Fig. 8.10 and the theoretical spectra presented in Figs. 8.1 and 8.2. The marked positive peak in the value of the ratio, H/V, is a feature of both. This feature of seismic behaviour has been commented on elsewhere. The similarity of the analytical and field spectra indicate that modelling could prove instructive in identifying the origin of the effects observed.

The infragravity band

The character of the spectra of Fig. 8.10 in the infragravity band is typical of the activity observed during the Southern Ocean measurements. Individual spectra showed considerable variation in level but the monthly averages tended to the comparatively stable form shown. The measurement site is exposed to much of the Southern Ocean and the observed response is obviously the result of the interaction of many active wave regimes. It seems likely, however, that at least some of the variation in the infragravity spectral levels arises from the difference-frequency interactions discussed in Sect. 6.3.

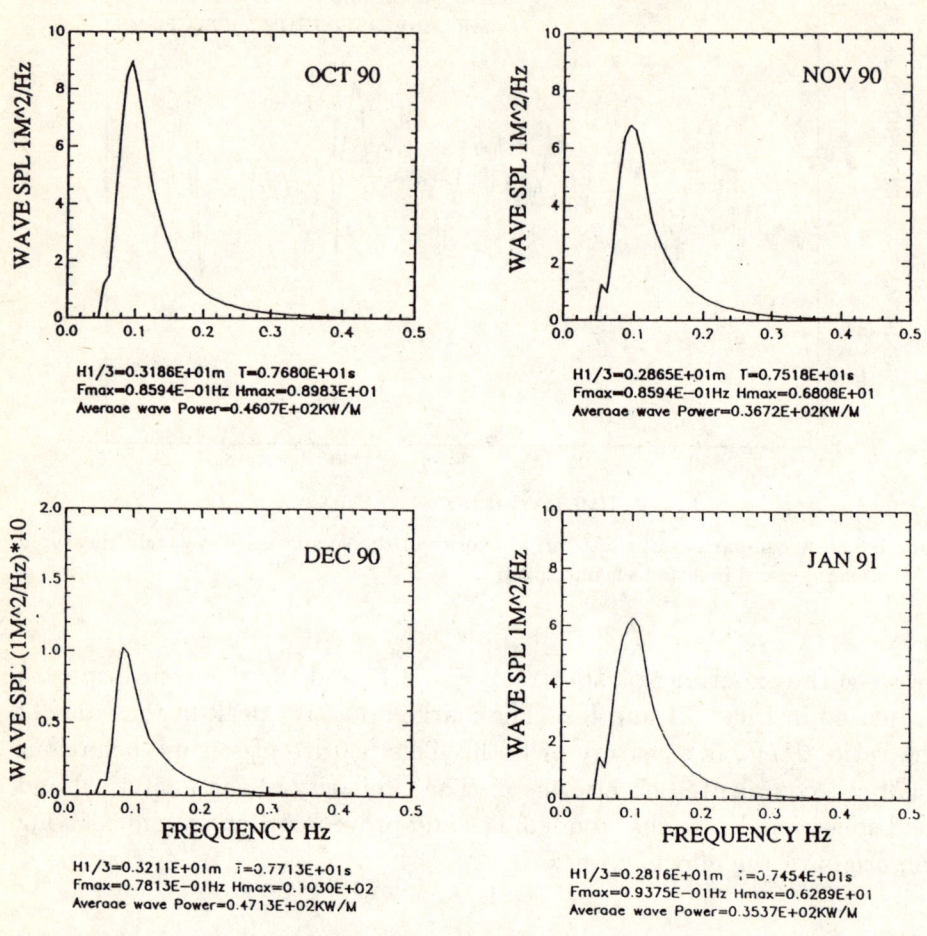

Figure 8.9: Average power density spectra of the sea-surface displacement for the months October 1990 (133 spectra), November 1990 (175), December 1990 (146) and January 1991 (184), established from 20-min records (taken every 4 hours) of the seismic response to the offshore wave field.

8.5 Range Dependence

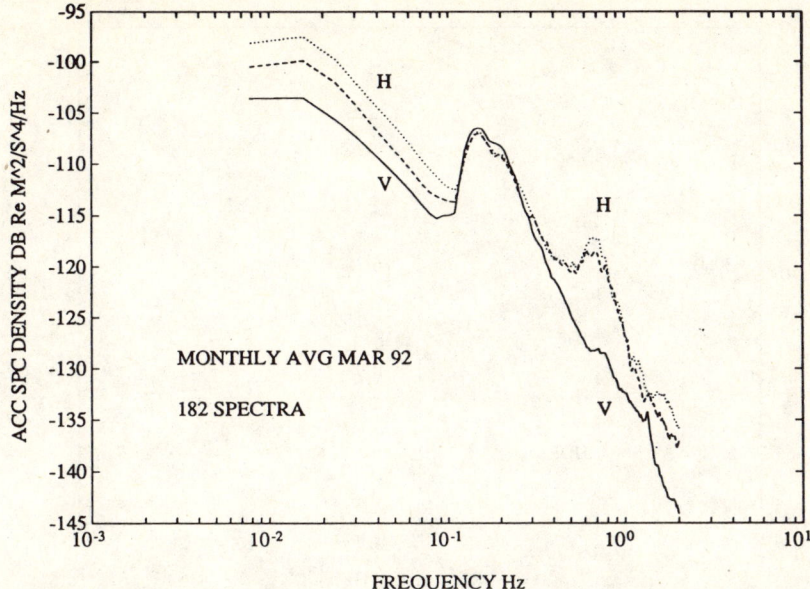

Figure 8.10: The average vertical (V) and horizontal (H,EW/NS) power-density acceleration spectra for March 1992, established from 182 20-min records (taken every 4 hours) of the seismic response (acceleration) to the offshore wavefield.

8.6.2 Inversion Studies of Seabed Structure

As was mentioned above, the pressure field generated by the wave-interaction process represents a unique and potentially useful source for the geophysical investigation of seabed structure. Figure 8.11 attempts to demonstrate this potential in a semi-quantitative way.

Fig. 8.11a indicates the range in the frequency-wavenumber domain over which the wave-induced pressure field represents an effective acoustic source (this figure is based on the sum-frequency interactions only). The dashed lines 5(J), 15(J), 30(J) indicate the lowest frequency for JONSWAP seas, under wind speeds of 5, 15 and 30 ms^{-1} respectively, at which the level is sufficiently high for the wave-induced pressure field to act as an effective acoustic source. Because of the sharp drop in level below the spectral peak, this frequency is taken to be that of the peak in the pressure spectrum at each wind speed. The line 15(PM) indicates the same limit for a PM sea and a wind speed of 15 ms^{-1}. The upper-frequency limit is taken to be 1 Hz for all sea states. It is defined by the line at 1 Hz and reflects the small wind dependence of spectral level at this frequency. The spectral level at both limits is approximately 0 dB re

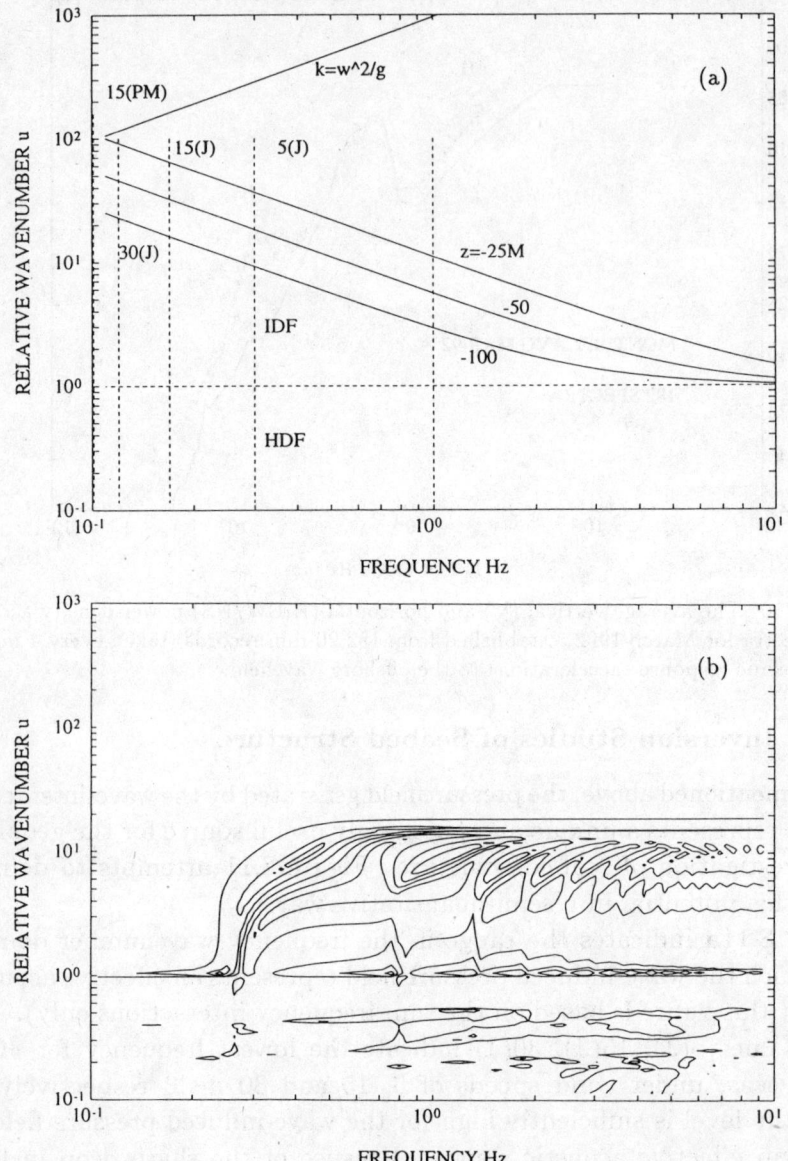

Figure 8.11: **a** The effective operating band (in the wavenumber-frequency domain) of the pressure field generated by the wave-interaction source, as a function of wind speed and water depth, for JONSWAP and PM seas; **b** Contour levels of the reflection coefficient for model MDL2 characterised by an unconsolidated, porous upper layer 100 m in thickness. The contour levels of 0.2, 0.5, 1, 2, 5, 10, and 20 are the same as in Fig.7.4.

8.5 Range Dependence

1 P_a^2/Hz, a level which is well above background noise level - see Fig.6.11.

The three curves indicate the range of wavenumbers at each frequency for which the inhomogeneous component of the pressure field still represents an effective source at depths 25, 50 and 100 m below the sea surface. It will be recalled that the inhomogeneous component of the pressure field decays with depth at a rate which is approximately $exp(-k|z|)$. The maximum value of u for each frequency has been defined, somewhat arbitrarily, as the relative wavenumber at which $exp(-2k|z|) = 0.1$. The spectral level so defined still exceeds 0 dB re 1 P_a^2/Hz even at the lower wind speeds. The horizontal line at u=1 defines, as usual, the boundary between the homogeneous and inhomogeneous components of the pressure field. The straight line at the top left-hand corner represents the dispersion relation for the primary-frequency (PF) component of the pressure field.

The areas under the curves for each depth and bounded by the dashed lines, thus provide a measure of the range in the frequency-wavenumber domain over which the second-order pressure field represents an effective source at each seastate. The wide band of wavenumbers present at each frequency in the DF field contrasts markedly with the single wavenumber associated with the PF field and emphasises the superiority of the former as a low-frequency source. The superiority of the second-order field is highlighted further by Fig. 8.11b. Here, by way of example, we present contours of the amplitude of the reflection coefficient for model MDL2, plotted as a function of frequency and wavenumber. This model, it will be recalled, is characterised by a porous unconsolidated sediment (see Fig.7.4 for the corresponding 3-D presentation). Comparison of Figs. 8.11a and b demonstrates that:

(i) Apart from the minor structure in the homogeneous region, associated with normal-mode effects, the most significant information about the bottom appears in the inhomogeneous region of the plot.

(ii) This character, which is a consequence of interface modes, occurs at frequencies and wavenumbers not available in conventional geoacoustical sources, but present in the second-order component of the wave-induced pressure field-see Fig. 8.11a.

(iii) The pressure levels produced by commonly experienced seastates are sufficient to allow insonification to depth of most of the structure on the continental shelf.

(iv) In contrast, the energy provided by the first-order field lies well away from the most significant frequencies and wavenumbers represented in the behaviour of the reflection coefficient and there is only one wavenumber associated with each frequency.

We can conclude from this comparison that the second-order pressure field is a more effective source than its first-order equivalent and offers the possibility of exploiting features in the seabed structure not readily accessible by current geophysical techniques. Its full potential remains to be realised.

Chapter 9

Summary

As ambient noise measurements were extended progressively to lower and lower frequencies, it became clear that the sea-noise spectrum was characterised by a large peak below 1 Hz, the level and peak frequency of which were dependent on sea state. While several possible sources were proposed it was generally accepted that the source of this peak was the mechanism responsible for microseisms, a description of which had been provided by Longuet-Higgins. As early as 1950, he recognised the role of nonlinear interactions between components of the gravity-wave field and, in an analysis which was based on a perturbation procedure, demonstrated the seismo-acoustic consequences of these. Other theoretical studies were to follow throughout the 1960's. While all these analyses involved a number of approximations, their description of the wave-interaction mechanism provided an adequate explanation for the seismo-acoustic effects observed at that time. However, the 1980's saw a renewed interest in the natural mechanisms of surface-generated noise and the initiation of extensive field programmes based on much superior technology. Experimental data from a wide range of environments is now beginning to appear, the proper interpretation of which has called for the removal of certain limitations of wave-interaction theory in its original form.

This contribution presents the results of an investigation aimed at removing the major restrictions of the earlier analyses. In particular the deep-water assumption has been removed and the formalism extended to cover environments of any depth. A geometric analysis of the wave-interaction process has provided a clearer physical picture of the mechanism and this, in turn, has led to a more complete description of the importance of the homogeneous and inhomogeneous components of the wave-induced pressure field. The contribution of the difference-frequency interactions to the source field has also been examined, and, its relationship to the sum-frequency component of the pressure field, on which the traditional treatments are based, has been evaluated. The overall result is a description of the wave-induced pressure field, which incorporates both the first- and second-order components of the wave-induced pressure field and describes the spectral and coherence properties of the source spectrum at any depth in the water column.

To provide a description of the response of the environment to the total wave-induced source field a parallel investigation has examined the reflection coefficient and transfer functions of multilayered seabeds as a function of both frequency and relative wavenumber. An additional refinement of this theoretical development has been the incorporation of porosity as a seabed property, to permit the calculation of the reflection coefficient and the related transfer functions for general structures.

Using the codes developed for the source spectrum and transfer functions, pressure and seismic spectra have been calculated for a number of typical environments, and their dependence on wind-speed, water depth, and bottom structure has been demonstrated. The spectra so calculated not only describe the influence of the first- and second-order components of the pressure field, but also demonstrate why the total wave-induced pressure field represents a low-frequency source with unique properties of potential value for the investigation of seabed parameters and structure by inversion techniques. It is believed that these results should be of interest not only to those concerned with the properties of the ULF noise field but also to those involved in using the wave-induced pressure field in geophysical investigations of the seafloor.

Appendix A

Equations Quoted in Chapter 2

The equation numbers in [ ] refer to the equation numbers in the original text.

Longuet-Higgins: See paper [25]

Standing-wave pressure field, [Eq.(31)]:

$$\frac{\bar{p}_h - p_s}{\rho} - gh = -2a_1 a_2 \sigma^2 \cos 2\sigma t \qquad (A.1)$$

where ρ, g, and p_s are respectively the density of the water, the gravitational acceleration and the pressure at the surface.

General dynamic equation, [Eq.(94)]:

$$\frac{\partial^2 \Phi}{\partial t^2} - c^2 \nabla^2 \Phi - g\frac{\partial \Phi}{\partial z} - \frac{\partial}{\partial t}(\frac{1}{2}\vec{u}^2) - \vec{u} \cdot \nabla(\frac{1}{2}\vec{u}^2) = 0 \qquad (A.2)$$

Boundary conditions, [Eq.(97)]:

$$\left.\frac{\partial \Phi}{\partial t} - \frac{1}{2}\vec{u}^2 + gz\right|_{z=\zeta} = 0 \qquad (A.3)$$

[Eq.(95)]:

$$\left.\nabla^2 \Phi\right|_{z=\zeta} = 0 \qquad (A.4)$$

[Eq.(99)]:

$$\left.\frac{\partial \Phi}{\partial z}\right|_{z=h} = 0 \tag{A.5}$$

where the particle velocity $\vec{u} = -\nabla\Phi$ and ζ is the surface displacement.

Perturbation series, [Eq.(109)]:

$$\left.\begin{aligned}\Phi &= \epsilon\Phi_1 + \epsilon^2\Phi_2 + \cdots \\ \vec{u} &= \epsilon\vec{u}_1 + \epsilon^2\vec{u}_2 + \cdots \\ \zeta &= \epsilon\zeta_1 + \epsilon^2\zeta_2 + \cdots \\ p - p_0 &= \epsilon p_1 + \epsilon^2 p_2 + \cdots \\ \rho - \rho_0 &= \epsilon \rho_1 + \epsilon^2 \rho_2 + \cdots\end{aligned}\right\} \tag{A.6}$$

First order field:

Field equation, [Eq.(110)]:

$$\frac{\partial^2 \Phi_1}{\partial t^2} - c^2 \nabla^2 \Phi_1 - g\frac{\partial \Phi_1}{\partial z} = 0 \tag{A.7}$$

Boundary conditions, [Eq.(111)]:

$$\left.\frac{\partial \Phi_1}{\partial z}\right|_{z=h} = 0 \tag{A.8}$$

$$\nabla^2 \Phi_1|_{z=0} = 0 \tag{A.9}$$

$$\vec{u}_1 = -\nabla \Phi_1 \tag{A.10}$$

$$g\zeta_1 = -\left(\frac{\partial \Phi_1}{\partial t}\right)_{z=0} \tag{A.11}$$

$$\frac{p_1}{\rho_s} = c^2\frac{\rho_1}{\rho_s} = \frac{\partial \Phi_1}{\partial t}e^{2\gamma z} \tag{A.12}$$

where $\gamma = g/2c^2$, c is used here for the sound speed in water.

Equations Quoted in Chapter 2

Second-order field:

Field equation, [Eq.(112)]:

$$\frac{\partial^2 \Phi_2}{\partial t^2} - c^2 \nabla^2 \Phi_2 - g\frac{\partial \Phi_2}{\partial z} = \frac{\partial}{\partial t}[u_1]^2 \tag{A.13}$$

Boundary conditions, [Eq.(113)]:

$$\frac{\partial \Phi_2}{\partial z}\bigg|_{z=h} = 0 \tag{A.14}$$

$$\nabla^2 \Phi_2|_{z=0} = -\zeta_2 \frac{\partial}{\partial z}\nabla^2 \Phi_1|_{z=0} \tag{A.15}$$

with

$$\vec{u}_2 = -\nabla \Phi_2 \tag{A.16}$$

$$g\zeta_2 = -\left(\frac{\partial \Phi_2}{\partial t} - \frac{1}{2}\vec{u}_1^2\right)\bigg|_{z=0} - \zeta_1 \left(\frac{\partial^2 \Phi_1}{\partial z \partial t}\right)\bigg|_{z=0} \tag{A.17}$$

$$\frac{p_2}{\rho_s} = c^2 \frac{\rho_2}{\rho_s} = \left[\frac{\partial \Phi_2}{\partial t} - \frac{1}{2}\vec{u}_1^2 + \frac{1}{2c^2}\left(\frac{\partial \Phi_1}{\partial t}\right)^2\right] e^{2\gamma z} \tag{A.18}$$

Hasselmann: See paper [26]

First-order potential:

Field equation (z-axis positive downwards), [Eq.(2.1)] to [Eq.(2.4)]:

$$\nabla^2 \Phi = 0 \quad -h \leq z \leq 0, \tag{A.19}$$

Boundary conditions, [Eqs.(2.3)-(2.4)]:

$$\frac{\partial \Phi}{\partial z} = 0 \quad \text{at} \quad z = -h \tag{A.20}$$

$$\frac{\partial \zeta}{\partial t} - \frac{\partial \Phi}{\partial z} = 0 \quad \text{at} \quad z = 0 \tag{A.21}$$

$$\frac{\partial \Phi}{\partial t} + g\zeta = 0 \quad \text{at} \quad z = 0 \qquad (A.22)$$

Second-order potential:

Field equation, [Eq.(2.8)]:

$$\nabla^2 \Phi - \frac{1}{\alpha_1^2}\left(\frac{\partial^2 \Phi}{\partial t^2} + g\frac{\partial \Phi}{\partial z}\right) = -\frac{1}{2\alpha_1^2}\frac{\partial}{\partial t}(\nabla \Phi)^2 + \cdots \quad -h \leq z \leq 0 \quad (A.23)$$

where α_1 is the sound speed in the water.

Boundary conditions, [Eqs.(2.9), (2.10)]:

$$\frac{\partial^2 \Phi}{\partial t^2} + g\frac{\partial \Phi}{\partial z} = -\frac{\partial}{\partial t}(\nabla \Phi)^2 + \cdots \quad \text{at} \quad z = 0 \qquad (A.24)$$

$$\frac{\partial \zeta}{\partial t} - \frac{\partial \Phi}{\partial z} = -\frac{\partial \Phi}{\partial x}\frac{\partial \zeta}{\partial x} - \frac{\partial \Phi}{\partial y}\frac{\partial \zeta}{\partial y} - \zeta\frac{\partial^2 \Phi}{\partial z^2} + \cdots \quad \text{at} \quad z = 0 \quad (A.25)$$

Fourier-Stieltjes integrals:

First-order potential, [Eq.(2.5)]:

$$\Phi(\vec{r}, z, t) = \int [d\Phi_+(\vec{k})e^{-i\sigma t} + d\Phi_-(\vec{k})e^{i\sigma t}]e^{i(\vec{k}\vec{r}+kz)} \qquad (A.26)$$

Sea-surface displacement, [Eq.(2.6)]:

$$\zeta(\vec{r}, t) = \int\int [dZ_+(\vec{k})e^{-i\sigma t} + dZ_-(\vec{k})e^{i\sigma t}]e^{i\vec{k}\vec{r}} \qquad (A.27)$$

with the deep-water dispersion relation, $\sigma = \sqrt{gk}$, applying.

Second-order pressure field, [Eq.(2.11)]:

$$p_2|_{z=0} = \rho_1 \frac{\partial}{\partial t}(\nabla \Phi_1)^2|_{z=0} \qquad (A.28)$$

which, on substitution for Φ_1 leads to, [Eq.(2.12)]:

$$p_2|_{z=0} = \rho_1 \sum_{s's''} \int \cdots \int d\Phi_{s'}(\vec{k}')d\Phi_{s''}(\vec{k}'')(k'k'' - \vec{k}'\vec{k}'')e^{i[(\vec{k}'+\vec{k}'')\vec{r}-(s'\sigma'+s''\sigma'')]t}$$

(A.29)

where s' and s'' denote sign indices.

Second-order power-spectral density function, [Eq.(2.13)]:

$$F_p^{(2)}(\vec{k},\omega) = \rho_1^2 g^4 \int\int F_\zeta(\vec{k}')F_\zeta(\vec{k}'')(\sigma',\sigma'')^{-2} \cdot (k'k'' + \vec{k}'\vec{k}'')^2 \delta(\vec{k}' + \vec{k}'' - \vec{k})$$
$$\delta(\sigma' - \sigma'' + \omega)(k'k'' - \vec{k}'\vec{k}'')^2 \delta(\vec{k}' + \vec{k}'' - \vec{k})\delta(\sigma' + \sigma'' - \omega)d\vec{k}'d\vec{k}''$$

(A.30)

"Standing-wave" approximation, [Eq.(2.15)]:

$$F_p^{(2)}(\vec{k},\omega) = \frac{\rho_1^2 g^2 \omega^2}{2} \int_\pi^\pi f_\zeta(\omega/2,\theta) f_\zeta(\omega/2,\theta+\pi)d\theta \qquad (A.31)$$

where

$$f_\zeta(\omega,\theta) = F_\zeta(\vec{k})kdk/d\omega \qquad (A.32)$$

Acoustic-analogue formulation, [Eq.(3.3)]:

$$\frac{1}{\alpha_1^2}\frac{\partial^2 p}{\partial t^2} - \nabla^2 p = \frac{\partial^2}{\partial x_\alpha \partial x_\beta}(\rho u_\alpha u_\beta) \qquad (A.33)$$

Brekhovskikh: See paper [28]

First-order potential, [Eq.(11)]:

$$c^2 \nabla^2 \Phi_1 = \frac{\partial^2 \Phi_1}{\partial t^2} - g\frac{\partial \Phi}{\partial t} \qquad (A.34)$$

Boundary conditions, [Eq.(12)]:

$$\left(\frac{\partial \Phi_1}{\partial t}\right)_{z=0} + g\zeta_1 = \gamma \nabla_{xy} \zeta_1 \tag{A.35}$$

$$\left.\left(\frac{\partial^2 \Phi_1}{\partial t^2} - g\frac{\partial \Phi_1}{\partial z} + \gamma \nabla_{xy}^2 \frac{\partial \Phi_1}{\partial z}\right)\right|_{z=0} = 0 \tag{A.36}$$

where $\gamma \equiv T/\rho_1$, T is the capillary constant, and $\nabla_{xy}^2 = \frac{\partial^2}{\partial x \partial y}$

Second-order potential, [Eq.(20)]:

$$D\Phi_2 = -\frac{\partial}{\partial t}(\nabla \Phi_1)^2, \quad \text{where} \quad D \equiv c^2 \nabla^2 - \frac{\partial^2}{\partial t^2} + g\frac{\partial}{\partial z} \tag{A.37}$$

Boundary conditions, [Eq.(21)]:

$$\left.(\nabla^2 \Phi_2)\right|_{z=0} = -\frac{1}{c^2}\zeta_1 \left.\left(\frac{\partial^2 \Phi_1}{\partial z \partial t} - g\frac{\partial^2 \Phi_1}{\partial z^2}\right)\right|_{z=0} \tag{A.38}$$

Solution of first-order field, [Eq.(38)]:

$$\zeta_1 = \int a(\vec{k}) \sin(\vec{k}\vec{r} - \omega t + \epsilon) d\vec{k}, \quad \phi_1 = \int b(\vec{k}) e^{-kz} \cos(\vec{k}\vec{r} - \omega t + \epsilon) d\vec{k} \tag{A.39}$$

where $d\vec{k} \equiv dk_x dk_y$ and $a(\vec{k}) = -(k/\omega) b(\vec{k})$.

Second-order solutions, [Eq.(41)]:

$$\Phi_2' = \frac{1}{2}\int\int e^{-(k_1+k_2)z} \Big\{ E \sin[(\vec{k}_1 - \vec{k}_2)\vec{r} - (\omega_1 - \omega_2)t + \epsilon_1 - \epsilon_2] $$
$$+ F \sin[(\vec{k}_1 + \vec{k}_2)\vec{r} - (\omega_1 + \omega_2)t + \epsilon_1 + \epsilon_2] \Big\} d\vec{k}_1 d\vec{k}_2 \tag{A.40}$$

and [Eq.(44)]:

$$\Phi_2'' = \frac{-1}{2}\int\int M(\vec{k}_1, \vec{k}_2) \sin[(\vec{k}_1 + \vec{k}_2)\vec{r} + \beta^2 - (\omega_1 + \omega_2)t + \epsilon_1 + \epsilon_2] d\vec{k}_1 d\vec{k}_2 \tag{A.41}$$

Equations Quoted in Chapter 2

Hughes: See paper [29]

First-order field equation, [Eq.(7)]:

$$\nabla^2 p_1 - \frac{1}{c^2}\frac{\partial^2 p_1}{\partial t} = 0 \qquad (A.42)$$

Boundary conditions, [Eqs.(8), (9)]:

$$p_1 = P_a + \mu \nabla_H^2 \zeta_1 + \rho_A g \zeta_1 \quad \text{at} \quad z = 0 \qquad (A.43)$$

First-order field equation, [Eq.(10)]:

$$\frac{\partial \zeta_1}{\partial t} = w_1 \quad \text{at} \quad z = 0 \qquad (A.44)$$

and

Boundary conditions, [Eqs.(11), (12)]:

$$\nabla^2 p_2 - \frac{1}{c^2}\frac{\partial^2 p_2}{\partial t} = -\nabla \rho_0 \vec{u}_1 \cdot \nabla \vec{u}_1 + \vec{u}_1 \nabla(\rho_0 \vec{u}_1) \qquad (A.45)$$

$$p_2 = \mu \nabla_H^2 \zeta_2 - \rho_A g \zeta_2 - \rho_A g \frac{\zeta_1^2}{2c^2} - \zeta_1 \frac{\partial p_1}{\partial z} \quad \text{at} \quad z = 0 \qquad (A.46)$$

$$\frac{\partial \zeta_2}{\partial t} = w_2 - \vec{u}_1 \cdot \nabla_H \zeta_1 + \zeta_1 \frac{\partial w_1}{\partial z} \quad \text{at} \quad z = 0 \qquad (A.47)$$

where $\nabla_H \equiv (\partial/\partial x, \partial/\partial y)$, ρ_A is the density of the water at the surface, and $\vec{u} \equiv (u, v, w)$.

Change of variable, [Eq.(14)]:

$$p_2' = p_2 + \frac{1}{2}\rho_0(\vec{u}_1 \cdot \vec{u}_1) \qquad (A.48)$$

and

[Eq.(15)]:
$$\frac{\partial \zeta_2'}{\partial t} = \frac{\partial \zeta_2}{\partial t} + \nabla_H(\vec{u}\zeta_1) \qquad (A.49)$$

New boundary conditions, [Eq.(17)]:

$$p'^2 = \mu\nabla_H^2\zeta_2' - \rho_A g\zeta_2' + \frac{1}{2}\rho_A(\vec{u}_1\vec{u}_1) - \zeta_1\frac{\partial p_1}{\partial z} + \rho_A g^2\frac{\zeta_1^2}{2c^2} \quad \text{at} \quad z=0$$
$$(A.50)$$

and [Eq.(18)]:

$$\frac{\partial \zeta_2'}{\partial t} = w_2 \quad \text{at} \quad z=0 \qquad (A.51)$$

External source field, [Eq.(20)]:

$$P_{2a} = \frac{1}{2}\rho_A(\vec{u}_1\vec{u}_1) - \zeta_1\frac{\partial p_1}{\partial z} = \frac{1}{2}\rho_A(\vec{u}_1\vec{u}_1) + \rho_A\zeta_1\frac{\partial^2\zeta_1}{\partial t^2} \qquad (A.52)$$

Ocean-wave spectrum, [Eq.(29)]:

$$\langle \zeta_1^2 \rangle = \int\int X(k_1)G(\theta)d\vec{k}_1 = \int_0^\infty X(k_1)k_1 dk_1 \qquad (A.53)$$

with [Eq.(31)]:

$$\int_0^{2\pi} G(\theta)d\theta = 1 \qquad (A.54)$$

Lloyd: See paper [30]

Preliminary form of Lighthill's equation, [Eq.(9)]:

$$\rho\left(\frac{D}{Dt} - \nu'\nabla^2\right)\frac{1}{\rho c^2}\frac{Dp}{Dt} - \rho\nabla\cdot\left(\frac{1}{\rho}\nabla p\right) = \Lambda \quad \text{in} \quad R_t \qquad (A.55)$$

where [Eq.(7)]:

$$\Lambda = \rho\frac{\partial v_s}{\partial x_r}\frac{\partial v_r}{\partial x_s} \qquad (A.56)$$

is the Reynolds stress tensor and R_t the volume.

Gravity-acoustic equation, [Eq.(11)]:

$$\frac{1}{c_0^2}\frac{\partial^2 p}{\partial t^2} - \nabla^2 p - \gamma\frac{\partial p}{\partial z} = \Lambda_0 \quad \text{in} \quad R_t \qquad (A.57)$$

where

$$\Lambda_0 = \rho_a\frac{\partial v_s \partial v_r}{\partial x_r \partial x_s}, \quad \text{and} \quad \gamma = \frac{N^2}{g} + \frac{g}{c_0^2}$$

and N is the Väisälä frequency.

Ribner-Lighthill equation, [Eq.(13)]:

$$\frac{1}{c_0^2}\frac{\partial^2 p_1}{\partial t^2} - \nabla^2 p_1 - \gamma\frac{\partial p_1}{\partial z} = -\frac{1}{c_0^2}\frac{\partial^2 p_0}{\partial t^2} \quad \text{in} \quad R_t \qquad (A.58)$$

or [Eq.(16)]:

$$\frac{1}{c_0^2}\frac{\partial^2 p_1}{\partial t^2} - \nabla^2 p_1 = -\frac{1}{c_0^2}\frac{\partial^2 p_0}{\partial t^2} \quad \text{in} \quad R_t \qquad (A.59)$$

with the pressure-release conditions,

$$\begin{aligned} p_1 &= 0 \quad \text{on} \quad S_t \\ p_1 &= 0(1) \quad \text{as} \quad z \to -\infty. \end{aligned} \qquad (A.60)$$

First-order equation, [Eq.(21)]:

$$\nabla^2 \Phi_1 = 0 \quad \text{in} \quad R_t$$

$$\frac{\partial \Phi_1}{\partial t} = -g(h - T\nabla_H^2 h) \quad \text{on} \quad z = 0 \qquad (A.61)$$

and $\Phi_1 \to 0$, as $z \to -\infty$.

Second-order equation,

$$\nabla^2 \Phi_2 = 0 \quad \text{in} \quad R_t$$

[Eq.(23)]:

$$\frac{\partial \Phi_2}{\partial t} = -h\frac{\partial^2 \Phi_1}{\partial z \partial t} - \frac{1}{2}\nabla\Phi_1 \cdot \nabla\Phi_1 \quad \text{on} \quad z = 0 \qquad (A.62)$$

Guo: See paper [54]

Acoustic analogue equation, [Eq.(2.1)]:

$$\left(\frac{\partial^2}{\partial \tau^2} - c_a^2 \nabla^2 - g\frac{\partial}{\partial y_3}\right)\rho'_a = \frac{\partial T_{ij}}{\partial y_i \partial y_j} \qquad (A.63)$$

where $T_{ij} = \rho u_i u_j + \delta_{ij}(p'_a - c_a^2 \rho'_a)$

Cato: See paper [51]

Doak's equation, [Eq.(4)]:

$$\frac{1}{c_0^2}\frac{\partial^2 p}{\partial t^2} - \frac{\partial^2 p}{\partial x_i^2} = \frac{\partial^2}{\partial x_i \partial x_j}(\rho u_i u_j + p_{ij} - p\delta_{ij}) + \frac{\partial^2}{\partial t^2}\left(\frac{p}{c_0^2} - \rho\right) \qquad (A.64)$$

Pressure field, [Eq.(9)]:

$$p(\vec{x},t) = \frac{1}{4\pi}\int_{V_w}\left[\frac{\partial^2}{\partial y_i \partial y_j}(\rho_w u_i u_j + p_{ij} - p\delta_{ij}) + \frac{\partial^2}{\partial t^2}\left(\frac{p}{c_0^2} - \rho_w\right)\right]\frac{d\vec{y}}{r} +$$

$$\frac{1}{4\pi}\int_{V_a}\left[\frac{\partial^2}{\partial y_i y_j}(\rho_a u_i u_j + p_{ij} - p\delta_{ij}) + \frac{\partial^2}{\partial t^2}\left(\frac{p}{c_0^2} - \rho_a\right)\right]\frac{d\vec{y}}{r}$$

$$(A.65)$$

Appendix B

Basic Equations Governing Wave Motion in a Viscoelastic Layer

Following Ewing[55], we denote $\vec{u} = (u_x, u_y, u_z)$ as the displacement vector of the particle movement in an isotropic elastic medium, and

$$\begin{pmatrix} e_{xx} & e_{xy} & e_{xz} \\ e_{yx} & e_{yy} & e_{yz} \\ e_{zx} & e_{zy} & e_{zz} \end{pmatrix} \text{ and } \begin{pmatrix} p_{xx} & p_{xy} & p_{xz} \\ p_{yx} & p_{yy} & p_{yz} \\ p_{zx} & p_{zy} & p_{zz} \end{pmatrix}$$

as the strain and stress tensors where

$$e_{xx} = \frac{\partial u_x}{\partial x}, \qquad e_{yy} = \frac{\partial u_y}{\partial y}, \qquad e_{zz} = \frac{\partial u_z}{\partial z}$$

$$e_{xy} = \frac{1}{2}\left(\frac{\partial u_x}{\partial y} + \frac{\partial u_y}{\partial x}\right), \quad e_{xz} = \frac{1}{2}\left(\frac{\partial u_x}{\partial z} + \frac{\partial u_z}{\partial x}\right), \quad e_{yz} = \frac{1}{2}\left(\frac{\partial u_y}{\partial z} + \frac{\partial u_z}{\partial y}\right)$$

Under small strains the linear strain-stress relations become (for convenience, the symbol $\tilde{p}$ used in the main text is written here as p.)

$$p_{xx} = \lambda\theta + 2\mu\frac{\partial u_x}{\partial x}$$

$$p_{yy} = \lambda\theta + 2\mu\frac{\partial u_y}{\partial y}$$

$$p_{zz} = \lambda\theta + 2\mu\frac{\partial u_z}{\partial z}$$

$$p_{xy} = \mu\left(\frac{\partial u_x}{\partial y} + \frac{\partial u_y}{\partial x}\right)$$

$$p_{xz} = \mu\left(\frac{\partial u_x}{\partial z} + \frac{\partial u_z}{\partial x}\right)$$

$$p_{yz} = \mu\left(\frac{\partial u_y}{\partial z} + \frac{\partial u_z}{\partial y}\right) \tag{B.1}$$

where

$$\theta = \nabla\vec{u} = \frac{\partial u_x}{\partial x} + \frac{\partial u_y}{\partial y} + \frac{\partial u_z}{\partial z} \tag{B.2}$$

and λ and μ are the Lamé constants of the isotropic medium. As usual here, p_{xy} (and similarly for p_{xz} and p_{yz}) means the stress component in the plane perpendicular to the x-axis and in the positive y direction. Because of mechanical equilibrium

$$p_{yx} = p_{xy}, \quad p_{zx} = p_{xz}, \quad p_{zy} = p_{yz} \tag{B.3}$$

Eqs.(B.1) can also be written as

$$p_{xx} = \left(K + \frac{4}{3}\mu\right)\theta - 2\mu\left(\frac{\partial u_y}{\partial y} + \frac{\partial u_z}{\partial z}\right)$$

$$p_{yy} = \left(K + \frac{4}{3}\mu\right)\theta - 2\mu\left(\frac{\partial u_x}{\partial x} + \frac{\partial u_z}{\partial z}\right)$$

$$p_{zz} = \left(K + \frac{4}{3}\mu\right)\theta - 2\mu\left(\frac{\partial u_x}{\partial x} + \frac{\partial u_y}{\partial y}\right)$$

$$p_{xy} = \mu\left(\frac{\partial u_x}{\partial y} + \frac{\partial u_y}{\partial x}\right)$$

$$p_{xz} = \mu\left(\frac{\partial u_x}{\partial z} + \frac{\partial u_z}{\partial x}\right)$$

$$p_{yz} = \mu\left(\frac{\partial u_y}{\partial z} + \frac{\partial u_z}{\partial y}\right) \tag{B.4}$$

where

$$K = \lambda + \frac{2}{3}\mu, \quad K + \frac{4}{3}\mu = \lambda + 2\mu \equiv H_0 \tag{B.5}$$

Equations for Elastic Media

Introducing Eqs.(B.1) into the dynamic equations

$$\rho \frac{d^2 u_x}{dt^2} = \rho X + \frac{\partial p_{xx}}{\partial x} + \frac{\partial p_{yx}}{\partial y} + \frac{\partial p_{zx}}{\partial z}$$

$$\rho \frac{d^2 u_y}{dt^2} = \rho Y + \frac{\partial p_{xy}}{\partial x} + \frac{\partial p_{yy}}{\partial y} + \frac{\partial p_{zy}}{\partial z}$$

$$\rho \frac{d^2 u_z}{dt^2} = \rho Z + \frac{\partial p_{xz}}{\partial x} + \frac{\partial p_{yz}}{\partial y} + \frac{\partial p_{zz}}{\partial z} \tag{B.6}$$

and neglecting the body force components ρX, ρY and ρZ leads to

$$\rho \frac{\partial^2 u_x}{\partial t^2} = (H_0 - \mu)\frac{\partial \theta}{\partial x} + \mu \nabla^2 u_x$$

$$\rho \frac{\partial^2 u_y}{\partial t^2} = (H_0 - \mu)\frac{\partial \theta}{\partial y} + \mu \nabla^2 u_y$$

$$\rho \frac{\partial^2 u_z}{\partial t^2} = (H_0 - \mu)\frac{\partial \theta}{\partial z} + \mu \nabla^2 u_z \tag{B.7}$$

in which the operator $\partial/\partial t$ has been used to replace $d/dt (\equiv \partial/\partial t + \vec{u}\cdot\nabla)$ because the second-order terms are neglected.

Since a vector field may be expressed as the sum of a gradient of a scalar field and the curl of a zero-divergence vector, we can here write the displacement vector field $\vec{u}$ as[58]

$$\vec{u} = \nabla \Phi + \nabla \times \vec{\Psi}, \quad \nabla \cdot \vec{\Psi} = 0 \tag{B.8}$$

where Φ is a scalar potential and $\vec{\Psi}$ a vector potential. Equation (B.2) then becomes

$$\theta = \nabla^2 \Phi \tag{B.9}$$

Substituting relation (B.8) for the displacement components into Eq.(B.7) it is clear that Eqs.(B.7) will be satisfied if both

$$\rho \frac{\partial^2}{\partial t^2}\Phi - H_0 \nabla^2 \Phi = 0 \tag{B.10}$$

and

$$\rho \frac{\partial^2}{\partial t^2}(\nabla \times \vec{\Psi}) - \mu \nabla^2 (\nabla \times \vec{\Psi}) = 0 \tag{B.11}$$

These relations define respectively the compressional and shear-wave motions in a medium with corresponding wave speeds

$$\alpha = \sqrt{\frac{H_0}{\rho}} \quad \left(= \sqrt{\frac{\lambda + 2\mu}{\rho}}\right) \tag{B.12}$$

and

$$\beta = \sqrt{\frac{\mu}{\rho}} \tag{B.13}$$

In the case of viscoelastic media, both H_0 and μ are complex, $H_0 = H_0' + iH_0''$ and $\mu = \mu' + i\mu''$. The sign of H_0'' and μ'' depends on the selection of the time factor $e^{i\omega t}$ or $e^{-i\omega t}$. If $e^{-i\omega t}$ is chosen H_0'' and μ'' must be negative, to ensure that the wavenumbers $k_\alpha = \omega/\alpha$ and $k_\beta = \omega/\beta$ have positive imaginary parts and that the wave motion decays at large distances from the source. As a matter of convenience we denote the symbol "sgn" as the sign of the time factor and write

$$H_0 = H_0' + i\operatorname{sgn}|H_0''| \quad \text{and} \quad \mu = \mu' + i\operatorname{sgn}|\mu''| \tag{B.14}$$

It is noted that under the influence of a plane-compressional wave with horizontal wavenumber vector, $\vec{k}_0$, the medium particles move backwards and forwards in the plane containing $\vec{k}_0$ and the z-axis, while under a shear wave the movement is in a direction perpendicular to this plane, as indicated by the curl operator $\nabla \times$. It is therefore convenient to set

$$\vec{\Psi} = (\vec{k}_0 \times \vec{n})\Psi \tag{B.15}$$

where $\vec{k}_0$ and $\vec{n}$ are respectively the unit vectors in the directions of the horizontal wavenumber vector, $\vec{k}$, and the positive vertical coordinate axis. By this means we can write the potential equations to be solved in the ith layer as

$$\nabla^2 \Phi_i + \frac{\omega^2}{\alpha_i^2}\Phi_i = 0 \tag{B.16}$$

$$\nabla^2 \Psi_i + \frac{\omega^2}{\beta_i^2}\Psi_i = 0 \tag{B.17}$$

with

$$\vec{u}_i = \nabla \Phi_i + \nabla \times (\vec{k}_0 \times \vec{n})\Psi_i \tag{B.18}$$

and the time factor $e^{\pm i\omega t}$ suppressed.

Appendix C

Basic Equations Governing Wave Motion in a Water-Saturated Porous Viscoelastic Layer

Stoll and his colleagues have developed a comprehensive mathematical model, based on Biot's theory, to predict the propagation characteristics of acoustic waves in marine sediments[46]. This enables us to extend our ocean model to include the effects of unconsolidated sedimentary layers. Since our purpose here is not so much the study of the physical properties of the porous medium itself but rather the effect of given medium parameters on wave propagation, we base the discussion to follow on Stoll's model and adopt his expressions and parameter values.

From the point of view of wave propagation, a water-saturated, porous-elastic medium differs from an ordinary elastic medium mainly because of a relative movement between the pore water and the solid frame. This relative motion produces a new kind of compressional wave in the medium, characterised by a relatively low wave speed and high attenuation, in addition to the compressional and shear-wave motion of the frame material. Denoting $\vec{u}$ as the displacement vector of the solid frame and $\vec{U}$ that of the pore fluid, a quantity "increment of fluid content", ζ_p, can be introduced to describe the volume of fluid flowing into a volume element attached to the frame,

$$\zeta_p = \beta \, \mathbf{div}(\vec{u} - \vec{U}) \tag{C.1}$$

where β, the porosity, is the ratio of the pore volume to the total volume of the element of porous material. The pressure in the pore water, p_f, will obviously be a function of ζ_p and the strain in the frame. Further, the stress (compressional) components will be influenced by the relative motion. Assuming that the amplitudes of all motions are small the stress-strain relation of such a motion can be written as (see Appendix B)

$$p_{xx} = \bar{H}\theta - 2\mu\left(\frac{\partial u_y}{\partial y} + \frac{\partial u_z}{\partial z}\right) - C\zeta_p$$

$$p_{yy} = \bar{H}\theta - 2\mu\left(\frac{\partial u_x}{\partial x} + \frac{\partial u_z}{\partial z}\right) - C\zeta_p$$

$$p_{zz} = \bar{H}\theta - 2\mu\left(\frac{\partial u_x}{\partial x} + \frac{\partial u_y}{\partial y}\right) - C\zeta_p$$

$$p_{xy} = \mu\left(\frac{\partial u_x}{\partial y} + \frac{\partial u_y}{\partial x}\right)$$

$$p_{xz} = \mu\left(\frac{\partial u_x}{\partial z} + \frac{\partial u_z}{\partial x}\right)$$

$$p_{yz} = \mu\left(\frac{\partial u_y}{\partial z} + \frac{\partial u_z}{\partial y}\right)$$

$$p_f = M\zeta_p - C\theta \tag{C.2}$$

where p_{xy} etc. again denote the stress components, $\vec{u} = (u_x, u_y, u_z)$, $\theta = \nabla \vec{u}$ and μ is the shear modulus of the frame. $\bar{H}$ is a new modulus equivalent to $H_0 (\equiv \lambda + 2\mu)$ of the elastic material and M and C are two new moduli.

Denoting ρ as an average density, $\rho = \beta\rho_f + (1-\beta)\rho_r$, with ρ_r the density of the frame material and ρ_f that of the fluid and $\vec{V} = \beta(\vec{u} - \vec{U})$, the dynamic equation for the x-component becomes

$$\frac{\partial p_{xx}}{\partial x} + \frac{\partial p_{yx}}{\partial y} + \frac{\partial p_{zx}}{\partial z} = \frac{d^2}{dt^2}[\beta\rho_f U_x + (1-\beta)\rho_r u_x]$$

$$= \frac{d^2}{dt^2}[\rho u_x - \beta\rho_f(u_x - U_x)]$$

Combining the expressions for all three components and replacing the differential d/dt by $\partial/\partial t$ leads to

$$\mu\nabla^2 \vec{u} + (\bar{H} - \mu)\nabla\theta - C\nabla\zeta_p = \rho\frac{\partial^2 \vec{u}}{\partial t^2} - \rho_f\frac{\partial^2 \vec{V}}{\partial t^2} \tag{C.3}$$

Equations for Porous Media

The equation governing the motion of pore fluid takes the form

$$-\nabla p_f = \rho_f \frac{\partial^2 \vec{U}}{\partial t^2} + \beta \frac{\eta}{k_p} \frac{\partial}{\partial t}(\vec{U} - \vec{u}) \qquad (C.4)$$

where the second term on the right-hand-side of Eq.(C.4) accounts for the frictional force, which is proportional to the velocity difference and the viscosity η but inversely proportional to the permeability of the medium, in accordance with Darcy's law. In the above $\partial/\partial t$ is used in place of d/dt in view of the restriction to small amplitude motion. The right-hand-side of Eq.(C.4) can be written as

$$\rho_f \frac{\partial^2 \vec{u}}{\partial t^2} - \frac{\rho_f}{\beta} \frac{\partial^2 \vec{V}}{\partial t^2} - \frac{\eta}{k_\rho} \frac{\partial \vec{V}}{\partial t}.$$

As Stoll has pointed out, however, under a given pressure gradient not all the fluid will necessarily move in the direction of the force, as some part may be forced in different directions by the multi-directional nature of the pores. To account for such a diversion a parameter $m = \alpha \rho_f/\beta$, where $\alpha > 1$, is used in place of ρ_f/β, so that from Eqs.(C.4) and (C.2),

$$C\nabla\theta - M\nabla\zeta_p = \rho_f \frac{\partial^2 \vec{u}}{\partial t^2} - m\frac{\partial^2 \vec{V}}{\partial t^2} - \frac{\eta}{k_\rho} \frac{\partial \vec{V}}{\partial t}. \qquad (C.5)$$

Equations (C.3) and (C.5) form coupled differential equations describing the wave motions in the medium. As in the case of the elastic layer (Appendix B) the velocity vectors $\vec{u}$ and $\vec{V}$ are expressed through potentials Φ_s, $\vec{\Psi}_s$, Φ_f and $\vec{\Psi}_f$

$$\vec{u} = \nabla\Phi_s + \nabla \times \vec{\Psi}_s \qquad \nabla\vec{\Psi}_s = 0 \qquad (C.6)$$

$$\vec{V} = \nabla\Phi_f + \nabla \times \vec{\Psi}_f \qquad \nabla\vec{\Psi}_f = 0 \qquad (C.7)$$

Their introduction into Eqs.(C.3) and (C.5) leads to two sets of equations coupling the scalar potentials Φ_s and Φ_f and the vector potentials $\vec{\Psi}_s$ and $\vec{\Psi}_f$ as

$$\bar{H}\nabla^2\Phi_s - C\nabla^2\Phi_f = \rho\frac{\partial^2\Phi_s}{\partial t^2} - \rho_f\frac{\partial^2\Phi_f}{\partial t^2}$$

$$C\nabla^2\Phi_s - M\nabla^2\Phi_f = \rho_f\frac{\partial^2\Phi_s}{\partial t^2} - m\frac{\partial^2\Phi_f}{\partial t^2} - \frac{\eta}{k_p}\frac{\partial\Phi_f}{\partial t}$$

and
$$\mu\nabla^2\vec{\Psi}_s - \rho\frac{\partial^2\vec{\Psi}_s}{\partial t^2} + \rho_f\frac{\partial^2\vec{\Psi}_f}{\partial t^2} = 0$$

$$\rho_f\frac{\partial^2\vec{\Psi}_s}{\partial t^2} - m\frac{\partial^2\vec{\Psi}_f}{\partial t^2} - \frac{\eta}{k_p}\frac{\partial\vec{\Psi}_f}{\partial t} = 0$$

We write the time factor of the wave motion as $\exp[\text{sgn}(i\omega t)]$ with sgn=+1 or -1 to be chosen as convenient (in the present contribution we take sgn=-1). Before proceeding, it is to be noted that the moduli and other parameters can be frequency dependent in a porous medium. Accordingly, when the time factor is introduced into the above equations, it should be borne in mind that these moduli are replaced by their Fourier transforms. Thus we have

$$(\bar{H}\nabla^2 + \omega^2\rho)\Phi_s - (C\nabla^2 + \omega^2\rho_f)\Phi_f = 0$$

$$(C\nabla^2 + \omega^2\rho_f)\Phi_s - \left(M\nabla^2 + m\omega^2 - i\,\text{sgn}\frac{\eta}{k_p}\omega\right)\Phi_f = 0 \qquad (C.8)$$

and

$$(\mu\nabla^2 + \rho\omega^2)\vec{\Psi}_s - \rho_f\omega^2\vec{\Phi}_f = 0$$

$$-\omega^2\rho_f\vec{\Psi}_s + \left(m\omega^2 - i\,\text{sgn}\frac{\eta\omega}{k_p}\right)\vec{\Psi}_f = 0 \qquad (C.9)$$

Substituting the wave forms

$$\Phi_{s,f} = A_\Phi \exp[i\,\text{sgn}(\omega t \pm \vec{l}_c\vec{r})]$$

and

$$\vec{\Psi}_{s,f} = A_\Psi \exp[i\,\text{sgn}(\omega t \pm \vec{l}_s\vec{r})]$$

into Eqs.(C.8) and (C.9) leads to the dispersion relations for two kinds of waves

$$\begin{vmatrix} \bar{H}l_c^2 - \omega^2\rho & \omega^2\rho_f - Cl_c^2 \\ Cl_c^2 - \omega^2\rho_f & m\omega^2 - Ml_c^2 - i\,\text{sgn}\,\eta F\omega/k_p \end{vmatrix} = 0 \qquad (C.10)$$

and

$$\begin{vmatrix} \mu l_s^2 - \omega^2\rho & \omega^2\rho_f \\ -\omega^2\rho_f & m\omega^2 - i\,\text{sgn}\,\eta F\omega/k_p \end{vmatrix} = 0 \qquad (C.11)$$

in which the frequency-correction factor, F, introduced to account for non-Poiseuille[1] friction (see Biot[44]), is given by:

$$F(k_\eta) = \frac{k_\eta T(k_\eta)}{4[1 - 2T(k_\eta)/i\,k_\eta]} \tag{C.12}$$

where

$$T(k_\eta) = \frac{ber'(k_\eta) + i\,bei'(k_\eta)}{ber(k_\eta) + i\,bei(k_\eta)}.$$

$k_\eta = a\sqrt{\mathrm{sgn}\,\omega \rho_f/\eta}$ and a is a pore size constant. $ber(x)$, $bei(x)$, $ber'(x)$ and $bei'(x)$ are respectively the real and imaginary parts of the Kelvin function,

$$J_0(i\sqrt{i}\,z) = ber(z) + i\,bei(z)$$

and their derivatives.

Solving Eqs.(C.10) and (C.11) leads to two independent solutions for the compressional wavenumbers, l_{c1} and l_{c2}, and one for the shear wave, l_s, as

$$l_{c1} = \pm\sqrt{-\frac{\epsilon_2}{2\epsilon_1}\left(1 - \sqrt{1 - \frac{4\epsilon_1\epsilon_3}{\epsilon_2^2}}\right)} \tag{C.13}$$

$$l_{c2} = \pm\sqrt{-\frac{\epsilon_2}{2\epsilon_1}\left(1 + \sqrt{1 - \frac{4\epsilon_1\epsilon_3}{\epsilon_2^2}}\right)} \tag{C.14}$$

and

$$l_s = \pm\omega\sqrt{\frac{\rho}{\mu}\left[1 - \frac{\rho_f^2}{\rho^2(\frac{m}{\rho} - i\,\mathrm{sgn}\,sF/\omega)}\right]} \tag{C.15}$$

where

$$\epsilon_1 = \bar{H}M - C^2 \tag{C.16}$$

$$\epsilon_2 = \omega^2\rho\left[2C\frac{\rho_f}{\rho} - M - \bar{H}\left(\frac{m}{\rho} - \frac{i\,\mathrm{sgn}\,sF}{\omega}\right)\right] \tag{C.17}$$

$$\epsilon_3 = \omega^4\rho^2\left[\frac{m}{\rho} - \left(\frac{\rho_f}{\rho}\right)^2 - \frac{i\,\mathrm{sgn}\,sF}{\omega}\right] \tag{C.18}$$

---

[1] At very low frequencies the friction between the flow of pore fluid and the wall of the pore itself can be regarded as linearly proportional to the viscosity coefficient.

and
$$s = \eta/(\rho k_p) \qquad (C.19)$$
The signs + and − correspond to the propagation direction of plane waves.

According to Stoll[46], moduli $\bar{H}$, C and M can be expressed in terms of K_b, the bulk modulus of the free-draining porous frame, K_f, the bulk modulus of the pore water and that of the solid frame, K_r, as

$$\bar{H} = [(K_r - K_b)^2/(D - K_b)] + K_b + 4\mu/3 \qquad (C.20)$$

$$C = K_r(K_r - K_b)/(D - K_b) \qquad (C.21)$$

$$M = K_r^2/(D - K_b) \qquad (C.22)$$

where
$$D = K_r[1 + \beta(K_r/K_f - 1)] \qquad (C.23)$$

The imaginary parts of the moduli, $\bar{H}$, C and M can be found by assigning the corresponding imaginary part $K_b'' = K_b' \Delta_c/\pi$ to the complex bulk modulus $K_b = K_b' + i \operatorname{sgn} K_b''$, and $\mu'' = \mu' \Delta_s/\pi$ to the complex shear modulus $\mu = \mu' + i \operatorname{sgn} \mu'''$.

In the case of an elastic medium $\beta = 0$, $\vec{V} = 0$, and $K_b = K_r \equiv K$, so that $\bar{H} = \lambda + 2\mu$, $\rho_f = 0$, $C = 0$ and $\zeta_p \equiv 0$. Equation (C.5) therefore disappears while Eq.(C.3) degenerates, as expected, to the dynamic equation for an ordinary elastic medium (see Eq.(B.7)).

Finally, the ratio of the amplitude of the compressional waves in the fluid to that in the frame can be determined by substituting
$\Phi_s = C_s e^{i \operatorname{sgn}(\omega t - \vec{l}_c \vec{r})}$ and $\Phi_f = C_f e^{i \operatorname{sgn}(\omega t - \vec{l}_c \vec{r})}$ into the first equation of (C.8), which leads to

$$\delta_c = \frac{C_f}{C_s} = \frac{\bar{H} l_c^2 - \omega^2 \rho}{C l_c^2 - \omega^2 \rho_f}$$

or more specifically, for the fast and slow waves

$$\delta_{fast} = \frac{\bar{H} l_{fast}^2 - \omega^2 \rho}{C l_{fast}^2 - \omega^2 \rho_f} \quad \text{and} \quad \delta_{slow} = \frac{\bar{H} l_{slow}^2 - \omega^2 \rho}{C l_{slow}^2 - \omega^2 \rho_f} \qquad (C.24)$$

In the same way Eq.(C.9) can be used to establish the ratio for the shear wave

$$\delta_{sh} = \frac{\rho}{\rho_f}\left(1 - \frac{\mu l_{sh}^2}{\omega^2 \rho}\right)$$

	Sand	Soft Sediment
K_r	3.6×10^{10}	3.6×10^{10}
K_f	2.0×10^9	2.0×10^9
ρ_s	2656	2656
ρ_f	1000	1000
μ	2.61×10^7	2.21×10^7
K_b	4.36×10^7	3.69×10^7
Δ_s	0.15	0.5
Δ_c	0.15	0.5
β	0.47	0.76
k_p	$5 \times 10^{-13} \to 10^{-10}$	1.6×10^{-15}

Table C.1: Parameters for an unconsolidated porous sediment layer.

For convenience of application the parameters for the two classic sediments discussed by Stoll are listed in Table C.1. All parameters have been converted to the MKS unit system with the values of water viscosity $\eta = 10^{-3}$(kg/ms), the pore-size parameter $a = 3 \times 10^{-5}$ and $\alpha = 1.25$ (in the expression for $m = \alpha \rho_f / \beta$) being assumed.

Appendix D

The Calculation of the Reflection Coefficient for a Multilayered Viscoelastic Half Space

Numerical procedures, which can be used to calculate the reflection coefficient, R_b, and transmission field in a stratified viscoelastic medium, were first described by Schmidt and Jensen[62]. We essentially follow this development here but extend it in places to provide a clarification of the methodology in certain anomalous situations.

We assume a plane wave of unit amplitude, $\Phi_{1inc} = e^{-i\gamma_1(z+H_1)+i(kx-\omega t)}$, is incident from water upon a seabed ($z = -H_1$) consisting of parallel, homogeneous viscoelastic layers, in which ρ_n, α_{rn}, β_{rn}, $Q_{\alpha n}$, $Q_{\beta n}$ and h_n are respectively the density, real part of the compressional and shear-wave velocities, Q-values and the thickness of the layers. Since the reflection and transmission coefficients are the same for velocity and displacement, for convenience the displacement potential is also used for the water medium.

As usual, by introducing the scalar and vector potentials Φ_n and $\vec{\Psi}_n$, the displacement vector in the n-th layer can be expressed as

$$\vec{u}_n = \nabla \Phi_n + \nabla \times \vec{\Psi}_n \quad \nabla \cdot \vec{\Psi}_n = 0 \tag{D.1}$$

With the $x - z$ plane chosen as the plane of incidence we can denote the vector potential

$$\vec{\Psi}_n = (\vec{k}_0 \times \vec{n})\Psi_n \tag{D.2}$$

where $\vec{k}_0$ and $\vec{n}$ are unit vectors in the directions of the horizontal wave vector (here the positive x direction) and positive z direction, and thus write

$$\nabla \times \vec{\Psi}_n = \left(\frac{\partial}{\partial z}, 0, -\frac{\partial}{\partial x}\right)\Psi_n. \tag{D.3}$$

The displacement vector then takes the form

$$u_{nx} = \frac{\partial \Phi_n}{\partial x} + \frac{\partial \Psi_n}{\partial z}$$

$$u_{nz} = \frac{\partial \Phi_n}{\partial z} - \frac{\partial \Psi_n}{\partial x} \qquad (D.4)$$

Since (see Appendix B)

$$p_{nxx} = \bar{H}_n \nabla \cdot \vec{u}_n - 2\mu_n \frac{\partial u_{nz}}{\partial z}$$

$$p_{nzz} = \bar{H}_n \nabla \cdot \vec{u}_n - 2\mu_n \frac{\partial u_{nx}}{\partial x}$$

$$p_{nxz} = p_{nzx} = \mu_n \left(\frac{\partial u_{nx}}{\partial z} + \frac{\partial u_{nz}}{\partial x} \right) \qquad (D.5)$$

where $\bar{H}_n \equiv (\lambda_n + 2\mu_n) = \rho_n \alpha_n^2$ and α_n is the complex compressional-wave velocity in the nth layer,

$$p_{nxx} = \bar{H}_n \nabla^2 \Phi_n - 2\mu_n \left(\frac{\partial^2 \Phi_n}{\partial z^2} - \frac{\partial^2 \Psi_n}{\partial x \partial z} \right)$$

$$p_{nzz} = \bar{H}_n \nabla^2 \Phi_n - 2\mu_n \left(\frac{\partial^2 \Phi_n}{\partial x^2} - \frac{\partial^2 \Psi_n}{\partial x \partial z} \right)$$

$$p_{nxz} = p_{nzx} = \mu_n \left(2 \frac{\partial^2 \Phi_n}{\partial x \partial z} - \frac{\partial^2 \Psi_n}{\partial x^2} + \frac{\partial^2 \Psi_n}{\partial z^2} \right) \qquad (D.6)$$

Denoting

$$\left. \begin{array}{l} \Phi_n = A_{n1} e^{-i\gamma_{n1}(z+H_{n-1})} + B_{n1} e^{i\gamma_{n1}(z+H_n)} \\ \Psi_n = A_{n3} e^{-i\gamma_{n3}(z+H_{n-1})} + B_{n3} e^{i\gamma_{n3}(z+H_n)} \end{array} \right\} e^{i(kx-\omega t)} \quad n \geq 2 \quad (D.7)$$

we have

$$u_{nx} = ik \left[A_{n1} e^{-i\gamma_{n1}(z+H_{n-1})} + B_{n1} e^{i\gamma_{n1}(z+H_n)} \right]$$
$$- i\gamma_{n3} \left[A_{n3} e^{-i\gamma_{n3}(z+H_{n-1})} - B_{n3} e^{i\gamma_{n3}(z+H_n)} \right]$$

$$u_{nz} = -i\gamma_{n1} \left[A_{n1} e^{-i\gamma_{n1}(z+H_{n-1})} - B_{n1} e^{i\gamma_{n1}(z+H_n)} \right]$$
$$- ik \left[A_{n3} e^{-i\gamma_{n3}(z+H_{n-1})} + B_{n3} e^{i\gamma_{n3}(z+H_n)} \right]$$

Reflection from Elastic Layers

$$p_{nzz} = -(\omega^2 \rho_n - 2\mu_n \gamma_{n1}^2)\left[A_{n1}e^{-i\gamma_{n1}(z+H_{n-1})} + B_{n1}e^{i\gamma_{n1}(z+H_n)}\right]$$
$$-2\mu_n k\gamma_{n3}\left[A_{n3}e^{-i\gamma_{n3}(z+H_{n-1})} - B_{n3}e^{i\gamma_{n3}(z+H_n)}\right]$$

$$p_{nxz} = 2\mu_n k\gamma_{n1}\left[A_{n1}e^{-i\gamma_{n1}(z+H_{n-1})} - B_{n1}e^{i\gamma_{n1}(z+H_n)}\right]$$
$$+\mu_n(k^2 - \gamma_{n3}^2)\left[A_{n3}e^{-i\gamma_{n3}(z+H_{n-1})} + B_{n3}e^{i\gamma_{n3}(z+H_n)}\right] \quad (D.8)$$

In the water ($z > -H_1$),

$$\Phi_1 = \left[e^{-i\gamma_1(z+H_1)} + R_b e^{i\gamma_1(z+H_1)}\right]e^{i(kx-\omega t)}$$

$$u_{1x} = \frac{\partial \Phi_1}{\partial x} = ik\left[e^{-i\gamma_1(z+H_1)} + R_b e^{i\gamma_1(z+H_1)}\right]e^{i(kx-\omega t)}$$

$$u_{1z} = \frac{\partial \Phi_1}{\partial z} = -i\gamma_1\left[e^{-i\gamma_1(z+H_1)} - R_b e^{i\gamma_1(z+H_1)}\right]e^{i(kx-\omega t)}$$

$$p_1 = -\rho_1 \frac{\partial^2 \Phi_1}{\partial t^2} = \rho_1 \omega^2 \left[e^{-i\gamma_1(z+H_1)} + R_b e^{i\gamma_1(z+H_1)}\right]e^{i(kx-\omega t)} \quad (D.9)$$

The boundary conditions at $z = -H_1$, ie., $u_{2z} = u_{1z}$, $p_{2zz} = -p_1$ and $p_{2zx} = 0$, require that

$$-i\gamma_1(1 - R_b) = -i\gamma_{21}\left[A_{21} - B_{21}e^{i\gamma_{21}(-H_1+H_2)}\right] - ik\left[A_{23} + B_{23}e^{i\gamma_{23}(-H_1+H_2)}\right]$$

$$-\rho_1\omega^2(1 + R_b) = -(\omega^2\rho_2 - 2\mu_2 k^2)\left[A_{21} + B_{21}e^{i\gamma_{21}(-H_1+H_2)}\right] - 2\mu_2 k\gamma_{23}\left[A_{23} - B_{23}e^{i\gamma_{23}(-H_1+H_2)}\right]$$

$$0 = 2\mu_2 k\gamma_{21}\left[A_{21} - B_{21}e^{i\gamma_{21}(-H_1+H_2)}\right] + \mu_2(k^2 - \gamma_{23}^2)\left[A_{23} + B_{23}e^{i\gamma_{23}(-H_1+H_2)}\right]$$

or by denoting

$$e_{n1} = e^{i\gamma_{n1}h_n}, \quad e_{n3} = e^{i\gamma_{n3}h_n}$$

(here the index "n2" is not used deliberately to be consistent with the expressions for porous media where "n2" is used for the slow compressional wave), $\mu_n = \rho_n \beta_n^2 = \rho_n \omega^2 s_{n3}^2$ and $m_{n,n+1} = \rho_n/\rho_{n+1}$,

$$\gamma_1 R_b + \gamma_{21} A_{21} - \gamma_{21} e_{21} B_{21} + k A_{23} + k e_{23} B_{23} = \gamma_1$$

$$\rho_1\omega^2 R_b - (\rho_2\omega^2 - 2\mu_2 k^2)A_{21} - (\rho_2\omega^2 - 2\mu_2 k^2)e_{21}B_{21}$$
$$-2k^2\mu_2\gamma_{23}A_{23} + 2k\mu_2\gamma_{23}e_{23}B_{23} = -\rho_1\omega^2$$

$$2k\gamma_{21}A_{21} - 2k\gamma_{21}e_{21}B_{21} + (k^2 - \gamma_{23}^2)A_{23} + (k^2 - \gamma_{23}^2)e_{23}B_{23} = 0$$
$$\text{(D.10)}$$

At the interface $z = -H_n$ ($n \geq 2$) the continuity conditions $u_{nz} = u_{n+1,z}$, $u_{nx} = u_{n+1,x}$, $p_{nzz} = p_{n+1,zz}$ and $p_{nzx} = p_{n+1,zx}$ lead to

$$\gamma_{n1}\left[A_{n1}e^{i\gamma_{n1}h_n} - B_{n1}\right] + k\left[A_{n3}e^{i\gamma_{n3}h_n} + B_{n3}\right] =$$
$$\gamma_{n+1,1}\left[A_{n+1,1} - B_{n+1,1}e^{i\gamma_{n+1,1}h_{n+1}}\right]$$
$$+k\left[A_{n+1,3} - B_{n+1,3}e^{i\gamma_{n+1,3}h_{n+1}}\right]$$

$$k\left[A_{n1}e^{i\gamma_{n1}h_n} + B_{n1}\right] - \gamma_{n3}\left[A_{n3}e^{i\gamma_{n3}h_n} - B_{n3}\right] =$$
$$k\left[A_{n+1,1} - B_{n+1,1}e^{i\gamma_{n+1,1}h_{n+1}}\right]$$
$$-\gamma_{n+1,3}\left[A_{n+1,3} - B_{n+1,3}e^{i\gamma_{n+1,3}h_{n+1}}\right]$$

$$-(\omega^2\rho_n - 2\mu_n k^2)\left[A_{n1}e^{i\gamma_{n1}h_n} + B_{n1}\right] - 2\mu_n k\gamma_{n3}\left[A_{n3}e^{i\gamma_{n3}h_n} - B_{n3}\right] =$$
$$-(\omega^2\rho_{n+1} - 2\mu_{n+1}k^2)\left[A_{n+1,1} + B_{n+1,1}e^{i\gamma_{n+1,1}h_{n+1}}\right]$$
$$-2\mu_{n+1}k\gamma_{n+1,3}\left[A_{n+1,3} - B_{n+1,3}e^{i\gamma_{n+1,3}h_{n+1}}\right]$$

$$2\mu_n k\gamma_{n1}\left[A_{n1}e^{i\gamma_{n1}h_n} - B_{n1}\right] + \mu_n(k^2 - \gamma_{n3}^2)\left[A_{n3}e^{i\gamma_{n3}h_n} + B_{n3}\right] =$$
$$2\mu_{n+1}k\gamma_{n+1,1}\left[A_{n+1,1} - B_{n+1,1}e^{i\gamma_{n+1,1}h_{n+1}}\right]$$
$$+\mu_{n+1}(k^2 - \gamma_{n+1,3}^2)\left[A_{n+1,3} + B_{n+1,3}e^{i\gamma_{n+1,3}h_{n+1}}\right]$$

which can be written more clearly as

$$\gamma_{n1}(A_{n1}e_{n1} - B_{n1}) + k(A_{n3}e_{n3} + B_{n3}) - \gamma_{n+1,1}(A_{n+1,1} - B_{n+1,1}e_{n+1,1})$$
$$-k(A_{n+1,3} + B_{n+1,3}e_{n+1,3}) = 0$$

$$k(A_{n1}e_{n1} + B_{n1}) - \gamma_{n3}(A_{n3}e_{n3} - B_{n3}) - k(A_{n+1,1} + B_{n+1,1}e_{n+1,1})$$
$$+\gamma_{n+1,3}(A_{n+1,3} - B_{n+1,3}e_{n+1,3}) = 0$$
$$\text{(D.11)}$$

$$-(\omega^2\rho_n - 2\mu_n k^2)(A_{n1}e_{n1} + B_{n1}) - 2\mu_n k\gamma_{n3}(A_{n3}e_{n3} - B_{n3})$$

$$+(\omega^2 \rho_{n+1} - 2\mu_{n+1}k^2)(A_{n+1,1} + B_{n+1,1}e_{n+1,1})$$
$$+2\mu_{n+1}k\gamma_{n+1,3}(A_{n+1,3} - B_{n+1,3}e_{n+1,3}) = 0$$

$$2\mu_n k\gamma_{n1}(A_{n1}e_{n1} - B_{n1}) + \mu_n(k^2 - \gamma_{n3}^2)(A_{n3}e_{n3} + B_{n3})$$
$$-2\mu_{n+1}k\gamma_{n+1,1}(A_{n+1,1} - B_{n+1,1}e_{n+1,1})$$
$$-\mu_{n+1}(k^2 - \gamma_{n+1,3}^2)(A_{n+1,3} + B_{n+1,3}e_{n+1,3}) = 0 \quad (\text{D.12})$$

The conditions at the last interface ($z = -H_N$) are of the same form as Eqs.(D.11)→(D.12) but with $B_{N+1,1} = 0$ and $B_{N+1,3} = 0$. For convenience, we add three equations, say $x_1 = 0$, $x_2 = 0$ and $x_3 = 0$, together with two equations about $B_{n+1,1}$ and $B_{n+1,3}$ to form a matrix

$$\tilde{C} \cdot \tilde{A} = \tilde{B} \quad (\text{D.13})$$

where $\tilde{C}$ is a $4(N+1) \times 4(N+1)$ matrix and $\tilde{A}$ and $\tilde{B}$ are $4(N+1) \times 1$ matrices:

$$\tilde{C} = \begin{pmatrix} \begin{smallmatrix} 1 & 0 & 0 & 0 \\ 0 & 0 & 0 & \gamma_1 \\ 0 & 0 & 0 & \rho_1\omega^2 \\ 0 & 0 & 0 & 0 \end{smallmatrix} & \tilde{C}_{12} & 0 & 0 & 0 & 0 & 0 \\ 0 & \tilde{C}_{22} & \tilde{C}_{23} & 0 & 0 & 0 & 0 \\ 0 & 0 & \ddots & \ddots & 0 & 0 & 0 \\ 0 & 0 & 0 & \tilde{C}_{nn} & \tilde{C}_{n,n+1} & 0 & 0 \\ 0 & 0 & 0 & 0 & \ddots & \ddots & 0 \\ 0 & 0 & 0 & 0 & 0 & \tilde{C}_{NN} & \tilde{C}_{N,N+1} \\ \begin{smallmatrix} 0 & 1 & 0 & 0 \\ 0 & 0 & 1 & 0 \\ 0 & 0 & 0 & 0 \\ 0 & 0 & 0 & 0 \end{smallmatrix} & 0 & 0 & 0 & 0 & 0 & \begin{smallmatrix} 0 & 0 & 0 & 0 \\ 0 & 0 & 0 & 0 \\ 0 & 1 & 0 & 0 \\ 0 & 0 & 0 & 1 \end{smallmatrix} \end{pmatrix}$$

$$\tilde{A} = \begin{pmatrix} x_1 \\ x_2 \\ x_3 \\ R_b \\ A_{21} \\ B_{21} \\ A_{23} \\ B_{23} \\ \vdots \\ \vdots \\ A_{n1} \\ B_{n1} \\ A_{n3} \\ B_{n3} \\ \vdots \\ \vdots \\ A_{N1} \\ B_{N1} \\ A_{N3} \\ B_{N3} \\ A_{N+1,1} \\ B_{N+1,1} \\ A_{N+1,3} \\ B_{N+1,3} \end{pmatrix} \qquad \tilde{B} = \begin{pmatrix} 0 \\ \gamma_1 \\ -\rho_1\omega^2 \\ 0 \\ \\ 0 \\ \\ \\ \\ 0 \\ \\ \\ \\ 0 \\ \\ \\ \\ 0 \\ \\ \\ \\ 0 \end{pmatrix}$$

$$\tilde{C}_{12} = \begin{pmatrix} 0 & 0 & 0 & 0 \\ \gamma_{21} & -\gamma_{21}e_{21} & k & ke_{23} \\ -(\rho_2\omega^2 - 2\mu_2 k^2) & -(\rho_2\omega^2 - 2\mu_2 k^2)e_{21} & -2k\mu_2\gamma_{23} & 2k\mu_2\gamma_{23}e_{23} \\ 2k\gamma_{21}\mu_2 & -2k\gamma_{21}\mu_2 e_{21} & (k^2-\gamma_{23}^2)\mu_2 & (k^2-\gamma_{23}^2)\mu_2 e_{23} \end{pmatrix},$$

$$\tilde{C}_{nn} = \begin{pmatrix} \gamma_{n1}e_{n1} & -\gamma_{n1} & ke_{n3} & k \\ ke_{n1} & k & -\gamma_{n3}e_{n3} & \gamma_{n3} \\ -(\omega^2\rho_n - 2\mu_n k^2)e_{n1} & -(\omega^2\rho_n - 2\mu_n k^2) & -2\mu_n k\gamma_{n3}e_{n3} & 2\mu_n k\gamma_{n3} \\ 2\mu_n k\gamma_{n1}e_{n1} & -2\mu_n k\gamma_{n1} & \mu_n(k^2-\gamma_{n3}^2)e_{n3} & \mu_n(k^2-\gamma_{n3}^2) \end{pmatrix},$$

and

$$\tilde{C}_{n,n+1} = \begin{pmatrix} -\gamma_{n+1,1} & \gamma_{n+1,1}e_{n+1,1} & -k & -ke_{n+1,3} \\ -k & -ke_{n+1,1} & \gamma_{n+1,3} & \gamma_{n+1,3}e_{n+1,3} \\ (\omega^2\rho_{n+1} - 2\mu_{n+1}k^2) & (\omega^2\rho_{n+1} - 2\mu_{n+1}k^2)e_{n+1,1} & 2\mu_{n+1}k\gamma_{n+1,3} & -2\mu_{n+1}k\gamma_{n+1,3}e_{n+1,3} \\ -2\mu_{n+1}k\gamma_{n+1,1} & 2\mu_{n+1}k\gamma_{n+1,1}e_{n+1,1} & -\mu_{n+1}(k^2-\gamma_{n+1,3}^2) & -\mu_{n+1}(k^2-\gamma_{n+1,3}^2)e_{n+1,3} \end{pmatrix}$$

By solving Eq.(D.13) we can find R_b, A_{n1}, B_{n1}, A_{n2}, B_{n2}, ... and therefore construct the wave field in the whole space when a plane wave is incident from the water,

$$\Phi_1 = [e^{-i\gamma_1(z+H_1)} + R_b e^{i\gamma_1(z+H_1)}]e^{ikx-i\omega t} \} \quad -H_1 \le z$$

$$\Phi_n = [A_{n1}e^{-i\gamma_{n1}(z+H_{n-1})} + B_{n1}e^{i\gamma_{n1}(z+H_n)}]e^{ikx-i\omega t}$$

$$\Psi_n = [A_{n2}e^{-i\gamma_{n2}(z+H_{n-1})} + B_{n2}e^{i\gamma_{n2}(z+H_n)}]e^{ikx-i\omega t}$$

$$-H_n \leq z \leq -H_{n-1}$$

$$\Phi_{N+1} = A_{N+1,1}e^{-i\gamma_{N+1,1}(z+H_N)}e^{ikx-i\omega t}$$

$$\Psi_{N+1} = A_{N+1,2}e^{-i\gamma_{N+1,2}(z+H_N)}e^{ikx-i\omega t}$$

$$z \leq -H_N \quad \text{(D.14)}$$

Because of the choice of the time factor, $e^{-i\omega t}$, the wavenumber, $k_c = \omega/c$, must have a positive imaginary part. According to the definition of the Q value and attenuation coefficient, η_c[64], we have the relations

$$e^{ik_c r} = e^{i\frac{\omega r}{c'+ic''}} = e^{i\frac{\omega}{c'}r\left(1+\frac{i}{2Q_c}\right)} = e^{i\frac{\omega}{c'}r - \eta_c r}$$

so that

$$\left(1 + i\frac{c''}{c'}\right)^{-1} = \left(1 + \frac{i}{2Q_c}\right) \quad \text{and} \quad \eta_c = \frac{\omega}{2c'Q_c} \quad \text{(D.15)}$$

If the attenuation is weak then

$$\frac{c''}{c'} \cong -\frac{1}{2Q_c}$$

and the complex velocity can be expressed through the Q-value as

$$c = c' + ic'' = c'\left(1 - \frac{i}{2Q_c}\right) \quad \text{(D.16)}$$

By using this expression the following relations can be established between the medium parameters.

$$\alpha_n = \alpha_{rn}\left(1 - \frac{i}{2Q_{\alpha n}}\right), \quad \beta_n = \beta_{rn}\left(1 - \frac{i}{2Q_{\beta n}}\right) \quad \text{(D.17)}$$

$$\gamma_{n1} = \sqrt{\frac{\omega^2}{\alpha_n^2} - k^2} = \frac{\omega}{\alpha_1}\sqrt{\left(\frac{\alpha_1}{\alpha_n}\right)^2 - u^2}, \quad \gamma_{n2} = \frac{\omega}{\alpha_1}\sqrt{\left(\frac{\alpha_1}{\alpha_n}\right)^2 - u^2}$$

$$\text{(D.18)}$$

$$\mu_n = \rho_n\beta_n^2 \cong \rho_n\beta_{rn}^2(1 - Q_{\beta n}^{-1}) = \mu'_n + i\mu''_n \quad \text{(D.19)}$$

$$\bar{H}_n = \rho_n\alpha_n^2 \cong \rho_n\alpha_{rn}^2(1 - Q_{\alpha n}^{-1}) = \bar{H}'_n + i\bar{H}''_n \quad \text{(D.20)}$$

where u is the normalised or relative horizontal wavenumber

$$u \equiv \frac{k\alpha_1}{\omega} \qquad (D.21)$$

The attenuation is often expressed in units of dB per meter or dB per wavelength, which relate to the Q values through

$$\xi_{m\alpha} \equiv 20 \log_{10} e^{\eta_\alpha} = \frac{8.686\pi f}{\alpha Q_\alpha} \quad \text{dB/m}$$

$$\xi_{m\beta} = \frac{8.686\pi f}{\beta Q_\beta} \quad \text{dB/m}$$

$$\xi_{\lambda\alpha} \equiv 20 \log_{10} e^{\eta_\alpha \lambda} = \frac{8.686\pi}{Q_\alpha} \quad \text{dB}/\lambda$$

$$\xi_{\lambda\beta} = \frac{8.686\pi}{Q_\beta} \quad \text{dB}/\lambda \qquad (D.22)$$

Appendix E

Frictional Force in a 3-D Duct

This appendix describes the friction force in a 3-D duct. The material presented is extracted from [45] but with the time factor $e^{i\omega t}$ changed to $e^{-i\omega t}$ and a few printing mistakes corrected. It is quoted in detail since the changes involved are not straight forward.

Consider a straight duct of circular cross section radius a. The equation of motion for the fluid in the x direction is

$$\rho_f \ddot{U} = -\frac{\partial p}{\partial x} + \mu \nabla^2 \dot{U} \tag{E.1}$$

We assume that $\dot{U}$ is independent of x and that the flow is axially symmetric so that

$$\nabla^2 = \frac{\partial^2}{\partial r^2} + \frac{1}{r}\frac{\partial}{\partial r} \tag{E.2}$$

Introducing the relative velocity $V = \dot{U} - \dot{u}$ we write

$$\rho_f \dot{V} = -\frac{\partial p}{\partial x} - \rho_f \ddot{u} + \mu \nabla^2 V \tag{E.3}$$

We may consider

$$X \rho_f = -\frac{\partial p}{\partial x} - \rho_f \ddot{u}$$

to be equivalent to an external volume force so that Eq.(E.3) becomes

$$\dot{V} = X + \nu \nabla^2 V, \quad \nu = \frac{\mu}{\rho_f}$$

or

$$\nu \left(\frac{\partial^2 V}{\partial r^2} + \frac{1}{r}\frac{\partial V}{\partial r} \right) - \frac{\partial V}{\partial t} = -X \tag{E.4}$$

Assuming a time factor $e^{i\,\mathrm{sgn}\,\omega t}$ where $\mathrm{sgn}=\pm 1$

$$\frac{d^2V}{dr^2} + \frac{1}{r}\frac{dV}{dr} - \frac{i\,\mathrm{sgn}\,\omega}{\nu}V = -\frac{X}{\nu} \qquad (E.5)$$

or

$$\frac{i\nu}{\mathrm{sgn}\,\omega}\frac{d^2V}{dr^2} + \frac{i\nu}{\mathrm{sgn}\,\omega}\frac{1}{r}\frac{dV}{dr} + V = \frac{-i}{\mathrm{sgn}\,\omega}X$$

Setting $r_1 = \sqrt{\frac{\mathrm{sgn}\,\omega}{i\nu}}\,r$, the above becomes

$$\frac{d^2V}{dr_1^2} + \frac{1}{r_1}\frac{dV}{dr_1} + V = \frac{-i}{\mathrm{sgn}\,\omega}X$$

Since the general form of the Bessel function is

$$Y'' + \frac{1}{x}Y' + \left(1 - \frac{m^2}{x^2}\right)Y = 0$$

the required solution becomes

$$V = CJ_0\left(\sqrt{\frac{\mathrm{sgn}\,\omega}{i\nu}}\,r\right) + \frac{X}{i\,\mathrm{sgn}\,\omega} = CJ_0\left(i^{\frac{3}{2}}\sqrt{\frac{\mathrm{sgn}\,\omega}{\nu}}\,r\right) + \frac{X}{i\,\mathrm{sgn}\,\omega}$$

By using the boundary condition $V = 0$ at $r = a$

$$C = -\frac{X}{i\,\mathrm{sgn}\,\omega}\frac{1}{J_0\left(i^{\frac{3}{2}}\sqrt{\frac{\mathrm{sgn}\,\omega}{\nu}}\,a\right)}$$

and so

$$\frac{i\,\mathrm{sgn}\,\omega V}{X} = 1 - \frac{J_0\left(i^{\frac{3}{2}}\sqrt{\frac{\mathrm{sgn}\,\omega}{\nu}}\,r\right)}{J_0\left(i^{\frac{3}{2}}\sqrt{\frac{\mathrm{sgn}\,\omega}{\nu}}\,a\right)} \qquad (E.6)$$

The average velocity over the section can be defined as

$$V_{av} = \frac{2}{a^2}\int_0^a V \cdot r\,dr$$

so

$$\frac{i\,\mathrm{sgn}\,\omega\,V_{av}}{X} = 1 - \frac{2}{k_\eta^2 J_0(i^{\frac{3}{3}}k_\eta)}\int_0^{k_\eta} J_0(i^{\frac{3}{2}}\zeta)\zeta\,d\zeta \qquad (E.7)$$

where

$$k_\eta = \sqrt{\frac{\mathrm{sgn}\,\omega}{\nu}}\,a \qquad (E.8)$$

The Frictional Force in a 3-D Duct

Since $J_0(i^{\frac{3}{2}}z) \equiv ber\, z + i\, bei\, z$ where $ber\, z$ and $bei\, z$, are the real and imaginary parts of the Kelvin function, we can use the relations,

$$\int x J_0(x) dx = x J_1(x), \quad J_1(x) = -J_0'(x) \quad \text{and} \quad k_\eta' \equiv i^{\frac{3}{2}} k_\eta$$

to establish the expression

$$\int_0^{k_\eta} J_0(i^{\frac{3}{2}}\zeta)\zeta d\zeta = i^{-3} \int_0^{k_\eta'} J_0(z) z\, dz = i^{-3} k_\eta' J_1(k_\eta') = -i^{-3} k_\eta' J_0'(k_\eta')$$

$$= -i\, i^{\frac{3}{2}} k_\eta J_0'(i^{\frac{3}{2}} k_\eta) = i^{\frac{1}{2}} k_\eta J_0'(i^{\frac{3}{2}} k_\eta) = i^{\frac{1}{2}} k_\eta i^{-\frac{3}{2}} \frac{d}{dk_\eta} J_0(i^{\frac{3}{2}} k_\eta)$$

$$= -i k_\eta [ber'(k_\eta) + i\, bei'(k_\eta)] = k_\eta [bei'(k_\eta) - i\, ber'(k_\eta)]$$

Hence, from Eq.(E.7)

$$\frac{i\, \text{sgn}\, \omega V_{av}}{X} = 1 - \frac{2(-i)}{k_\eta} \frac{ber'(k_\eta) + i\, bei'(k_\eta)}{ber(k_\eta) + i\, bei(k_\eta)}$$

$$= 1 - \frac{2}{i k_\eta} \frac{ber'(k_\eta) + i\, bei'(k_\eta)}{ber(k_\eta) + i\, bei(k_\eta)}$$

We can also calculate the friction between the the fluid and the wall. From Eq.(E.6), the stress at the wall is

$$\tau = -\mu \left(\frac{dV}{dr}\right)_{r=a}$$

$$= \frac{-X\mu}{i\, \text{sgn}\, \omega} \left[-\sqrt{\frac{\text{sgn}\, \omega}{\nu}} \frac{\frac{d}{dk_\eta} J_0(i^{\frac{3}{2}} k_\eta)}{J_0(i^{\frac{3}{2}} k_\eta)} \right]$$

$$= \frac{-X\mu}{i\, \text{sgn}\, \omega} \sqrt{\frac{\text{sgn}\, \omega}{\nu}} T(k_\eta)$$

and the ratio of the total frictional force, $2\pi a \tau$, to the average velocity is [1]

$$\frac{2\pi a \tau}{V_{av}} = 2\pi \mu k_\eta \frac{T(k_\eta)}{1 - \frac{2}{ik_\eta}T(k_\eta)} \qquad (E.9)$$

where

$$T(k_\eta) = \frac{ber'(k_\eta) + i\, bei'(k_\eta)}{ber(k_\eta) + i\, bei(k_\eta)} \qquad (E.10)$$

From the expressions

$$ber(k_\eta) + i\, bei(k_\eta) = J_0(i^{\frac{3}{2}} k_\eta) = 1 + \frac{i}{2^2} k_\eta^2 - \frac{1}{2^2 2^4} k_\eta^4 + \cdots$$

$$ber'(k_\eta) + i\, bei'(k_\eta) = \frac{d}{dk_\eta} J_0(i^{\frac{3}{2}} k_\eta) = \frac{ik_\eta}{2} - \frac{k_\eta^3}{16} + \cdots$$

we have, when $k_\eta \to 0$,

$$1 - \frac{2}{ik_\eta} T(k_\eta) \to 1 - \frac{2}{ik_\eta} \left[\frac{\frac{ik_\eta}{2} - \frac{k_\eta^3}{16}}{1 + \frac{i}{4} k_\eta^2} \right] \to 1 - \frac{1 + i\frac{k_\eta^2}{8}}{1 + \frac{i}{4} k_\eta^2} \cong 1 - \left(1 + i\frac{k_\eta^2}{8} - i\frac{k_\eta^2}{4} \right)$$

$$\to i\frac{k_\eta^2}{8}, \qquad \frac{2\pi a \tau}{V_{av}} \to -2\pi \mu k_\eta \frac{4}{k_\eta} \to 8\pi \mu,$$

and

$$\frac{T(k_\eta)}{1 - \frac{2}{ik_\eta} T(k_\eta)} \to \frac{\frac{ik_\eta}{2}\left(1 - i\frac{k_\eta^2}{8}\right)}{i\frac{k_\eta^2}{8}} \to \frac{4}{k_\eta}$$

We can therefore express Eq.(E.9) as

$$\frac{2\pi a \tau}{V_{av}} = 8\pi \mu F(k_\eta) \qquad (E.11)$$

with

$$F(k_\eta) = \frac{1}{4} \frac{k_\eta T(k_\eta)}{\left[1 - \frac{2}{ik_\eta} T(k_\eta)\right]}, \qquad F(0) = 1 \qquad (E.12)$$

---

[1]

$$\frac{2\pi a \tau}{V_{av}} = \frac{2\pi a \frac{X\mu}{i \operatorname{sgn}\omega} \sqrt{\frac{\operatorname{sgn}\omega}{\nu}} T(k_\eta)}{\frac{X}{i \operatorname{sgn}\omega}\left(1 - \frac{2}{ik_\eta} T(k_\eta)\right)}$$

$$= 2\pi \mu a \sqrt{\frac{\operatorname{sgn}\omega}{\nu}} \frac{T(k_\eta)}{1 - \frac{2}{ik_\eta} T(k_\eta)} = 2\pi \mu k_\eta \frac{T(k_\eta)}{1 - \frac{2}{ik_\eta} T(k_\eta)}$$

$$J_0(x) = \sum_{m=0}^{\infty} \frac{(-1)^m x^{2m}}{2^{2m} m! \Gamma(m+1)} = 1 - \frac{x^2}{2^2} + \frac{x^4}{2^2 2^4} - \cdots$$

At large k_η

$$ber(k_\eta) + i\,bei(k_\eta) \to \frac{1}{\sqrt{2\pi k_\eta}} e^{\frac{k_\eta}{\sqrt{2}} + i\frac{k_\eta}{\sqrt{2}}} = \frac{1}{\sqrt{2\pi k_\eta}} e^{\frac{1}{\sqrt{2}}(1+i)k_\eta}$$

$$ber'(k_\eta) + i\,bei'(k_\eta) \to \frac{1}{\sqrt{2}}(1+i)\frac{1}{\sqrt{2\pi k_\eta}} e^{\frac{1}{\sqrt{2}}(1+i)k_\eta}$$

$$T(k_\eta) \to \frac{1}{\sqrt{2}}(1+i) \Rightarrow F(k_\eta) \to \frac{1}{4}\frac{k_\eta}{\sqrt{2}}(1+i)$$

It is found experimentally[45] that, at high frequency and constant velocity, the friction is proportional to the square root of the frequency and is 45 degrees out of phase with the velocity. The behaviour is as if the static viscosity coefficient, μ, were replaced by a dynamic complex value, $\mu F(k_\eta)$. We therefore write

$$F(k_\eta) = F_r(k_\eta) + i\,F_i(k_\eta) \tag{E.13}$$

The functions $F_r(k_\eta)$ and $F_i(k_\eta)$ are described in [45]. For large (k_η)

$$ber(z) \to \frac{e^{a(z)}}{\sqrt{2\pi z}} \cos\beta(z)$$

$$ber(z) \to \frac{e^{a(z)}}{\sqrt{2\pi z}} \sin\beta(z)$$

$$ber(z) + i\,bei(z) \to \frac{e^{a(z)+i\beta(z)}}{\sqrt{2\pi z}}$$

$$a(z) \cong \frac{z}{\sqrt{2}} + \frac{1}{8z\sqrt{2}} - \cdots$$

$$\beta(z) \cong \frac{z}{\sqrt{2}} - \frac{\pi}{8} - \frac{1}{8z\sqrt{2}} - \cdots$$

Finally, it is of interest to examine the difference between the values of $T(x)$ and $T(ix)$ and those of $F(x)$ and $F(ix)$ when x is real. From its definition[2]

$$T(x) = \frac{\frac{d}{dx} J_0(i^{\frac{3}{2}}x)}{J_0(i^{\frac{3}{2}}x)} = \frac{-i^3\frac{1}{2}x + \frac{i^6}{16}x^3 - \cdots}{1 - \frac{1}{4}i^3 x^2 + \frac{1}{64}i^6 x^4 - \cdots}$$

---

[2] $-i^{-\frac{3}{2}} = -e^{-i\frac{3}{4}\pi} = e^{i\frac{\pi}{4}}$

$$T(ix) \;=\; \frac{-\frac{1}{16}x^3 + i\frac{1}{2}x - \cdots}{1 - \frac{x^4}{64} + \frac{i}{4}x^2 - \cdots} = T_r(x) + i\,T_i(x)$$

$$T(ix) \;=\; \frac{\frac{i}{16}x^3 - \frac{1}{2}x + \cdots}{1 - \frac{x^4}{64} - i\frac{x^2}{4} + \cdots} = -i\frac{\frac{-1}{16}x^3 - i\frac{x}{2} + \cdots}{1 - \frac{x^4}{64} - i\frac{x^2}{4} + \cdots} = (-i)\,[T_r(x) - i\,T_i(x)]$$

$$F(ix) \;=\; \frac{1}{4}\frac{ixT(ix)}{[1 - \frac{2}{i(ix)}T(ix)]} = \frac{x}{4}\frac{[T_r(x) - i\,T_i(x)]}{\{1 + \frac{2}{ix}[T_r(x) - i\,T_i(x)]\}} = F_r(x) - i\,F_i(x)$$

or

$$F(\sqrt{sgn}\,x) = \frac{x}{4}\frac{[T_r(x) + i\,\mathrm{sgn}\,T_i(x)]}{\left\{1 - \mathrm{sgn} \cdot \frac{2}{ix}[T_r(x) + i\,\mathrm{sgn}\,T_i(x)]\right\}} \tag{E.14}$$

Equation (E.14) is the expression used to calculate the three wave velocities and reflection coefficient of a porous layer - see Eq.(C.12). It involves only Kelvin functions with real arguments. This is fortunate from the computational viewpoint.

Appendix F

Numerical Procedure for the Calculation of the Reflection Coefficient for a Multilayered Porous Viscoelastic Medium

We assume a plane wave of unit amplitude $\Phi_{1inc} = e^{-i\gamma_1(z+H_1)+i(kx-\omega t)}$ is incident from water upon the seabed ($z = -H_1$) composed of parallel porous/viscoelastic layers overlying an elastic half space. For convenience, however, all layers will be modelled as porous. The redundant equations can then be discarded by setting the amplitudes of the slow wave to zero in any layers which are purely elastic.

By defining the displacement vector of the solid frame and that of the pore fluid relative to the frame (in the nth layer) as

$$\vec{u}_n = \nabla \Phi_{ns} + \nabla \times \vec{\Psi}_{ns}$$

$$\vec{V}_n = \nabla \Phi_{nf} + \nabla \times \vec{\Psi}_{nf} \qquad \text{(F.1)}$$

where $\vec{\Psi}_{ns} = (\vec{k}_0 \times \vec{n})\Psi_{ns}$, $\vec{k}_0$ and $\vec{n}$ being unit vectors in the direction of the horizontal wavenumber vector (positive x-direction in the present case) and the normal to the interfaces (positive z-direction here). We can write the displacement components as

$$u_{nx} = \frac{\partial \Phi_{ns}}{\partial x} + \frac{\partial \Psi_{ns}}{\partial z}$$

$$u_{nz} = \frac{\partial \Phi_{ns}}{\partial z} - \frac{\partial \Psi_{ns}}{\partial x}$$

$$V_{nx} = \frac{\partial \Phi_{nf}}{\partial x} + \frac{\partial \Psi_{nf}}{\partial z}$$

$$V_{nz} = \frac{\partial \Phi_{nf}}{\partial z} - \frac{\partial \Psi_{nf}}{\partial x} \qquad (F.2)$$

and the stress and pressure components as,

$$p_{nzz} = \bar{H}_n \theta_n - 2\mu_n \frac{\partial u_{nx}}{\partial x} - C_n \zeta_{pn}$$

$$p_{nzx} = \mu_n \left(\frac{\partial u_{nx}}{\partial z} + \frac{\partial u_{nz}}{\partial x} \right)$$

$$p_{fn} = M_n \zeta_{pn} - C_n \theta_n, \qquad (F.3)$$

in which $\bar{H}_n$, C_n, M_n and D_n are constants of the layer (see Appendix C).

$$\zeta_{pn} = \nabla \cdot \beta(\vec{u}_n - \vec{U}_n) \equiv \nabla \cdot \vec{V}_n = \nabla^2 \Phi_{nf} \qquad (F.4)$$

and

$$\theta_n = \nabla \cdot \vec{u}_n = \nabla^2 \Phi_{ns} \qquad (F.5)$$

with β_n defining the porosity.

It has been shown (Appendix C) that in a porous layer three kinds of waves can propagate, the fast- and slow-compressional waves and the shear wave, with wave numbers l_{c1n}, l_{c2n} and l_{sn} respectively. The amplitude of the three waves is different in the solid frame and the pore water. The ratios of the two amplitudes for the three waves, defined as δ_{1n}, δ_{2n} and δ_{3n}, all functions of the medium constants -see Eq.(C.24)- are:

$$\delta_{1n} = \frac{\bar{H}_n l_{c1n} - \omega^2 \rho_n}{C_n l_{c1n} - \omega^2 \rho_{fn}}$$

$$\delta_{2n} = \frac{\bar{H}_n l_{c2n} - \omega^2 \rho_n}{C_n l_{c2n} - \omega^2 \rho_{fn}}$$

$$\delta_{sn} = \frac{\rho_n}{\rho_{fn}} \left(1 - \frac{\mu_n l_{sn}^2}{\omega^2 \rho_n} \right) \qquad (F.6)$$

Here, ρ_f is the density of the pore fluid, $\rho_n = \beta_n \rho_{fn} + (1 - \beta_n) \rho_{rn}$ with ρ_{rn} being the density of the frame, all referred to the n_{th} layer.

Calculation of the Reflection Coefficient

We can therefore express the displacement potentials in each layer as

$$\Phi_{ns} = \Phi_{ns1} + \Phi_{ns2}$$

$$\Phi_{nf} = \Phi_{nf1} + \Phi_{nf2} = \delta_1 \Phi_{ns1} + \delta_2 \Phi_{ns2}$$

$$\Psi_{nf} = \delta_{3n} \Psi_{ns} \tag{F.7}$$

and

$$\left.\begin{array}{l} \Phi_{ns1} = A_{n1} e^{-i\gamma_{n1}(z+H_{n-1})} + B_{n1} e^{i\gamma_{n1}(z+H_n)} \\ \Phi_{ns2} = A_{n2} e^{-i\gamma_{n2}(z+H_{n-1})} + B_{n2} e^{i\gamma_{n2}(z+H_n)} \\ \Psi_{ns} = A_{n3} e^{-i\gamma_{n3}(z+H_{n-1})} + B_{n3} e^{i\gamma_{n3}(z+H_n)} \end{array}\right\} e^{ikx-i\omega t} \tag{F.8}$$

where

$$\gamma_{n1} = \sqrt{l_{c1n}^2 - k^2}, \qquad \gamma_{n2} = \sqrt{l_{c2n}^2 - k^2}, \qquad \gamma_{n3} = \sqrt{l_{sn}^2 - k^2}$$

Denoting the potential in the water as

$$\Phi_1 = \left[e^{-i\gamma_1(z+H_1)} + R_b e^{i\gamma_1(z+H_1)} \right] e^{ikx-i\omega t}$$

the displacement vector and pressure field become

$$\vec{U}_1 = \nabla \Phi_1, \qquad p_w = -\rho_1 \frac{\partial^2 \Phi_1}{\partial t^2} \tag{F.9}$$

So, at $z = -H_1$,

$$U_{1z} = -i\gamma_1 (1 - R_b) e^{ikx-i\omega t}$$

$$p_w = \rho_1 \omega^2 (1 + R_b) e^{ikx-i\omega t} \tag{F.10}$$

The boundary conditions at the water-sediment interface, $z = -H_1$, require

$$U_{1z} = u_{2z} - V_{2z}$$
$$-p_w = p_{2zz}$$
$$0 = p_{2zx}$$
$$p_w = p_{2f} \tag{F.11}$$

while at the interface of the porous layers, $z = -H_n$, the requirements are

$$u_{nz} = u_{n+1,z}$$
$$u_{nx} = u_{n+1,x}$$
$$u_{nz} - V_{nz} = u_{n+1,z} - V_{n+1,z}$$
$$p_{nzz} = p_{n+1,zz}$$
$$p_{nzx} = p_{n+1,zx}$$
$$p_{fn} = p_{f,n+1} \tag{F.12}$$

By using Eqs. (F.8), (F.2) and (F.3), and noting that at $z = -H_n$ (for $n \geq 2$)

$$e^{-i\gamma_{n1}(-H_n+H_{n-1})} = e^{i\gamma_{n1}h_n} \equiv e_{n1}, \quad e^{i\gamma_{n1}(-H_n+H_n)} = 1$$

it follows that

$$u_{nz} = \frac{\partial \Phi_{ns}}{\partial z} - \frac{\partial \Psi_{ns}}{\partial x}$$
$$= -i\gamma_{n1}(A_{n1}e_{n1} - B_{n1}) - i\gamma_{n2}(A_{n2}e_{n2} - B_{n2})$$
$$\quad -ik(A_{n3}e_{n3} + B_{n3}) \tag{F.13}$$

$$u_{nx} = \frac{\partial \Phi_{ns}}{\partial x} + \frac{\partial \Psi_{ns}}{\partial z}$$
$$= ik(A_{n1}e_{n1} + B_{n1}) + ik(A_{n2}e_{n2} + B_{n2}) - i\gamma_{n3}(A_{n3}e_{n3} - B_{n3}) \tag{F.14}$$

$$u_{nz} - v_{nz} = \frac{\partial}{\partial z}(\Phi_{ns} - \Phi_{nf}) - \frac{\partial}{\partial x}(\Psi_{ns} - \Psi_{nf})$$
$$= (1-\delta_{1n})\frac{\partial \Phi_{ns1}}{\partial z} + (1-\delta_{2n})\frac{\partial \Phi_{ns2}}{\partial z} - (1-\delta_{3n})\frac{\partial \Psi_{ns}}{\partial x}$$
$$= -i\gamma_{n1}(1-\delta_{1n})(A_{n1}e_{n1} - B_{n1}) - i\gamma_{n2}(1-\delta_{2n})(A_{n2}e_{n2} - B_{n2})$$
$$\quad -ik(1-\delta_{3n})(A_{n3}e_{n3} + B_{n3}) \tag{F.15}$$

$$p_{nzz} = \bar{H}_n \nabla^2 \Phi_{ns} - 2\mu_n \left(\frac{\partial^2 \Phi_{ns1}}{\partial x^2} + \frac{\partial^2 \Phi_{ns2}}{\partial x^2} + \frac{\partial^2 \Psi_{ns}}{\partial x \partial z}\right) - C_n \nabla^2 \Phi_{nf}$$
$$= \left[(C_n \delta_{1n} - \bar{H}_n)l_{c1n}^2 + 2\mu_n k^2\right](A_{n1}e_{n1} + B_{n1})$$
$$\quad + \left[(C_n \delta_{2n} - \bar{H}_n)l_{c2n}^2 + 2\mu_n k^2\right](A_{n2}e_{n2} + B_{n2})$$
$$\quad -2\mu_n k\gamma_{n3}(A_{n3}e_{n3} - B_{n3}) \tag{F.16}$$

Calculation of the Reflection Coefficient

$$p_{nzx} = \mu_n \frac{\partial}{\partial z}\left(\frac{\partial \Phi_{ns}}{\partial x} + \frac{\partial \Psi_{ns}}{\partial z}\right) + \mu_n \frac{\partial}{\partial x}\left(\frac{\partial \Phi_{ns}}{\partial z} - \frac{\partial \Psi_{ns}}{\partial x}\right)$$

$$= 2\mu_n \frac{\partial^2 \Phi_{ns}}{\partial z \partial x} - \mu_n \left(\frac{\partial^2}{\partial x^2} - \frac{\partial^2}{\partial z^2}\right)\Psi_{ns}$$

$$= 2\mu_n k \gamma_{n1}(A_{n1}e_{n1} - B_{n1}) + 2\mu_n k \gamma_{n2}(A_{n2}e_{n2} - B_{n2})$$

$$+ \mu_n(k^2 - \gamma_{n3}^2)(A_{n3}e_{n3} + B_{n3}) \qquad (F.17)$$

$$p_{nf} = M_n \nabla^2 \Phi_{nf} - C_n \nabla^2 \Phi_{ns}$$

$$= -(C_n - M_n \delta_{1n})\nabla^2 \Phi_{ns1} - (C_n - M_n \delta_{2n})\nabla^2 \Phi_{ns2}$$

$$= l_{c1n}^2 (C_n - M_n \delta_{1n})(A_{n1}e_{n1} + B_{n1})$$

$$+ l_{c2n}^2 (C_n - M_n \delta_{2n})(A_{n2}e_{n2} + B_{n2}) \qquad (F.18)$$

in which

$$e_{n1} = e^{i\gamma_{n1}h_n}, \qquad e_{n2} = e^{i\gamma_{n2}h_n}, \qquad e_{n3} = e^{i\gamma_{n3}h_n} \qquad (F.19)$$

We can write the boundary conditions at $z = -H_1$ as (compare Eq.(D.10))

$$\gamma_1 R_b + \gamma_{21}(1 - \delta_{12})(A_{21} - B_{21}e_{21}) + \gamma_{22}(1 - \delta_{22})(A_{22} - B_{22}e_{22})$$
$$+ k(1 - \delta_{32})(A_{23} + B_{23}e_{23}) = \gamma_1 \qquad (F.20)$$

$$\rho_1 \omega^2 R_b + \left[(C_2 \delta_{21} - \bar{H}_2)l_{c12}^2 + 2\mu_2 k^2\right](A_{21} + B_{21}e_{21})$$
$$+ \left[(C_2 \delta_{22} - \bar{H}_2)l_{c22}^2 + 2\mu_2 k^2\right](A_{22} + B_{22}e_{22})$$
$$- 2\mu_2 k \gamma_{23}(A_{23} - B_{23}e_{23}) = -\rho_1 \omega^2 \qquad (F.21)$$

$$2k\gamma_{21}(A_{21} - B_{21}e_{21}) + 2k\gamma_{22}(A_{22} - B_{22}e_{22})$$
$$+ (k^2 - \gamma_{23}^2)(A_{23} + B_{23}e_{23}) = 0 \qquad (F.22)$$

$$-\rho_1 \omega^2 R_b + (C_2 - M_2 \delta_{21})l_{c12}^2 (A_{21} + B_{21}e_{21})$$
$$+ (C_2 - M_2 \delta_{22})l_{c22}^2 (A_{22} + B_{22}e_{22}) = \rho_1 \omega^2 \qquad (F.23)$$

Corresponding equations for each of the intimate interfaces can be easily derived from Eqs. (F.13) to (F.18). At the last interface, $z = -H_N$, it is only necessary to set $B_{N+1,1} = B_{N+1,2} = B_{N+1,3} = 0$. As an example, consider the equations at $z = -H_n$, $n \geq 2$:

$$\gamma_{n1}(A_{n1}e_{n1} - B_{n1}) + \gamma_{n2}(A_{n2}e_{n2} - B_{n2}) + k(A_{n3}e_{n3} + B_{n3}) =$$
$$\gamma_{n+1,1}(A_{n+1,1} - B_{n+1,1}e_{n+1,1}) + \gamma_{n+1,2}(A_{n+1,2} - B_{n+1,2}e_{n+1,2})$$
$$+k(A_{n+1,3} + B_{n+1,3}e_{n+1,3}) \tag{F.24}$$

$$k(A_{n1}e_{n1} + B_{n1}) + k(A_{n2}e_{n2} + B_{n2}) - \gamma_{n3}(A_{n3}e_{n3} - B_{n3}) =$$
$$k(A_{n+1,1} + B_{n+1,1}e_{n+1,1}) + k(A_{n+1,2} + B_{n+1,2}e_{n+1,2})$$
$$-\gamma_{n+1,3}(A_{n+1,3} - B_{n+1,3}e_{n+1,3}) \tag{F.25}$$

$$\gamma_{n1}(1 - \delta_{1n})(A_{n1}e_{n1} - B_{n1}) + \gamma_{n2}(1 - \delta_{2n})(A_{n2}e_{n2} - B_{n2})$$
$$+k(1 - \delta_{3n})(A_{n3}e_{n3} + B_{n3}) =$$
$$\gamma_{n+1,1}(1 - \delta_{1,n+1})(A_{n+1,1} - B_{n+1,1}e_{n+1,1})$$
$$+\gamma_{n+1,2}(1 - \delta_{2,n+1})(A_{n+1,2} - B_{n+1,2}e_{n+1,2})$$
$$+k(1 - \delta_{3,n+1})(A_{n+1,3} + B_{n+1,3}e_{n+1,3}) \tag{F.26}$$

$$[\ ]_{n1}(A_{n1}e_{n1} + B_{n1}) + [\ ]_{n2}(A_{n2}e_{n2} + B_{n2}) - 2\mu_n k \gamma_{n3}$$
$$(A_{n3}e_{n3} - B_{n3}) =$$
$$[\ ]_{n+1,1}(A_{n+1,1} + B_{n+1,1}e_{n+1,1}) + [\ ]_{n+1,2}(A_{n+1,2} + B_{n+1,2}e_{n+1,2})$$
$$-2\mu_{n+1} k \gamma_{n+1,3}(A_{n+1,3}e_{n+1,3} - B_{n+1,3}) \tag{F.27}$$

where

$$[\ ]_{ni} = \left[C_n \delta_{in} - \bar{H}_n\right]l^2_{cin} + 2\mu_n k^2\right]$$

$$[\ ]_{n+1,i} = \left[(C_{n+1}\delta_{i,n+1} - \bar{H}_{n+1})l^2_{ci,n+1} + 2\mu_{n+1} k^2\right]$$

$$2\mu_n k\gamma_{n1}(A_{n1}e_{n1} - B_{n1}) + 2\mu_n k\gamma_{n2}(A_{n2}e_{n2} - B_{n2})$$
$$+\mu_n(k^2 - \gamma_{n3}^2)(A_{n3}e_{n3} + B_{n3}) = 2\mu_{n+1}k\gamma_{n+1,1}(A_{n+1,1} - B_{n+1,1}e_{n+1,1}) +$$
$$2\mu_{n+1}k\gamma_{n+1,2}(A_{n+1,2} - B_{n+1,2}e_{n+1,2})$$
$$+\mu_{n+1}(k^2 - \gamma_{n+1,3}^2)(A_{n+1,3} + B_{n+1,3}e_{n+1,3})$$
(F.28)

$$(C_n - M_n\delta_{1n})l_{c1n}^2(A_{n1}e_{n1} + B_{n1}) + (C_n - M_n\delta_{2n})l_{c2n}^2(A_{n2}e_{n2} + B_{n2}) =$$
$$(C_{n+1} - M_{n+1}\delta_{1,n+1})l_{c1,n+1}^2(A_{n+1,1} + B_{n+1,1}e_{n+1,1})$$
$$+(C_{n+1} - M_{n+1}\delta_{2,n+1})l_{c2,n+1}^2(A_{n+1,2} + B_{n+1,2}e_{n+1,2})$$
(F.29)

For a N-layer model, there are $6 \times (N-1)$ parameters involved for the intimate layers plus R_b and 3 amplitudes in the basement, altogether $6N - 2$ unknowns, a number compatible with the number of equations. For convenience, however, we add eight more equations into the system, say $x_i = 0$ ($i = 1, 5$) and $B_{N+1,1} = 0$, $B_{N+1,2} = 0$ and $B_{N+1,3} = 0$ with eight unknowns x_i and $B_{N+1,j}$, so that we can write these equations in the form of matrices as

$$\tilde{C} \cdot \tilde{A} = \tilde{B} \tag{F.30}$$

where $\tilde{A}$ and $\tilde{B}$ are both $6(N+1) \times 1$ matrices as indicated in Table F.1, and $\tilde{C}$ is a $6(N+1) \times 6(N+1)$ matrix as given in Table F.2.

$$\tilde{A} = \begin{pmatrix} x_1 \\ x_2 \\ x_3 \\ x_4 \\ x_5 \\ R_b \\ \\ A_{21} \\ B_{21} \\ A_{22} \\ B_{22} \\ A_{23} \\ B_{23} \\ \vdots \\ \\ A_{n1} \\ B_{n1} \\ A_{n2} \\ B_{n2} \\ A_{n3} \\ B_{n3} \\ A_{n+1,1} \\ B_{n+1,1} \\ A_{n+1,2} \\ B_{n+1,2} \\ A_{n+1,3} \\ B_{n+1,3} \\ \vdots \\ \\ A_{N1} \\ B_{N1} \\ A_{N2} \\ B_{N2} \\ A_{N3} \\ B_{N3} \\ A_{N+1,1} \\ B_{N+1,1} \\ A_{N+1,2} \\ B_{N+1,2} \\ A_{N+1,3} \\ B_{N+1,3} \end{pmatrix} \qquad \tilde{B} = \begin{pmatrix} 0 \\ 0 \\ \gamma_1 \\ -\rho_1\omega^2 \\ 0 \\ \rho_1\omega^2 \\ \\ \\ \\ 0 \\ \\ \\ \\ \\ 0 \\ \\ \\ \\ \\ 0 \\ \\ \\ \\ \\ 0 \\ \\ \\ \\ \\ 0 \\ \\ \\ \\ \\ 0 \end{pmatrix}$$

Table F.1: The matrices $\tilde{A}$ and $\tilde{B}$.

$$\tilde{C} = \begin{pmatrix} \begin{smallmatrix}1&0&0&0&0&0\\0&1&0&0&0&0\\0&0&0&0&0&\gamma_1\\0&0&0&0&0&\rho_1\omega^2\\0&0&0&0&0&0\\0&0&0&0&0&\rho_1\omega^2\end{smallmatrix} & \tilde{C}_{12} & 0 & 0 & 0 & 0 & 0 & 0 \\[1em] 0 & \tilde{C}_{22} & \tilde{C}_{23} & 0 & 0 & 0 & 0 & 0 \\[1em] 0 & 0 & \ddots & \ddots & 0 & 0 & 0 & 0 \\[1em] 0 & 0 & 0 & \tilde{C}_{nn} & \tilde{C}_{n,n+1} & 0 & 0 & 0 \\[1em] 0 & 0 & 0 & 0 & \ddots & \ddots & 0 & 0 \\[1em] 0 & 0 & 0 & 0 & 0 & \ddots & \ddots & 0 \\[1em] 0 & 0 & 0 & 0 & 0 & 0 & \tilde{C}_{NN} & \tilde{C}_{N,N+1} \\[1em] \begin{smallmatrix}0&0&1&0&0&0\\0&0&0&1&0&0\\0&0&0&0&1&0\\0&0&0&0&0&0\\0&0&0&0&0&0\\0&0&0&0&0&0\end{smallmatrix} & 0 & 0 & 0 & 0 & 0 & 0 & \begin{smallmatrix}0&0&0&0&0&0\\0&0&0&0&0&0\\0&0&0&0&0&0\\0&1&0&0&0&0\\0&0&0&1&0&0\\0&0&0&0&0&1\end{smallmatrix} \end{pmatrix}$$

In the case where one of the layers, say the $(n+1)$th layer, is elastic while both the nth and the $(n+2)$th layers are porous, the boundary conditions at $z = -H_n$ become

$$\begin{aligned} u_{nz} &= u_{n+1,z} \\ u_{nx} &= u_{n+1,x} \\ u_{nz} - V_{nz} &= u_{n+1,z} \quad (\text{ie.}, V_{nz} = 0) \\ p_{nzz} &= p_{n+1,zz} \\ p_{nzx} &= p_{n+1,zx} \end{aligned} \qquad (\text{F.31})$$

and those at $z = -H_{n+1}$

$$\begin{aligned}
u_{n+1,z} &= u_{n+2,z} \\
u_{n+1,x} &= u_{n+2,x} \\
u_{n+1,z} &= u_{n+2,z} - V_{n+2,z} \quad (\text{ie., } V_{n+2,z} = 0) \\
p_{n+1,zz} &= p_{n+2,zz} \\
p_{n+1,zx} &= p_{n+2,zx}
\end{aligned} \tag{F.32}$$

The number of equations is reduced by 2 because of the removal of the two unknowns $A_{n+1,2}$ and $B_{n+1,2}$ (the amplitudes of the slow wave in the nth-layer). To keep the matrix of the same dimension, we can introduce two equations $A_{n+1,2} = 0$ and $B_{n+1,2} = 0$ in this layer. If the $(n+1)$th layer is a homogeneous elastic half space it is only necessary to add one equation, $B_{n+1,2} = 0$. Since this is a case of interest we show the non-zero sub-matrices at $z = -H_{N-1}$ and $z = -H_N$, assuming the nth layer and the halfspace (the $(n+1)$th layer) are both elastic.

Since the top layer of the sediment can be porous-elastic or solid-elastic, the non-zero submatrix $[\tilde{C}_{11}\ \tilde{C}_{12}]$ for these two types of interface, water-porous (WP) and water-elastic (WE) can be written explicitly as in Tables F.3 and F.4. It is to be noted that, on the elastic side of the interface, the stress components are $p_{nzz} = (-\omega^2 \rho_n + 2\mu_n k^2)$.

The interface between two layers can be of four different types,

1. porous-elastic (PE),

2. porous-porous (PP),

3. elastic-porous (EP) and

4. elastic-elastic (EE).

The non-zero sub-matrices representing the 4 types of interface conditions take the form shown in Tables F.5-F.8.

$$\tilde{C}_{11} = \begin{bmatrix} 1 & 0 & 0 & 0 & 0 & 0 \\ 0 & 1 & 0 & 0 & 0 & 0 \\ 0 & 0 & 0 & 0 & 0 & \gamma_1 \\ 0 & 0 & 0 & 0 & 0 & \rho_1\omega^2 \\ 0 & 0 & 0 & 0 & 0 & 0 \\ 0 & 0 & 0 & 0 & 0 & -\rho_1\omega^2 \end{bmatrix}$$

and

$$\tilde{C}_{12} = \begin{bmatrix} 0 & 0 & 0 & 0 & 0 & 0 \\ 0 & 0 & 0 & 0 & 0 & 0 \\ \gamma_{21}(1-\delta_{12}) & -\gamma_{21}(1-\delta_{12})e_{21} & \gamma_{22}(1-\delta_{22}) & -\gamma_{22}(1-\delta_{22})e_{22} & k(1-\delta_{32}) & k(1-\delta_{32})e_{23} \\ [\,]_{21} & [\,]_{21}e_{21} & [\,]_{22} & [\,]_{22}e_{22} & -2\mu_2 k\gamma_{23} & 2\mu_2 k\gamma_{23}e_{23} \\ 2k\gamma_{21}\mu_2 & -2k\gamma_{21}\mu_2 e_{21} & 2k\gamma_{22}\mu_2 & -2k\gamma_{22}\mu_2 e_{22} & (k^2-\gamma_{23}^2)\mu_2 & (k^2-\gamma_{23}^2)\mu_2 e_{23} \\ (C_2-M_2\delta_{21})l_{c12}^2 & (C_2-M_2\delta_{21})l_{c12}^2 e_{21} & (C_2-M_2\delta_{22})l_{c22}^2 & (C_2-M_2\delta_{22})l_{c22}^2 e_{22} & 0 & 0 \end{bmatrix}$$

in which

$$[\,]_{21} = [(C_2\delta_{21} - \bar{H}_2)l_{c12}^2 + 2\mu_2 k^2]$$

$$[\,]_{22} = [(C_2\delta_{22} - \bar{H}_2)l_{c22}^2 + 2\mu_2 k^2]$$

Table F.3: Non-zero submatrix (WP) at $z = -H_1$.

$$\tilde{C}_{11} = \begin{bmatrix} 1 & 0 & 0 & 0 & 0 & 0 \\ 0 & 1 & 0 & 0 & 0 & 0 \\ 0 & 0 & 0 & 0 & 0 & \gamma_1 \\ 0 & 0 & 0 & 0 & 0 & \rho_1\omega^2 \\ 0 & 0 & 0 & 0 & 0 & 0 \\ 0 & 0 & 0 & 0 & 0 & 0 \end{bmatrix}$$

and

$$\tilde{C}_{12} = \begin{bmatrix} 0 & 0 & 0 & 0 & 0 & 0 \\ 0 & 0 & 0 & 0 & 0 & 0 \\ \gamma_{21} & -\gamma_{21}e_{21} & 0 & 0 & k & ke_{23} \\ -(\omega^2\rho_2 - 2\mu_2 k^2) & -(\omega^2\rho_2 - 2\mu_2 k^2)e_{21} & 0 & 0 & -2k\mu_2\gamma_{23} & 2k\mu_2\gamma_{23}e_{23} \\ 2k\gamma_{21}\mu_2 & -2k\gamma_{21}\mu_2 e_{21} & 0 & 0 & (k^2-\gamma_{23}^2)\mu_2 & (k^2-\gamma_{23}^2)\mu_2 e_{23} \\ 0 & 0 & 1 & 0 & 0 & 0 \end{bmatrix}$$

Table F.4: Non-zero submatrix (WE) at $z = -H_1$.

$$\tilde{C}_{nn} = \begin{bmatrix} \gamma_{n1}e_{n1} & -\gamma_{n1} & \gamma_{n2}e_{n2} & -\gamma_{n2} & ke_{n3} & k \\ ke_{n1} & k & ke_{n2} & k & -\gamma_{n3}e_{n3} & \gamma_{n3} \\ \gamma_{n1}(1-\delta_{1n})e_{n1} & -\gamma_{n1}(1-\delta_{1n}) & \gamma_{n2}(1-\delta_{2n})e_{n2} & -\gamma_{n2}(1-\delta_{2n}) & k(1-\delta_{3n})e_{n3} & k(1-\delta_{3n}) \\ [\,]_{n1}\,e_{21} & [\,]_{n1} & [\,]_{n2}\,e_{n2} & [\,]_{n2} & -2\mu_n k\gamma_{n3}e_{n3} & 2\mu_n k\gamma_{n3} \\ 2\mu_n k\gamma_{n1}e_{n1} & -2\mu_n k\gamma_{n1} & 2\mu_n k\gamma_{n2}e_{n2} & -2\mu_n k\gamma_{n2} & \mu_n(k^2-\gamma_{n3}^2)e_{n3} & \mu_n(k^2-\gamma_{n3}^2) \\ (\,)_{n1}e_{n1} & (\,)_{n1} & (\,)_{n2}e_{n2} & (\,)_{n2} & 0 & 0 \end{bmatrix}$$

and

$$\tilde{C}_{n,n+1} = \begin{bmatrix} -\gamma_{n+1,1} & \gamma_{n+1,1}e_{n+1,1} & -\gamma_{n+1,2} & \gamma_{n+1,2}e_{n+1,2} & -k & -ke_{n+1,3} \\ -k & -ke_{n+1,1} & -k & -ke_{n+1,2} & \gamma_{n+1,3} & -\gamma_{n+1,3}e_{n+1,3} \\ -(1-\delta_{1,n+1})\gamma_{n+1,1} & (1-\delta_{1,n+1})\gamma_{n+1,1}e_{n+1,1} & -(1-\delta_{2,n+1})\gamma_{n+1,2} & (1-\delta_{2,n+1})\gamma_{n+1,2}e_{n+1,2} & -k(1-\delta_{3,n+1}) & -k(1-\delta_{3,n+1})e_{n+1,3} \\ -[\,]_{n+1,1} & -[\,]_{n+1,1}e_{n+1,1} & -[\,]_{n+1,2} & -[\,]_{n+1,2}e_{n+1,2} & 2\mu_{n+1}k\gamma_{n+1,3} & -2\mu_{n+1}k\gamma_{n+1,3}e_{n+1,3} \\ -2\mu_{n+1}k\gamma_{n+1,1} & 2\mu_{n+1}k\gamma_{n+1,1}e_{n+1,1} & -2\mu_{n+1}k\gamma_{n+1,2} & 2\mu_{n+1}k\gamma_{n+1,2}e_{n+1,2} & -(k^2-\gamma_{n+1,3}^2)\mu_{n+1} & -(k^2-\gamma_{n+1,3}^2)\mu_{n+1}e_{n+1,3} \\ -(\,)_{n+1,1}e_{n+1,1} & -(\,)_{n+1,1} & -(\,)_{n+1,2}e_{n+1,2} & -(\,)_{n+1,2} & 0 & 0 \end{bmatrix}$$

where

$$[\,]_{n1} = (C_n\delta_{1n} - \bar{H}_n)l_{c1n}^2 + 2\mu_n k^2$$

$$[\,]_{2n} = (C_n\delta_{2n} - \bar{H}_n)l_{c2n}^2 + 2\mu_n k^2$$

$$(\,)_{n1} = (C_n - M_n\delta_{1n})l_{c1n}^2$$

$$(\,)_{n2} = (C_n - M_n\delta_{2n})l_{c2n}^2$$

Table F.5: The non-zero submatrix for the interface PP at $z = -H_n$.

Calculation of the Reflection Coefficient

$$\tilde{C}_{nn} = \begin{bmatrix} \gamma_{n1}e_{n1} & -\gamma_{n1} & \gamma_{n2}e_{n2} & -\gamma_{n2} & ke_{n3} & k \\ ke_{n1} & k & ke_{n2} & k & -\gamma_{n3}e_{n3} & \gamma_{n3} \\ \gamma_{n1}(1-\delta_{1n})e_{n1} & -\gamma_{n1}(1-\delta_{1n}) & \gamma_{n2}(1-\delta_{2n})e_{n2} & -\gamma_{n2}(1-\delta_{2n}) & k(1-\delta_{3n})e_{n3} & k(1-\delta_{3n}) \\ [\,]_{n1}e_{n1} & [\,]_{n1} & [\,]_{n2}e_{n2} & [\,]_{n2} & -2\mu_n k\gamma_{n3}e_{n3} & 2\mu_n k\gamma_{n3} \\ 2\mu_n k\gamma_{n1}e_{n1} & -2\mu_n k\gamma_{n1} & 2\mu_n k\gamma_{n2}e_{n2} & -2\mu_n k\gamma_{n2} & \mu_n(k^2-\gamma_{n3}^2)e_{n3} & \mu_n(k^2-\gamma_{n3}^2) \\ 0 & 0 & 0 & 0 & 0 & 0 \end{bmatrix}$$

and

$$\tilde{C}_{n,n+1} = \begin{bmatrix} -\gamma_{n+1,1} & \gamma_{n+1,1}e_{n+1,1} & 0 & 0 & -k & -ke_{n+1,3} \\ -k & -ke_{n+1,1} & 0 & 0 & \gamma_{n+1,3} & -\gamma_{n+1,3}e_{n+1,3} \\ -\gamma_{n+1,1} & \gamma_{n+1,1}e_{n+1,1} & 0 & 0 & -k & -ke_{n+1,3} \\ (\rho_{n+1}\omega^2 - 2\mu_{n+1}k^2) & (\rho_{n+1}\omega^2 - 2\mu_{n+1}k^2)e_{n+1,1} & 0 & 0 & 2\mu_{n+1}k\gamma_{n+1,3} & -2\mu_{n+1}k\gamma_{n+1,3}e_{n+1,3} \\ -2\mu_{n+1}k\gamma_{n+1,1} & 2\mu_{n+1}k\gamma_{n+1,1}e_{n+1,1} & 0 & 0 & -\mu_{n+1}(k^2-\gamma_{n+1,3}^2) & -\mu_{n+1}(k^2-\gamma_{n+1,3}^2)e_{n+1,3} \\ 0 & 0 & 1 & 0 & 0 & 0 \end{bmatrix}$$

where

$$[\,]_{n1} = (C_n \delta_{1n} - \bar{H}_n)l_{c1n}^2 + 2\mu_n k^2$$

$$[\,]_{2n} = (C_n \delta_{2n} - \bar{H}_n)l_{c2n}^2 + 2\mu_n k^2$$

Table F.6: The non-zero submatrix for the interface PE at $z = -H_n$.

$$\tilde{C}_{nn} = \begin{bmatrix} \gamma_{n1}e_{n1} & -\gamma_{n1} & 0 & 0 & ke_{n3} & k \\ ke_{n1} & k & 0 & 0 & -\gamma_{n3}e_{n3} & \gamma_{n3} \\ \gamma_{n1}e_{n1} & -\gamma_{n1} & 0 & 0 & ke_{n3} & k \\ -(\rho_n\omega^2 - 2\mu_n k^2)e_{n1} & -(\rho_n\omega^2 - 2\mu_n k^2) & 0 & 0 & -2\mu_n k\gamma_{n3}e_{n3} & 2\mu_n k\gamma_{n3} \\ 2\mu_n k\gamma_{n1}e_{n1} & -2\mu_n k\gamma_{n1} & 0 & 0 & \mu_2(k^2 - \gamma_{n3}^2)e_{n3} & \mu_n(k^2 - \gamma_{n3}^2) \\ 0 & 0 & 0 & 1 & 0 & 0 \end{bmatrix}$$

and

$$\tilde{C}_{n,n+1} = \begin{bmatrix} -\gamma_{n+1,1} & \gamma_{n+1,1}e_{n+1,1} & -\gamma_{n+1,2} & \gamma_{n+1,2}e_{n+1,2} & -k & -ke_{n+1,3} \\ -k & -ke_{n+1,1} & -k & -ke_{n+1,2} & \gamma_{n+1,3} & -\gamma_{n+1,3}e_{n+1,3} \\ -(1-\delta_{1,n+1})\gamma_{n+1,1} & (1-\delta_{1,n+1})\gamma_{n+1,1}e_{n+1,1} & -(1-\delta_{2,n+1})\gamma_{n+1,2} & -(1-\delta_{2,n+1})\gamma_{n+1,2}e_{n+1,2} & -k(1-\delta_{3,n+1}) & -k(1-\delta_{3,n+1})e_{n+1,3} \\ -[\,]_{n+1,1} & -[\,]_{n+1,1}e_{n+1,1} & -[\,]_{n+1,2} & -[\,]_{n+1,2}e_{n+1,2} & 2\mu_{n+1}k\gamma_{n+1,3} & -2\mu_{n+1}k\gamma_{n+1,3}e_{n+1,3} \\ -2\mu_{n+1}k\gamma_{n+1,1} & 2\mu_{n+1}k\gamma_{n+1,1}e_{n+1,1} & -2\mu_{n+1}k\gamma_{n+1,2} & 2\mu_{n+1}k\gamma_{n+1,2}e_{n+1,2} & -(k^2-\gamma_{n+1,3}^2)\mu_{n+1} & -(k^2-\gamma_{n+1,3}^2)\mu_{n+1}e_{n+1,3} \\ 0 & 0 & 0 & 0 & 0 & 0 \end{bmatrix}$$

where

$$[\,]_{n+1,1} = (C_{n+1}\delta_{1,n+1} - \bar{H}_{n+1})l_{c1,n+1}^2 + 2\mu_{n+1}k^2$$

$$[\,]_{n+1,2} = (C_{n+1}\delta_{2,n+1} - \bar{H}_{n+1})l_{c2,n+1}^2 + 2\mu_{n+1}k^2$$

Table F.7: The non-zero submatrix for the interface EP at $z = -H_n$.

Calculation of the Reflection Coefficient

$$\tilde{C}_{nn} = \begin{bmatrix} \gamma_{n1}e_{n1} & -\gamma_{n1} & 0 & 0 & ke_{n3} & k \\ ke_{n1} & k & 0 & 0 & -\gamma_{n3}e_{n3} & \gamma_{n3} \\ 0 & 0 & 0 & 1 & 0 & 0 \\ -(\rho_n\omega^2 - 2\mu_n k^2)e_{n1} & -(\rho_n\omega^2 - 2\mu_n k^2) & 0 & 0 & -2\mu_n k\gamma_{n3}e_{n3} & 2\mu_n k\gamma_{n3} \\ 2\mu_n k\gamma_{n1}e_{n1} & -2\mu_n k\gamma_{n1} & 0 & 0 & \mu_2(k^2 - \gamma_{n3}^2)e_{n3} & \mu_n(k^2 - \gamma_{n3}^2) \\ 0 & 0 & 0 & 0 & 0 & 0 \end{bmatrix}$$

and

$$\tilde{C}_{n,n+1} = \begin{bmatrix} -\gamma_{n+1,1} & \gamma_{n+1,1}e_{n+1,1} & 0 & 0 & -k & -ke_{n+1,3} \\ -k & -ke_{n+1,1} & 0 & 0 & \gamma_{n+1,3} & -\gamma_{n+1,3}e_{n+1,3} \\ 0 & 0 & 0 & 0 & 0 & 0 \\ (\rho_{n+1}\omega^2 - 2\mu_{n+1}k^2) & (\rho_{n+1}\omega^2 - 2\mu_{n+1}k^2)e_{n+1,1} & 0 & 0 & 2\mu_{n+1}k\gamma_{n+1,3} & -2\mu_{n+1}k\gamma_{n+1,3}e_{n+1,3} \\ -2\mu_{n+1}k\gamma_{n+1,1} & 2\mu_{n+1}k\gamma_{n+1,1}e_{n+1,1} & 0 & 0 & -\mu_{n+1}(k^2 - \gamma_{n+1,3}^2) & -\mu_{n+1}(k^2 - \gamma_{n+1,3}^2)e_{n+1,3} \\ 0 & 0 & 1 & 0 & 0 & 0 \end{bmatrix}$$

Table F.8: The non-zero submatrix for the interface EE at $z = -H_n$.

Appendix G

Spectral Representation of a Stationary Process

This Appendix summarizes, without derivation, those spectral properties of a stationary stochastic process, which are most relevant to this analysis, viz. the power-density spectrum and the correlation functions. For more detail readers are referred to Chap.I and Chap.IV of [57].

A periodic function can be expressed as the sum of sine and cosine terms over a discrete set of frequencies, or in the form of complex variables as

$$f(t) = \sum_{n=-\infty}^{\infty} A_n e^{-i\omega_n t} \qquad (G.1)$$

while a non-periodic function, if it is absolutely integrable, i.e. if

$$\int_{-\infty}^{\infty} |f(t)| dt < \infty$$

can be expressed as the sum of these terms over a continuous set of frequencies,

$$f(t) = \frac{1}{2\pi} \int_{-\infty}^{\infty} p(\omega) e^{-i\omega t} d\omega \qquad (G.2)$$

Both types of function can be represented as a Fourier-Stieltjes integral

$$f(t) = \int_{-\infty}^{\infty} e^{-i\omega t} dP(\omega) \qquad (G.3)$$

where the function $P(\omega)$ (generally complex) is stepwise for periodic functions and differentiable ($dP(\omega) = p(\omega)d\omega/(2\pi)$) for non-periodic, absolutely integrable functions.

As described in [57] the form of integration in (G.3) can also be used to describe a stationary random process, which is neither periodic nor absolutely integrable. In this case the amplitude, $dP(\omega)$, is proportional to the square root of the frequency interval, i.e

$$dP(\omega) = O(\sqrt{d\omega}) \tag{G.4}$$

It is also true that if $\{X(t)\}$ is a zero-mean, continuous parameter stationary process, $\{X(t)\}$, $-\infty < t < \infty$, then there exists an orthogonal process, $\{Z(\omega)\}$, such that for all t, $X(t)$ can be written in the form

$$X(t) = \int_{-\infty}^{\infty} e^{-i\omega t} dZ(\omega) \tag{G.5}$$

in the mean square sense[1]. The process $z(\omega)$ has the properties,

(i) $\langle dZ(\omega) \rangle = 0$

(ii) $\langle |dZ(\omega)|^2 \rangle = dH(\omega)$ (G.6)

(iii) $\langle dZ^*(\omega) \, dZ(\omega') \rangle = 0 \quad \omega \neq \omega'$

where $H(\omega)$ is the integrated power spectrum of $X(t)$, and relates to the power-density spectrum $h(\omega)$ through

$$h(\omega) = \frac{dH(\omega)}{d\omega} = \frac{\langle |dZ(\omega)|^2 \rangle}{d\omega} \tag{G.7}$$

The relation between the power-density spectrum, $h(\omega)$, and the auto-covariance function, $R(\tau)$,

$$R(\tau) = \langle X(t) \, X^*(t-\tau) \rangle \tag{G.8}$$

then takes the form

$$R(\tau) = \int_{-\infty}^{\infty} h(\omega) e^{-i\omega\tau} \frac{d\omega}{2\pi} \tag{G.9}$$

with

$$h(\omega) = \int_{-\infty}^{\infty} R(\tau) e^{i\omega\tau} d\tau \tag{G.10}$$

Similar definitions can be made for the spatial correlation function and related power-density spectrum

$$R(\vec{\rho}) = \int h(\vec{k}) e^{i\vec{k}\vec{\rho}} \frac{d\vec{k}}{(2\pi)^2} \tag{G.11}$$

---

[1] Let $z_1, z_2, \ldots, z_i$ be a sequence of random variables. The sequence $\{z_i\}$ is said to converge in mean square if and only if there exists a random variable, z, such that $\lim_{n\to\infty} \langle |z_n - z|^2 \rangle = 0$

with
$$h(\vec{k}) = \int R(\vec{\rho})e^{-i\vec{k}\vec{\rho}}d\vec{\rho} \tag{G.12}$$

When two processes $X_1(t)$ and $X_2(t)$ defined by
$$X_{1,2}(t) = \int_{-\infty}^{\infty} e^{-i\omega t}dZ_{1,2}(\omega) \tag{G.13}$$

are both stationary and the covariance function, $\langle |X_1(t_1)X_2^*(t_1-\tau)|\rangle$, is a function of τ only, we may define a covariance matrix $\mathbf{R}(\tau)$
$$\mathbf{R}(\tau) = \begin{bmatrix} R_{11}(\tau) & R_{12}(\tau) \\ R_{21}(\tau) & R_{22}(\tau) \end{bmatrix} \tag{G.14}$$

and a spectral matrix $\mathbf{h}(\omega)$
$$\mathbf{h}(\omega) = \begin{bmatrix} h_{11}(\omega) & h_{12}(\omega) \\ h_{21}(\omega) & h_{22}(\omega) \end{bmatrix} \tag{G.15}$$

to describe the interrelations between the two processes. In the above, the matrix elements are
$$R_{ij}(\tau) = \int_{-\infty}^{\infty} h_{ij}(\omega)e^{-i\omega\tau}\frac{d\omega}{2\pi} \tag{G.16}$$

and
$$h_{ij}(\omega) = \int_{-\infty}^{\infty} R_{ij}(\tau)e^{i\omega\tau}d\tau = \frac{\langle dZ_i(\omega)dZ_j^*(\omega)\rangle}{d\omega} \tag{G.17}$$

where (i and $j = 1,2$) are respectively the correlation function and coherence function(at frequency ω). The normalized coherence function is then defined as
$$C_{12}(\omega) = \frac{h_{12}(\omega)}{[h_{11}(\omega)h_{22}(\omega)]^{\frac{1}{2}}} \tag{G.18}$$

or
$$C_{12}(\omega) = \frac{\langle dZ_1(\omega), dZ_2^*(\omega)\rangle}{[\langle |dZ_1(\omega)|^2\rangle \langle |dZ_2(\omega)|^2\rangle]^{\frac{1}{2}}} \tag{G.19}$$

The two processes $dZ_1(\omega)$ and $dZ_2(\omega)$ are orthogonal, i.e.,
$$\langle dZ_1(\omega)dZ_1^*(\omega')\rangle = \langle dZ_2(\omega)dZ_2^*(\omega')\rangle = 0 \tag{G.20}$$

and
$$\langle dZ_1(\omega)dZ_2^*(\omega')\rangle = \langle dZ_2(\omega)dZ_1^*(\omega')\rangle = 0 \tag{G.21}$$

if $\omega \neq \omega'$.

Though, as mentioned in the text, only a stationary process can strictly be expressed as a Fourier-Stieltjes integration, in this analysis we still "formally" introduce an ordinary Fourier transform of the displacement potential field

$$\tilde{\Phi}(\vec{R}, t) = \frac{1}{2\pi} \int_{-\infty}^{\infty} \Phi_\omega(\vec{R}) e^{-i\omega t} d\omega \qquad (G.22)$$

to be able to treat the problem in hand as a steady-state boundary-value problem, but keep in mind that the function $\Phi_\omega(\vec{R})$ is regarded as the limit of

$$\Phi_\omega(\vec{R}) = \int_{-T}^{T} \tilde{\Phi}(\vec{R}, t) e^{-i\omega t} dt \qquad (G.23)$$

as $T \to \infty$.

As a spatial function $\Phi_\omega(\vec{R})$ can also be expanded in wavenumber space as

$$\Phi_\omega(\vec{R}) = \frac{1}{(2\pi)^2} \int_{-\infty}^{\infty} \Phi_\omega(\vec{k}, z) e^{i\vec{k}\cdot\vec{r}} d\vec{k} \qquad (G.24)$$

Accordingly the power-density spectrum in the frequency-wavenumber domain then takes the form

$$f_\Phi(\vec{k}, \omega, z) = \frac{1}{(2\pi)^6} \langle |\Phi_\omega(\vec{k}, z)|^2 \rangle d\vec{k} d\omega \qquad (G.25)$$

Finally, as with the function $h_{ij}(\omega)$ of Eq.(G.17), we can define the non-normalized horizontal coherence function of the potential field as

$$C_\Phi(\vec{r}_1, \vec{r}_2, z, \omega) = \frac{1}{(2\pi)^2} \langle \Phi_\omega(\vec{r}_1, z) \Phi_\omega^*(\vec{r}_2, z) \rangle d\omega \qquad (G.26)$$

and the vertical coherence function as

$$C_\Phi(\vec{r}, z_1, z_2, \omega) = \frac{1}{(2\pi)^2} \langle \Phi_\omega(\vec{r}, z_1) \Phi_\omega^*(\vec{r}, z_2) \rangle d\omega \qquad (G.27)$$

Appendix H

General Solution of Adjoint Boundary Value Problems

We first review the normal procedure for solving a Helmholtz equation,

$$L\psi(\vec{R}) = -4\pi\rho(\vec{R}) \tag{H.1}$$

where L is the Helmholtz operator, $L \equiv \nabla^2 + \omega^2/\alpha_1^2$, and the boundary conditions are defined. (The material of this appendix is based on Chap.7 of [58]).

To construct the required solution, $\psi(\vec{R})$, we introduce a Green's function satisfying

$$LG(\vec{R}, \vec{R}_0) = -4\pi\delta(\vec{R} - \vec{R}_0) \tag{H.2}$$

where $\vec{R}$ and $\vec{R}_0$ respectively define the field and the source points, and certain boundary conditions. These conditions are designed so that the solution $\psi(\vec{R})$ can be expressed in terms of this Green's function and the given boundary values of $\psi(\vec{R})$. To establish such conditions we multiply Eq.(H.1) by $G(\vec{R}, \vec{R}_0)$ and Eq.(H.2) by $\psi(\vec{R})$ and subtract to obtain

$$4\pi\psi(\vec{R})\delta(\vec{R}-\vec{R}_0) - 4\pi\rho(\vec{R})G(\vec{R},\vec{R}_0) = G(\vec{R},\vec{R}_0)L\psi(\vec{R}) - \psi(\vec{R})LG(\vec{R},\vec{R}_0) \tag{H.3}$$

Integration of (H.3) over space V, which includes $\vec{R}_0$, gives

$$\psi(\vec{R}_0) = \int_V \rho(\vec{R})G(\vec{R},\vec{R}_0)d\vec{R} + \frac{1}{4\pi}\int_V \left[G(\vec{R},\vec{R}_0)L\psi(\vec{R}) - \psi(\vec{R})LG(\vec{R},\vec{R}_0)\right]d\vec{R} \tag{H.4}$$

Since for the operator $L = \nabla^2 + \omega^2/\alpha_1^2$ the Green's theorem

$$uLv - vLu = \nabla \cdot (u\nabla v - v\nabla u) \tag{H.5}$$

applies, where u and v are two arbitrary scalar functions, the second integration over V in (H.4) can be transformed, using Gauss's Theorem, to one over the surface S surrounding the volume V, so that

$$\psi(\vec{R}_0) = \int_V \rho(\vec{R}) G(\vec{R},\vec{R}_0) d\vec{R} + \frac{1}{4\pi} \int_S \left[G(\vec{R},\vec{R}_0) \frac{\partial}{\partial n} \psi(\vec{R}) - \psi(\vec{R}) \frac{\partial}{\partial n} G(\vec{R},\vec{R}_0) \right] d\vec{R} \quad \text{(H.6)}$$

where $\partial/\partial n$ is the derivative along the outward normal direction to the surface.

Further, the Green's function of the Helmholtz operator is reciprocal in space, i.e.,

$$G(\vec{R},\vec{R}_0) = G(\vec{R}_0,\vec{R}) \quad \text{(H.7)}$$

We demonstrate this here (because of its importance later in the analysis) by assuming another solution of (H.2), $G(\vec{R},\vec{R}_1)$, under the same boundary conditions, satisfying

$$LG(\vec{R},\vec{R}_1) = -4\pi \delta(\vec{R} - \vec{R}_1) \quad \text{(H.8)}$$

Multiplying Eq.(H.2) by $G(\vec{R},\vec{R}_1)$ and Eq.(H.8) by $G(\vec{R},\vec{R}_0)$ and subtracting gives

$$G(\vec{R},\vec{R}_1) LG(\vec{R},\vec{R}_0) - G(\vec{R},\vec{R}_0) LG(\vec{R},\vec{R}_1) = -4\pi \left[G(\vec{R},\vec{R}_1) \delta(\vec{R} - \vec{R}_0) - G(\vec{R},\vec{R}_0) \delta(\vec{R} - \vec{R}_1) \right]$$

which leads after integration to

$$G(\vec{R}_0,\vec{R}_1) - G(\vec{R}_1,\vec{R}_0) =$$
$$\frac{-1}{4\pi} \int_V \left[G(\vec{R},\vec{R}_1) LG(\vec{R},\vec{R}_0) - G(\vec{R},\vec{R}_0) LG(\vec{R},\vec{R}_1) \right] d\vec{R} =$$
$$\frac{-1}{4\pi} \int_S \left[G(\vec{R},\vec{R}_1) \frac{\partial}{\partial n} G(\vec{R},\vec{R}_0) - G(\vec{R},\vec{R}_0) \frac{\partial}{\partial n} G(\vec{R},\vec{R}_1) \right] d\vec{R} \equiv 0$$
(H.9)

Eq.(H.9) vanishes because the functions, $G(\vec{R},\vec{R}_1)$ and $G(\vec{R},\vec{R}_0)$ both satisfy the same boundary conditions on S, i.e. at each point $\vec{R}$ on S

$$\frac{G(\vec{R},\vec{R}_0)}{\frac{\partial}{\partial n} G(\vec{R},\vec{R}_0)} = \frac{G(\vec{R},\vec{R}_1)}{\frac{\partial}{\partial n} G(\vec{R},\vec{R}_1)}$$

By using the reciprocal relation, (H.7), we can therefore interchange $\vec{R}_0$ and $\vec{R}$ and establish from (H.6) that

Solution of Adjoint Problems

$$\psi(\vec{R}) = \int_V \rho(\vec{R}_0) G(\vec{R}, \vec{R}_0) d\vec{R}_0 + \frac{1}{4\pi} \int_S \left[G(\vec{R}, \vec{R}_0) \frac{\partial}{\partial n_0} \psi(\vec{R}_0) \right.$$
$$\left. - \psi(\vec{R}_0) \frac{\partial}{\partial n_0} G(\vec{R}, \vec{R}_0) \right] d\vec{R}_0 \quad \text{(H.10)}$$

where $\partial/\partial n_0$ and $\partial/\partial n$ denote the normal derivatives taken at the source- and field-point respectively.

The boundary conditions for the Green's function can now be found in terms of the boundary conditions assigned to the problem. For example, if the value of the normal derivative is given on the boundary surface, viz

$$\frac{\partial \psi(\vec{R})}{\partial n} = F(\vec{R}) \quad \vec{R} \text{ at surface} \quad \text{(H.11)}$$

we then require the Green's function to satisfy

$$\frac{\partial G(\vec{R}, \vec{R}_0)}{\partial n} = 0 \quad \vec{R} \text{ at surface} \quad \text{(H.12)}$$

so that

$$\psi(\vec{R}) = \int_V \rho(\vec{R}_0) G(\vec{R}, \vec{R}_0) d\vec{R}_0 + \frac{1}{4\pi} \int_S F(\vec{R}_0) G(\vec{R}, \vec{R}_0) d\vec{R}_0 \quad \text{(H.13)}$$

On the other hand, if the value of $\psi(\vec{R})$ on the surface S is given, we require that the Green's function satisfies the homogeneous condition, $G(\vec{R}, \vec{R}_0) = 0$, whereupon the solution is determined by

$$\psi(\vec{R}) = \int_V \rho(\vec{R}_0) G(\vec{R}, \vec{R}_0) d\vec{R}_0 - \frac{1}{4\pi} \int_S F(\vec{R}_0) \frac{\partial}{\partial n_0} G(\vec{R}, \vec{R}_0) d\vec{R}_0 \quad \text{(H.14)}$$

When L is not a Helmholtz operator, however, the Green's theorem, (H.5), and the simple reciprocal relation, (H.7), are not satisfied. To establish the required solution for this situation the above discussion requires modification. As an extension of Eq.(H.5) we first introduce an "adjoint" operator, $\hat{L}$, so that a pair of scalar functions, u and v, can be related to each other through the generalised Green's theorem[58].

$$uLv - v\hat{L}u = \nabla \cdot \vec{P}(u, v) \quad \text{(H.15)}$$

As is shown in [58] the adjoint operator of a differential operator,

$$L = p(x_1, x_2, \cdots, x_s) \frac{\partial^n}{\partial x_1^a \partial x_2^b \cdots \partial x_s^k}, \quad a + b + \cdots k = n \quad \text{(H.16)}$$

takes the form,

$$\hat{L} = (-1)^n \frac{\partial^n}{\partial x_1^a \partial x_2^b \cdots \partial x_s^k}[p(x_1, x_2, \cdots, x_s)] \qquad (H.17)$$

The corresponding vector function, $\vec{P}$, known as the bilinear concomitant, is

$$\vec{P}(u,v) = \vec{a}_1 \left[pu \left(\frac{\partial^{(n-1)}v}{\partial x_1^{a-1}\partial x_2^b \cdots \partial x_s^k} \right) - \frac{\partial(pu)}{\partial x_1} \left(\frac{\partial^{(n-2)}v}{\partial x_1^{a-2}\partial x_2^b \cdots \partial x_s^k} \right) \right.$$
$$\left. + \cdots + (-1)^{a-1}\frac{\partial^{(a-1)}(pu)}{\partial x_1^{a-1}} \left(\frac{\partial^{(n-a)}v}{\partial x_2^b \cdots \partial x_s^k} \right) \right]$$
$$+ (-1)^a \vec{a}_2 \left[\left(\frac{\partial^a(pu)}{\partial x_1^a} \right) \left(\frac{\partial^{(n-a-1)}v}{\partial x_2^{b-1} \cdots \partial x_s^k} \right) - \left(\frac{\partial^{a+1}(pu)}{\partial x_1^a \partial x_2} \right) \left(\frac{\partial^{(n-a-2)}v}{\partial x_2^{b-2} \cdots \partial x_s^k} \right) \right.$$
$$\left. + \cdots + (-1)^{b-1} \left(\frac{\partial^{(a+b-1)}(pu)}{\partial x_1^a \partial x_2^{b-1}} \right) \left(\frac{\partial^{(n-a-b)}v}{\cdots \partial x_s^k} \right) \right]$$
$$+ \cdots + (-1)^{n-k}\vec{a}_s \left[\left(\frac{\partial^{n-k}(pu)}{\partial x_1^a \partial x_2^b \cdots} \right) \left(\frac{\partial^{k-1}v}{\partial x_s^{k-1}} \right) - \left(\frac{\partial^{n-k+1}(pu)}{\partial x_1^a \partial x_2^b \cdots \partial x_s} \right) \left(\frac{\partial^{k-2}v}{\partial x_s^{k-2}} \right) \right.$$
$$\left. + \cdots + (-1)^{k-1} \left(\frac{\partial^{n-1}(pu)}{\partial x_1^a \partial x_2^b \cdots \partial x_s^{k-1}} \right) v \right]$$

$$(H.18)$$

with $\vec{a}_i$ being the unit vector in the $\vec{x}_i$ direction (a few typing mistakes in (H.18) as it appears in [58] have been corrected here).

To solve Eq.(H.1) when L has its general form, we introduce the Green's function $\hat{G}(\vec{R}, \vec{R}_0)$ satisfying

$$\hat{L}\hat{G}(\vec{R}, \vec{R}_0) = -4\pi\delta(\vec{R} - \vec{R}_0) \qquad (H.19)$$

Multiplying (H.1) and (H.19) by $\hat{G}(\vec{R}, \vec{R}_0)$ and $\psi(\vec{R})$ respectively and integrating, we can establish a relation similar to (H.4) as

$$\psi(\vec{R}_0) =$$
$$\int_V \rho(\vec{R})\hat{G}(\vec{R}, \vec{R}_0)d\vec{R} + \frac{1}{4\pi}\int_V \left[\hat{G}(\vec{R}, \vec{R}_0)L\psi(\vec{R}) - \psi(\vec{R})\hat{L}\hat{G}(\vec{R}, \vec{R}_0) \right] d\vec{R}$$
$$= \int_V \rho(\vec{R})\hat{G}(\vec{R}, \vec{R}_0)d\vec{R} + \frac{1}{4\pi}\int_S \vec{n} \cdot \vec{P}\left[\hat{G}(\vec{R}, \vec{R}_0), \psi(\vec{R}) \right] d\vec{R} \qquad (H.20)$$

Solution of Adjoint Problems

Before proceeding further, we need to find the condition under which a reciprocal relationship exists between the Green's function, $\hat{G}(\vec{R}, \vec{R}_0)$, and $G(\vec{R}, \vec{R}_1)$ of the original operator L, satisfying

$$LG(\vec{R}, \vec{R}_1) = -4\pi\delta(\vec{R} - \vec{R}_1). \tag{H.21}$$

Multiplying (H.19) and (H.21) by $G(\vec{R}, \vec{R}_1)$ and $\hat{G}(\vec{R}, \vec{R}_0)$ respectively, using the subtraction and integration procedures applied to obtain (H.9), and introducing the relation, (H.15), establishes

$$G(\vec{R}_0, \vec{R}_1) - \hat{G}(\vec{R}_1, \vec{R}_0) = \frac{1}{4\pi}\int_V \left[\hat{G}(\vec{R}, \vec{R}_0)LG(\vec{R}, \vec{R}_1) - G(\vec{R}, \vec{R}_1)\hat{L}\hat{G}(\vec{R}, \vec{R}_0)\right]d\vec{R} = \frac{1}{4\pi}\int_S \vec{n}\cdot\vec{P}\left[\hat{G}(\vec{R}, \vec{R}_0), G(\vec{R}, \vec{R}_1)\right]d\vec{R} \tag{H.22}$$

Clearly, if the the right-hand-side of this equation is zero (demonstrated below) the following reciprocal relation exists,

$$G(\vec{R}_0, \vec{R}) = \hat{G}(\vec{R}, \vec{R}_0) \quad \text{or} \quad G(\vec{R}, \vec{R}_0) = \hat{G}(\vec{R}_0, \vec{R}) \tag{H.23}$$

Assuming the validity of (H.23), applying it to (H.20) and interchanging the symbols $\vec{R}$ and $\vec{R}_0$ we have

$$\psi(\vec{R}) = \int_V \rho(\vec{R}_0)G(\vec{R}, \vec{R}_0)d\vec{R}_0 + \frac{1}{4\pi}\int_S \vec{n}_0\cdot\vec{P}\left[G(\vec{R}, \vec{R}_0), \psi(\vec{R}_0)\right]d\vec{R}_0 \tag{H.24}$$

This expression corresponds to (H.10). In a way similar to that applied in the case of (H.10), we can use (H.24) to find the boundary conditions to be imposed upon the Green's functions.

As an example consider the solution of the equation

$$L\psi(\vec{R}) \equiv \left(\nabla^2 - 2a\frac{\partial}{\partial z} + \frac{\omega^2}{\alpha_1^2}\right)\psi(\vec{R}) = 0 \tag{H.25}$$

From Eqs.(H.16) to (H.18) we see that for an operator, $L = \partial^2/\partial x_i^2$, $\hat{L} = \partial^2/\partial x_i^2$ and $\vec{P} = (u\partial v/\partial x_i - v\partial u/\partial x_i)\vec{a}_i$; while for operator $L = \partial/\partial x_i$, $\hat{L} = -\partial/\partial x_i$ and $\vec{P} = uv\vec{a}_i$. Therefore in this case the adjoint operator $\hat{L}$ and the vector $\vec{P}(u, v)$ take the form

$$\hat{L} = \nabla^2 + 2a\frac{\partial}{\partial z} + \frac{\omega^2}{\alpha_1^2} \tag{H.26}$$

and
$$\vec{P}(u,v) = (u\nabla v - v\nabla u - 2auv\vec{z}_0) \tag{H.27}$$

where $\vec{z}_0$ is the unit vector in the z direction.

In the present case the validity of (H.23), requires that

$$\hat{G}(\vec{R},\vec{R}_0)\frac{\partial}{\partial n}G(\vec{R},\vec{R}_1) - G(\vec{R},\vec{R}_1)\frac{\partial}{\partial n}\hat{G}(\vec{R},\vec{R}_0)$$
$$-2a\hat{G}(\vec{R},\vec{R}_0)G(\vec{R},\vec{R}_1) = 0, \quad \vec{R} \text{ at surface} \tag{H.28}$$

or equivalently,

$$\frac{\hat{G}(\vec{R},\vec{R}_0)}{\frac{\partial}{\partial n}\hat{G}(\vec{R},\vec{R}_0) + a\hat{G}(\vec{R},\vec{R}_0)} = \frac{G(\vec{R},\vec{R}_1)}{\frac{\partial}{\partial n}G(\vec{R},\vec{R}_1) - aG(\vec{R},\vec{R}_1)} \tag{H.29}$$

It is easy to see that this condition is satisfied. In fact, if we change the sign before parameter "a" in the operator L and in the boundary condition (H.29), then the function G becomes $\hat{G}$ and both sides of (H.29) identical (see also Appendix J).

Let us now consider only the boundary effect, i.e., we assume $\rho(\vec{R}_0) = 0$ in the volume integral. (H.20) then becomes

$$\psi(\vec{R}_0) = \frac{1}{4\pi}\int_S \left[\hat{G}(\vec{R},\vec{R}_0)\frac{\partial}{\partial n}\psi(\vec{R}) - \psi(\vec{R})\frac{\partial}{\partial n}\hat{G}(\vec{R},\vec{R}_0) \right.$$
$$\left. - 2a\psi(\vec{R})\hat{G}(\vec{R},\vec{R}_0)\right]d\vec{R}$$

Therefore if the derivative of $\psi(\vec{R}_0)$ at the surface is given as the boundary value, i.e.,

$$\frac{\partial \psi(\vec{R})}{\partial n} = F(\vec{R}^s) \quad \text{at the surface } \vec{R} = \vec{R}^s.$$

we can then impose the condition,

$$\frac{\partial}{\partial n}\hat{G}(\vec{R},\vec{R}_0) + 2a\hat{G}(\vec{R},\vec{R}_0) = 0 \quad \vec{R} \text{ at the surface}$$

when deriving the function $\hat{G}$. Further, applying the reciprocal relation we finally establish the solution,

$$\psi(\vec{R}) = \frac{1}{4\pi}\int_S G(\vec{R},\vec{R}_0)\frac{\partial}{\partial n_0}\psi(\vec{R}_0)d\vec{R}_0. \tag{H.30}$$

Solution of Adjoint Problems

On the other hand, if the function $\psi(\vec{R})$ itself is given at the surface, the condition required is

$$\hat{G}(\vec{R}, \vec{R}_0) = 0 \quad \vec{R} \text{ at the surface}$$

and the solution takes the form

$$\psi(\vec{R}) = -\frac{1}{4\pi} \int_s \psi(\vec{R}_0) [\frac{\partial}{\partial n_0} G(\vec{R}, \vec{R}_0) + 2aG(\vec{R}, \vec{R}_0)] d\vec{R}_0 \qquad (H.31)$$

Finally, if only the volume source is involved both the above conditions must be met and the solution becomes

$$\psi(\vec{R}) = \int_V \rho(\vec{r}_0) G(\vec{R}, \vec{R}_0) d\vec{R}_0 \qquad (H.32)$$

Appendix I

Solutions to the First-Order Potential Field

The first-order potential, $\Phi_{1\omega}^{(1)}(\vec{R})$, satisfies

$$L\Phi_{1\omega}^{(1)} = 0$$

$$\left(\omega^2 - g\frac{\partial}{\partial z}\right)\Phi_{1\omega}^{(1)} = 0 \quad \text{at} \quad z = 0$$

$$\frac{\partial}{\partial z}\Phi_{1\omega}^{(1)} = -i\omega\zeta_\omega^{(1)} \quad \text{at} \quad z = 0$$

$$L_b\Phi_{1\omega}^{(1)} = 0 \quad \text{at} \quad z = -H_1 \quad \text{(I.1)}$$

where the linear differential operator, L, is

$$L \equiv \nabla^2 - \frac{g}{\alpha_1^2}\frac{\partial}{\partial z} + \frac{\omega^2}{\alpha_1^2} \quad \text{(I.2)}$$

and L_b represents the effect of the seabed on the field. Since L is not self-adjoint, the Green's theorem takes the form

$$\hat{G}^{(1)}(\vec{R}, \vec{R}_0)L\Phi_{1\omega}^{(1)}(\vec{R}) - \Phi_{1\omega}^{(1)}(\vec{R})\hat{L}\hat{G}^{(1)}(\vec{R}, \vec{R}_0) = \nabla \cdot \vec{P}[\hat{G}^{(1)}(\vec{R}, \vec{R}_0), \Phi_{1\omega}^{(1)}(\vec{R})]$$
$$\text{(I.3)}$$

In the above, $\hat{L}$ is the adjoint operator of L,

$$\hat{L} \equiv \nabla^2 + \frac{g}{\alpha_1^2}\frac{\partial}{\partial z} + \frac{\omega^2}{\alpha_1^2} \quad \text{(I.4)}$$

$\hat{G}^{(1)}(\vec{R}, \vec{R}_0)$ is the solution of

$$\hat{L}\hat{G}^{(1)}(\vec{R}, \vec{R}_0) = -4\pi\delta(\vec{R} - \vec{R}_0) \quad \text{(I.5)}$$

and the bilinear concomitant, $\vec{P}(u,v)$, takes the form (for more detail about the adjoint operation see the previous Appendix)

$$\vec{P}(u,v) = \left(u\frac{\partial v}{\partial x} - v\frac{\partial u}{\partial x}\right)\vec{x}_0 + \left(u\frac{\partial v}{\partial y} - v\frac{\partial u}{\partial y}\right)\vec{y}_0 + \left(u\frac{\partial v}{\partial z} - v\frac{\partial u}{\partial z} - 2auv\right)\vec{z}_0 \tag{I.6}$$

where $a \equiv g/2\alpha_1^2$, as before, and $\vec{x}_0$, $\vec{y}_0$, $\vec{z}_0$ are unit vectors along the coordinate axes.

Multiplying (I.5) by $\Phi_{1\omega}^{(1)}$ and the first equation of (I.1) by $\hat{G}^{(1)}$ and subtracting one from the other we can establish

$$\Phi_{1\omega}^{(1)}(\vec{R}_0) = \frac{1}{4\pi} \int_{S+\Sigma} \vec{n} \cdot \vec{P}[\hat{G}^{(1)}(\vec{R},\vec{R}_0), \Phi_{1\omega}^{(1)}(\vec{R})]d\vec{r} \tag{I.7}$$

where S and Σ are respectively the average surface plane, $z = 0$, and the seafloor plane, $z = -H_1$, $d\vec{r}$ defines an elemental area, and $\vec{n}$ is the unit vector in the outward normal direction. More explicitly we can write

$$\begin{aligned}\Phi_{1\omega}^{(1)}(\vec{R}_0) &= \frac{1}{4\pi}\int_S \left\{\hat{G}^{(1)}(\vec{R},\vec{R}_0)\frac{\partial}{\partial z}\Phi_{1\omega}^{(1)}(\vec{R})\right.\\ &\quad \left.- \Phi_{1\omega}^{(1)}(\vec{R})\left[\frac{\partial \hat{G}^{(1)}(\vec{R},\vec{R}_0)}{\partial z} + 2a\hat{G}^{(1)}(\vec{R},\vec{R}_0)\right]\right\}_{z=0} d\vec{r} \\ &\quad -\frac{1}{4\pi}\int_\Sigma \left\{\hat{G}^{(1)}(\vec{R},\vec{R}_0)\frac{\partial}{\partial z}\Phi_{1\omega}^{(1)}(\vec{R})\right.\\ &\quad \left.- \Phi_{1\omega}^{(1)}(\vec{R})\left[\frac{\partial \hat{G}^{(1)}(\vec{R},\vec{R}_0)}{\partial z} - 2a\hat{G}^{(1)}(\vec{R},\vec{R}_0)\right]\right\}_{z=-H_1} d\vec{r}\end{aligned} \tag{I.8}$$

If we require that the function $\hat{G}^{(1)}(\vec{R},\vec{R}_0)$ satisfies the conditions

$$\frac{\partial \hat{G}^{(1)}(\vec{R},\vec{R}_0)}{\partial z} + 2a\hat{G}^{(1)}(\vec{R},\vec{R}_0) = 0 \text{ at } z = 0 \tag{I.9}$$

and

$$\frac{\hat{G}^{(1)}(\vec{R},\vec{R}_0)}{\frac{\partial \hat{G}^{(1)}(\vec{R},\vec{R}_0)}{\partial z}} = \frac{\Phi_{1\omega}^{(1)}(\vec{R})}{\frac{\partial \Phi_{1\omega}^{(1)}(\vec{R})}{\partial z} + 2a\Phi_{1\omega}^{(1)}(\vec{R})} \text{ at } z = -H_1 \tag{I.10}$$

then the first-order potential will have a simpler form,

$$\begin{aligned}\Phi_{1\omega}^{(1)}(\vec{R}) &= \frac{1}{4\pi}\int_S G^{(1)}(\vec{R},\vec{R}_0)\frac{\partial}{\partial z_0}\Phi_{1\omega}^{(1)}(\vec{R}_0)\bigg|_{z_0=0} d\vec{r}_0 \\ &= \frac{-i\omega}{4\pi}\int_S G^{(1)}(\vec{R},\vec{R}_0)\zeta_\omega^{(1)}(\vec{r}_0)\,d\vec{r}_0\end{aligned} \tag{I.11}$$

Solution of First Order Potential

in which $\vec{R}$ and $\vec{R}_0$ have been interchanged and we have used the reciprocal relation $\hat{G}^{(1)}(\vec{R}, \vec{R}_0) = G^{(1)}(\vec{R}_0, \vec{R})$.

The required Green's function $\hat{G}^{(1)}(\vec{R}, \vec{R}_0)$ can be found by solving

$$\hat{L}\hat{G}^{(1)}(\vec{R}, \vec{R}_0) = -4\pi\delta(\vec{R} - \vec{R}_0) \tag{I.12}$$

using the boundary conditions,

$$\frac{\partial \hat{G}^{(1)}(\vec{R}, \vec{R}_0)}{\partial z} + 2a\hat{G}^{(1)}(\vec{R}, \vec{R}_0) = 0 \qquad \text{at} \quad z = 0 \tag{I.13}$$

$$\frac{\hat{G}^{(1)}(\vec{R}, \vec{R}_0)}{\frac{\partial \hat{G}^{(1)}(\vec{R}, \vec{R}_0)}{\partial z}} = \frac{\Phi_{1\omega}^{(1)}(\vec{R})}{\frac{\partial \Phi_{1\omega}^{(1)}(\vec{R})}{\partial z} + 2a\Phi_{1\omega}^{(1)}(\vec{R})} \qquad \text{at} \quad z = -H_1 \tag{I.14}$$

Condition (I.14) shows that when a is small the adjoint Green's function satisfies the same field-continuity conditions as the potential $\Phi_{1\omega}^{(1)}$ itself, i.e., the impedance is constant when passing through the sea-floor. As a reasonable approximation we can therefore replace it by introducing the bottom reflection coefficient, R_b.

To find $\hat{G}^{(1)}$ we introduce the Hankel transform,

$$\hat{G}^{(1)}(\vec{R}, \vec{R}_0) = \int_0^\infty \hat{g}_\omega^{(1)}(k, z, z_0) J_0(k\rho) k \, dk \tag{I.15}$$

where $\rho = |\vec{r} - \vec{r}_0|$. Using the relation

$$\delta(\vec{R} - \vec{R}_0) = \frac{1}{(2\pi)^2}\delta(z - z_0)\int e^{i\vec{k}|\vec{r}-\vec{r}_0|} d\vec{k} = \frac{1}{2\pi}\delta(z - z_0)\int_0^\infty J_0(k\rho) k \, dk \tag{I.16}$$

we obtain

$$\left(\frac{d^2}{dz^2} + 2a\frac{d}{dz} + \gamma_1^2\right)\hat{g}_\omega^{(1)}(k, z, z_0) = -2\delta(z - z_0) \tag{I.17}$$

$$\frac{d}{dz}\hat{g}_\omega^{(1)}(k, z, z_0) = -2a\hat{g}_\omega^{(1)}(k, z, z_0) \qquad \text{at} \quad z = 0 \tag{I.18}$$

$$L_b \hat{g}_\omega^{(1)}(k, z, z_0) = 0 \qquad \text{at} \quad z = -H_1 \tag{I.19}$$

where $\gamma_1^2 = \omega^2/\alpha_1^2 - k^2$. The inhomogeneous equation, (I.17), can be replaced by a homogeneous one (left-hand-side = 0) by the introduction of two additional field-continuity conditions at z_0 $(-H_1 < z_0 < 0)$, viz

$$\hat{g}_\omega^{(1)}(k, z_0^+, z_0) - \hat{g}_\omega^{(1)}(k, z_0^-, z_0) = 0$$

and

$$\frac{d}{dz}\hat{g}^{(1)}_\omega(k, z_0^+, z_0) - \frac{d}{dz}\hat{g}^{(1)}_\omega(k, z_0^-, z_0) = -2,$$

Here, $z_0^+ = z_0 + 0$ and $z_0^- = z_0 - 0$.

Further, by introducing

$$\hat{g}^{(1)}_\omega(k, z, z_0) = \begin{cases} A[e^{s_+(z-z_0)} + Be^{s_-(z-z_0)}] & z_0 \le z \le 0 \\ C[e^{s_+(z+H_1)} + R_b e^{s_-(z+H_1)}] & -H_1 \le z \le z_0 \end{cases} \quad (I.20)$$

where $s_\pm = -a \mp \sqrt{a^2 - \gamma_1^2} = -a \mp i\sqrt{\gamma_1^2 - a^2}$ are the solutions of the characteristic equation, $s^2 + 2as + \gamma_1^2 = 0$, and the constants A, B and C are to be determined by the conditions at $z = 0$ and the two continuity conditions at $z = z_0$, we can, after some algebraic manipulation (Appendix J), establish that for $z_0 < z \le 0$,

$$\hat{g}^{(1)}_\omega(k, z, z_0) = \frac{2\left[e^{\gamma_+(z_0+H_1)} + R_b e^{\gamma_-(z_0+H_1)}\right]\left(e^{\gamma_+ z} - \frac{1}{\epsilon}e^{\gamma_- z}\right)e^{-2az}}{\gamma_+\left(1 - \frac{1}{\epsilon}\right)[e^{\gamma_+ H_1} + \epsilon R_b e^{\gamma_- H_1}]}$$

(I.21)

and for $-H_1 \le z < z_0$,

$$\hat{g}^{(1)}_\omega(k, z, z_0) = \frac{2\left[e^{\gamma_+(z+H_1)} + R_b e^{\gamma_-(z+H_1)}\right]\left(e^{\gamma_+ z_0} - \frac{1}{\epsilon}e^{\gamma_- z_0}\right)e^{-2az}}{\gamma_+\left(1 - \frac{1}{\epsilon}\right)[e^{\gamma_+ H_1} + \epsilon R_b e^{\gamma_- H_1}]}.$$

(I.22)

The function $\hat{G}^{(1)}$ is thus established by bringing Eqs. (I.21) and (I.22) back into (I.15).

Further, by interchanging z and z_0 in (I.21) (it is noted that after interchanging z and z_0 (I.21) becomes the required solution in the region $-H_1 \le z < z_0$) and letting z_0 tend to zero, we establish the function $g^{(1)}_\omega$ as

$$g^{(1)}_\omega(k, z) = \frac{2}{\gamma_+}\frac{e^{\gamma_+(z+H_1)} + R_b e^{\gamma_-(z+H_1)}}{e^{\gamma_+ H_1} + \epsilon R_b e^{\gamma_- H_1}} \quad (I.23)$$

where $\gamma_\pm = 2a + s_\pm = a \mp \sqrt{a^2 - \gamma_1^2} = a \mp i\sqrt{\gamma_1^2 - a^2}$.

Thus

$$G^{(1)}(\vec{R}, \vec{R}_0) = \int_0^\infty g^{(1)}_\omega(k, z) J_o(k\rho) k\, dk$$

or

Solution of First Order Potential

$$G^{(1)}(\vec{R}, \vec{R}_0) = \frac{1}{2\pi} \int_{\vec{k}} g_\omega^{(1)}(k, z) e^{i\vec{k}(\vec{r} - \vec{r}_0)} d\vec{k} \tag{I.24}$$

Substituting (I.24) into (I.11) finally establishes the solution

$$\Phi_{1\omega}^{(1)}(\vec{R}) = -\frac{i\omega}{2\pi} \int_S \zeta_\omega^{(1)}(\vec{r}_0) \int_0^\infty \frac{1}{\gamma_+} \frac{e^{\gamma_+(z+H_1)} + R_b e^{\gamma_-(z+H_1)}}{e^{\gamma_+ H_1} + \epsilon R_b e^{\gamma_- H_1}} J_0(k\rho) k \, dk \, d\vec{r}_0 \tag{I.25}$$

or

$$\Phi_{1\omega}^{(1)}(\vec{R}) = \frac{-i\omega}{(2\pi)^2} \int_{\vec{k}} \int_S \frac{1}{\gamma_+} \zeta_\omega^{(1)}(\vec{r}_0) \frac{e^{\gamma_+(z+H_1)} + R_b e^{\gamma_-(z+H_1)}}{e^{\gamma_+ H_1} + \epsilon R_b e^{\gamma_- H_1}} e^{i\vec{k}(\vec{r} - \vec{r}_0)} d\vec{r}_0 d\vec{k} \tag{I.26}$$

To examine the reciprocity we can, for instance, introduce the function $\hat{g}_\omega^{(1)}(k, z, z_0)$ of Eq.(I.21) and its partner, $g_\omega^{(1)}(k, z, z_0)$, derived by interchanging z and z_0 in Eq.(I.22), i.e.,

$$g_\omega^{(1)}(k, z, z_0) = \frac{2\left[e^{\gamma_+(z_0+H_1)} + R_b e^{\gamma_-(z_0+H_1)}\right] \left(e^{\gamma_+ z} - \frac{1}{\epsilon} e^{\gamma_- z}\right) e^{-2az_0}}{\gamma_+ \left(1 - \frac{1}{\epsilon}\right) \left[e^{\gamma_+ H_1} + \epsilon R_b e^{\gamma_- H_1}\right]} \tag{I.27}$$

both into Eq.(H.28) of Appendix H. Taking $\partial/\partial n$ as $\partial/\partial z$ we see that the equation, i.e., the condition for the validity of the reciprocity is apparently satisfied. This can also be directly checked by interchanging the source and field points, z_0 and z, in (I.27), which leads to an expression identical to that for $\hat{g}_\omega^{(1)}(k, z, z_0)$, Eq.(I.21).

Appendix J

Calculation of the First-Order Green's Function

We write the solution $\hat{g}^{(1)}(k, z, z_0)$ in the form

$$\hat{g}^{(1)}_\omega(k, z, z_0) = \begin{cases} A\left[e^{s_+(z-z_0)} + Be^{s_-(z-z_0)}\right] & \text{for } z_0 \leq z \leq 0 \\ C\left[e^{s_+(z+H_1)} + R_b e^{s_-(z+H_1)}\right] & \text{for } -H_1 \leq z \leq z_0 \end{cases} \quad (\text{J.1})$$

where $s_\pm = -a \mp \sqrt{a^2 - \gamma_1^2}$ and R_b is the plane-wave reflection coefficient for the seafloor. Coefficients A, B and C can be determined by the conditions at $z = 0$ and $z = z_0$,

$$\left. \begin{array}{r} \frac{\partial}{\partial z}\hat{g}^{(1)}_\omega(k, z, z_0) + 2a\hat{g}^{(1)}_\omega(k, z, z_0) = 0 \quad \text{at } z = 0 \\[4pt] \frac{\partial}{\partial z}\hat{g}^{(1)}_\omega(k, z, z_0)\Big|_{z_0-0}^{z_0+0} = -2 \quad \text{at } z = z_0 \\[4pt] \hat{g}^{(1)}_\omega(k, z, z_0)\Big|_{z_0-0}^{z_0+0} = 0 \quad \text{at } z = z_0 \end{array} \right\} \quad (\text{J.2})$$

Substituting (J.1) into (J.2) leads to

$$\left. \begin{array}{r} (s_+ + 2a)e^{-s_+ z_0} + B(s_- + 2a)e^{-s_- z_0} = 0 \\[4pt] A(s_+ + Bs_-) - C\left(s_+ e^{s_+(z_0+H_1)} + R_b s_- e^{s_-(z_0+H_1)}\right) = -2 \\[4pt] A(1 + B) - C\left(e^{s_+(z_0+H_1)} + R_b e^{s_-(z_0+H_1)}\right) = 0 \end{array} \right\} \quad (\text{J.3})$$

from which we find

$$B = -\frac{s_+ + 2a}{s_- + 2a} e^{-(s_+ - s_-)z_0}$$

Denoting the last two equations of (J.3) as

$$AQ_1 + CQ_2 = -2$$

$$AQ_3 + CQ_4 = 0$$

establishes

$$A = -\frac{2Q_4}{\Delta} \quad \text{and} \quad C = \frac{2Q_3}{\Delta}$$

where

$$Q_1 = (s_+ + Bs_-)$$

$$Q_2 = -\left(s_+ e^{s_+(z_0+H_1)} + R_b s_- e^{s_-(z_0+H_1)}\right)$$

$$Q_3 = (1+B)$$

$$Q_4 = -\left(e^{s_+(z_0+H_1)} + R_b e^{s_-(z_0+H_1)}\right)$$

$$\begin{aligned}\Delta &= Q_1 Q_4 - Q_2 Q_3 \\ &= (1+B)\left(s_+ e^{s_+(z_0+H_1)} + R_b s_- e^{s_-(z_0+H_1)}\right) - (s_+ + Bs_-)\left(e^{s_+(z_0+H_1)} + R_b e^{s_-(z_0+H_1)}\right) = B(s_+ - s_-)e^{s_+(z_0+H_1)} + (s_- - s_+)R_b e^{s_-(z_0+H_1)}\end{aligned}$$

(J.4)

For convenience we define

$$\gamma_\pm \equiv s_\pm + 2a = a \mp \sqrt{a^2 - \gamma_1^2} = -s_\mp$$

and

$$\epsilon \equiv \frac{\gamma_-}{\gamma_+} = \frac{s_+}{s_-}$$

so that

$$B = -\frac{1}{\epsilon} e^{2\sqrt{a^2-\gamma_1^2} z_0}$$

and

$$\Delta = \gamma_+ \left(1 - \frac{1}{\epsilon}\right) \left[e^{\gamma_+ H_1} + \epsilon R_b e^{\gamma_- H_1}\right] e^{-2a(z_0+H_1)+\gamma_- z_0}$$

Calculation of First Order Potential

$$\Delta = \gamma_+ \left(1 - \frac{1}{\epsilon}\right) \left[e^{\gamma_+ H_1} + \epsilon R_b e^{\gamma_- H_1}\right] e^{-\gamma_- z_0 - 2a(z_0 + H_1)} \tag{J.5}$$

Substituting the above expressions into (J.1) establishes the solution:
for $z_0 < z \leq 0$,

$$\hat{g}_\omega^{(1)}(k, z, z_0) = \frac{2 \left[e^{\gamma_+(z_0 + H_1)} + R_b e^{\gamma_-(z_0 + H_1)}\right] \left(e^{\gamma_+ z} - \frac{1}{\epsilon} e^{\gamma_- z}\right) e^{-2az}}{\gamma_+ \left(1 - \frac{1}{\epsilon}\right) \left[e^{\gamma_+ H_1} + \epsilon R_b e^{\gamma_- H_1}\right]}, \tag{J.6}$$

and for $-H_1 \leq z < z_0$,

$$\hat{g}_\omega^{(1)}(k, z, z_0) = \frac{2 \left[e^{\gamma_+(z + H_1)} + R_b e^{\gamma_-(z + H_1)}\right] \left(e^{\gamma_+ z_0} - \frac{1}{\epsilon} e^{\gamma_- z_0}\right) e^{-2az}}{\gamma_+ \left(1 - \frac{1}{\epsilon}\right) \left[e^{\gamma_+ H_1} + \epsilon R_b e^{\gamma_- H_1}\right]}. \tag{J.7}$$

Appendix K

Solution of the Second-Order Potential Field

To establish solutions to the equations

$$L\tilde{\Phi}_{1a}^{(2)}(\vec{R},t) = \frac{1}{\alpha_1^2}\frac{\partial}{\partial t}\left[\nabla\tilde{\Phi}_1^{(1)}\right]^2 \tag{K.1}$$

$$\left(\frac{\partial^2}{\partial t^2} + g\frac{\partial}{\partial z}\right)\tilde{\Phi}_{1a}^{(2)}(\vec{R},t) = 0 \quad \text{at} \quad z = 0 \tag{K.2}$$

$$L_b\tilde{\Phi}_{1a}^{(2)}(\vec{R},t) = 0 \quad \text{at} \quad z = -H_1, \tag{K.3}$$

and

$$L\tilde{\Phi}_{1b}^{(2)}(\vec{R},t) = 0 \tag{K.4}$$

$$\left(\frac{\partial^2}{\partial t^2} + g\frac{\partial}{\partial z}\right)\tilde{\Phi}_{1b}^{(2)}(\vec{R},t) = -\frac{\partial}{\partial t}\left[\nabla\tilde{\Phi}_1^{(1)}\right]^2 \quad \text{at} \quad z = 0 \tag{K.5}$$

$$L_b\tilde{\Phi}_{1b}^{(2)}(\vec{R},t) = 0 \quad \text{at} \quad z = -H_1, \tag{K.6}$$

where

$$L = \nabla^2 - 2a\frac{\partial}{\partial z} - \frac{1}{\alpha_1^2}\frac{\partial^2}{\partial t^2},$$

and L_b is an operator representing the field continuity conditions at the bottom, we first perform the Fourier transform

$$\tilde{\Phi}_{1a}^{(2)}(\vec{R},t) = \frac{1}{2\pi}\int \Phi_{1a\omega}^{(2)}(\vec{R})e^{-i\omega t}d\omega \tag{K.7}$$

$$\tilde{\Phi}_{1b}^{(2)}(\vec{R},t) = \frac{1}{2\pi} \int \Phi_{1b\omega}^{(2)}(\vec{R}) e^{-i\omega t} d\omega \tag{K.8}$$

and substitute into (K.1) to (K.6). This leads to a set of new equations,

$$L\Phi_{1a\omega}^{(2)}(\vec{R}) = \bar{S}_a(\omega, \vec{r}, z) \tag{K.9}$$

$$\left(\frac{\partial}{\partial z} - \frac{\omega^2}{g}\right) \Phi_{1a\omega}^{(2)}(\vec{R}) = 0 \quad \text{at} \quad z = 0 \tag{K.10}$$

$$L_b \Phi_{1a\omega}^{(2)}(\vec{R}) = 0 \quad \text{at} \quad z = -H_1 \tag{K.11}$$

and

$$L\Phi_{1b\omega}^{(2)}(\vec{R}) = 0 \tag{K.12}$$

$$\left(\frac{\partial}{\partial z} - \frac{\omega^2}{g}\right) \Phi_{1b\omega}^{(2)}(\vec{R}) = \frac{1}{g} \bar{S}_b(\omega, \vec{r}, 0) \quad \text{at} \quad z = 0 \tag{K.13}$$

$$L_b \Phi_{1b\omega}^{(2)}(\vec{R}) = 0 \quad \text{at} \quad z = -H_1 \tag{K.14}$$

In the above

$$L = \nabla^2 - 2a\frac{\partial}{\partial z} + \frac{\omega^2}{\alpha_1^2}, \quad \text{where} \quad a = g/\alpha_1^2 \tag{K.15}$$

$$\bar{S}_a(\omega, \vec{r}, z) = \frac{1}{\alpha_1^2} \int_{-\infty}^{\infty} \left\{ \frac{\partial}{\partial t} \left[\nabla \tilde{\Phi}_1^{(1)}(\vec{R}, t) \right]^2 \right\} e^{i\omega t} dt \tag{K.16}$$

$$\bar{S}_b(\omega, \vec{r}, 0) = -\int_{-\infty}^{\infty} \left\{ \frac{\partial}{\partial t} \left[\nabla \tilde{\Phi}_1^{(1)}(\vec{R}, t) \right]^2 \right\} e^{i\omega t} dt \tag{K.17}$$

and $\vec{R}$ and $\vec{r}$ are respectively the position vectors in space and in the horizontal plane.

Following the analysis in Appendix H, we introduce the adjoint Green's functions, $G_{a,b}^{(2)}(\vec{R}, \vec{R}_0)$ and $\hat{G}_{a,b}^{(2)}(\vec{R}, \vec{R}_0)$, satisfying

$$LG_{a,b}^{(2)}(\vec{R}, \vec{R}_0) = -4\pi\delta(\vec{R} - \vec{R}_0), \quad L = \nabla^2 - 2a\frac{\partial}{\partial z} + \frac{\omega^2}{\alpha_1^2} \tag{K.18}$$

$$\hat{L}\hat{G}_{a,b}^{(2)}(\vec{R}, \vec{R}_0) = -4\pi\delta(\vec{R} - \vec{R}_0), \quad \hat{L} = \nabla^2 + 2a\frac{\partial}{\partial z} + \frac{\omega^2}{\alpha_1^2} \tag{K.19}$$

Solution of Second Order Potential

and note that by virtue of their reciprocity

$$\hat{G}^{(2)}_{a,b}(\vec{R}, \vec{R}_0) = G^{(2)}_{a,b}(\vec{R}_0, \vec{R}) \tag{K.20}$$

Multiplying Eq.(K.19) by $\Phi^{(2)}_{1a\omega}$ and the equation $L\Phi^{(2)}_{1a\omega}(\vec{R}) = -4\pi\rho(\vec{R})$ by $\hat{G}^{(2)}_a(\vec{R}, \vec{R}_0)$, subtracting, integrating, and applying Gauss' theorem, as was done in Chap. 5 to relate Eqs.(5.34) to (5.35), allows us to express the required solution, $\Phi^{(2)}_{1a\omega}(\vec{R})$, as

$$\Phi^{(2)}_{1a\omega}(\vec{R}_0) = \int_V \rho(\vec{R})\hat{G}^{(2)}_a(\vec{R}, \vec{R}_0)d\vec{R} + \frac{1}{4\pi}\int_S \vec{n}\cdot\vec{P}\left[\hat{G}^{(2)}_a(\vec{R}, \vec{R}_0), \Phi^{(2)}_{1a\omega}(\vec{R})\right]d\vec{R} \tag{K.21}$$

Similarly, multiplying Eq.(K.19) by $\Phi^{(2)}_{1b\omega}$ and $L\Phi^{(2)}_{1b\omega}(\vec{R}) = 0$ by $\hat{G}^{(2)}_b(\vec{R}, \vec{R}_0)$ leads to

$$\Phi^{(2)}_{1b\omega}(\vec{R}_0) = \frac{1}{4\pi}\int_S \vec{n}\cdot\vec{P}\left[\hat{G}^{(2)}_b(\vec{R}, \vec{R}_0), \Phi^{(2)}_{1b\omega}(\vec{R})\right]d\vec{R} \tag{K.22}$$

In the above,

$$\rho(\vec{R}) = -\frac{1}{4\pi}\bar{S}_a(\omega, \vec{r}, 0) \tag{K.23}$$

and

$$\vec{P}(u,v) = u\nabla v - v\nabla u - 2auv\vec{z}_0$$

is the bilinear concomitant vector, where $\vec{z}_0$ is the unit vector in the z direction.

We first derive the Green's function for the volume source. From (K.20) it is seen that to establish the solution,

$$\Phi^{(2)}_{1a\omega}(\vec{R}) = \int_V \rho(\vec{R}_0)G^{(2)}_a(\vec{R}, \vec{R}_0)d\vec{R}_0 \tag{K.24}$$

requires that the function, $\hat{G}^{(2)}_a$, must satisfy the boundary condition

$$\vec{n}\cdot\vec{P}[\hat{G}^{(2)}_a(\vec{R}, \vec{R}_0), \Phi^{(2)}_{1a\omega}(\vec{R})] = 0$$

which is equivalent to two conditions being met

$$\hat{G}^{(2)}_a(\vec{R}, \vec{R}_0)\frac{\partial}{\partial z}\Phi^{(2)}_{1a\omega}(\vec{R}) - \Phi^{(2)}_{1a\omega}(\vec{R})\frac{\partial}{\partial z}\hat{G}^{(2)}_a(\vec{R}, \vec{R}_0) -$$
$$-2a\hat{G}^{(2)}_a(\vec{R}, \vec{R}_0)\Phi^{(2)}_{1a\omega}(\vec{R}) = 0 \text{ at } z = 0 \tag{K.25}$$

and

$$\hat{G}^{(2)}_a(\vec{R}, \vec{R}_0)\frac{\partial}{\partial z}\Phi^{(2)}_{1a\omega}(\vec{R}) - \Phi^{(2)}_{1a\omega}(\vec{R})\frac{\partial}{\partial z}\hat{G}^{(2)}_a(\vec{R}, \vec{R}_0) +$$
$$+2a\hat{G}^{(2)}_a(\vec{R}, \vec{R}_0)\Phi^{(2)}_{1a\omega}(\vec{R}) = 0 \text{ at } z = -H_1 \tag{K.26}$$

As in Appendix I, we replace the condition (K.26) by introducing the plane-wave reflection coefficient, R_b, and apply (K.10) to rewrite condition (K.25) in the form,

$$\frac{\partial}{\partial z}\hat{G}_a^{(2)}(\vec{R},\vec{R}_0) + (2a - \frac{\omega^2}{g})\hat{G}_a^{(2)}(\vec{R},\vec{R}_0) = 0 \tag{K.27}$$

Performing the Hankel transforms

$$\hat{G}_a^{(2)}(\vec{R},\vec{R}_0) = \int_0^\infty \hat{g}_\omega^{(2)}(k,z,z_0) J_o(k\rho) k\, dk$$

and

$$G_a^{(2)}(\vec{R},\vec{R}_0) = \int_0^\infty g_\omega^{(2)}(k,z,z_0) J_o(k\rho) k\, dk \tag{K.28}$$

where $\rho \equiv |\vec{r}-\vec{r}_0|$, $\vec{R} = (z,\vec{r})$, $\vec{R}_0 = (z_0,\vec{r}_0)$ and applying the relation

$$\delta(\vec{R}-\vec{R}_0) = \frac{1}{2\pi}\delta(z-z_0) \int_0^\infty J_0(k\rho) k\, dk \tag{K.29}$$

allows us to establish the equation for $\hat{g}_\omega^{(2)}(k,z,z_0)$,

$$\left(\frac{d^2}{dz^2} + 2a\frac{d}{dz} + \gamma_1^2\right)\hat{g}_\omega^{(2)}(k,z,z_0) = -2\delta(z-z_0) \tag{K.30}$$

and the surface condition,

$$\left(\frac{d}{dz} - \frac{\omega^2}{g} + 2a\right)\hat{g}_\omega^{(2)}(k,z,z_0) = 0 \quad \text{at} \quad z=0. \tag{K.31}$$

As in the case of Eq.(I.17), Eq.(K.30) can be replaced by a homogeneous equation

$$\left(\frac{d^2}{dz^2} + 2a\frac{d}{dz} + \gamma_1^2\right)\hat{g}_\omega^{(2)}(k,z,z_0) = 0 \tag{K.32}$$

by the introduction of two additional continuity conditions at the point $z = z_0$, given by

$$\hat{g}_\omega^{(2)}(k,z,z_0)\Big|_{z_0-0}^{z_0+0} = 0 \quad \text{and} \quad \frac{d}{dz}\hat{g}_\omega^{(2)}(k,z,z_0)\Big|_{z_0-0}^{z_0+0} = -2 \tag{K.33}$$

As in Appendix J, we write the solution in the form

$$\hat{g}_\omega^{(2)}(k,z,z_0) = \begin{cases} A\left[e^{s_+(z-z_0)} + Be^{s_-(z-z_0)}\right] & \text{for } z_0 \le z \le 0 \\ C\left[e^{s_+(z+H_1)} + R_b e^{s_-(z+H_1)}\right] & \text{for } -H_1 \le z < z_0 \end{cases} \tag{K.34}$$

Solution of Second Order Potential

where
$$s_\pm = -a \mp \sqrt{a^2 - \gamma_1^2} = \gamma_\pm - 2a = -\gamma_\mp \tag{K.35}$$
and
$$\gamma_\pm = a \mp \sqrt{a^2 - \gamma_1^2} \tag{K.36}$$

Substituting (K.34) into conditions (K.31) and (K.33) leads to

$$\left(s_+ + 2a - \frac{\omega^2}{g}\right) e^{-s_+ z_0} + B\left(s_- + 2a - \frac{\omega^2}{g}\right) e^{-s_- z_0} = 0 \tag{K.37}$$

$$A(s_+ + Bs_-) - C\left(s_+ e^{s_+(z_0+H_1)} + R_b s_- e^{s_-(z_0+H_1)}\right) = -2 \tag{K.38}$$

$$A(1+B) - C\left(e^{s_+(z_0+H_1)} + R_b e^{s_-(z_0+H_1)}\right) = 0 \tag{K.39}$$

It is to be noted that Eqs.(K.37) to (K.39) are identical to the corresponding equations Eqs.(J.3) in Appendix J, except that in Eq.(K.37) the term $2a - \omega^2/g$ is replaced by the term $2a$. Accordingly in this case the coefficient B and the determinant of the coefficient of Eqs.(K.38) and (K.39) become

$$B = -\frac{\gamma_+ - \omega^2/g}{\gamma_- - \omega^2/g} e^{-(\gamma_+ - \gamma_-)z_0} = -\frac{b}{\epsilon} e^{2\sqrt{a^2 - \gamma_1^2} z_0} \tag{K.40}$$

where
$$b = \frac{1 - \omega^2/g\gamma_+}{1 - \omega^2/g\gamma_-} \tag{K.41}$$

and
$$\Delta = \gamma_+ \left(1 - \frac{1}{\epsilon}\right) \left[be^{\gamma_+ H_1} + \epsilon R_b e^{\gamma_- H_1}\right] e^{\gamma_- z_0 - 2a(z_0+H_1)}$$

This leads to the solutions:
for $z_0 < z \le 0$,

$$\hat{g}^{(2)}_\omega(k, z, z_0) = \frac{2 \left[e^{\gamma_+(z_0+H_1)} + R_b e^{\gamma_-(z_0+H_1)}\right] \left(e^{\gamma_+ z} - \frac{b}{\epsilon} e^{\gamma_- z}\right) e^{-2az}}{\gamma_+ \left(1 - \frac{1}{\epsilon}\right) \left[be^{\gamma_+ H_1} + \epsilon R_b e^{\gamma_- H_1}\right]}, \tag{K.42}$$

and for $-H_1 \le z < z_0$,

$$\hat{g}^{(2)}_\omega(k, z, z_0) = \frac{2 \left[e^{\gamma_+(z+H_1)} + R_b e^{\gamma_-(z+H_1)}\right] \left(e^{\gamma_+ z_0} - \frac{b}{\epsilon} e^{\gamma_- z_0}\right) e^{-2az}}{\gamma_+ \left(1 - \frac{1}{\epsilon}\right) \left[be^{\gamma_+ H_1} + \epsilon R_b e^{\gamma_- H_1}\right]} \tag{K.43}$$

According to the reciprocal relation, (K.20),

$$g^{(2)}_\omega(k, z_0, z)|_{z<z_0} = \hat{g}^{(2)}_\omega(k, z, z_0)|_{z_0<z}$$

and the required functions become:
for $z_0 < z \leq 0$,

$$g^{(2)}_\omega(k, z, z_0) = \frac{2\left[e^{\gamma_+(z_0+H_1)} + R_b e^{\gamma_-(z_0+H_1)}\right]\left(e^{\gamma_+ z} - \frac{b}{\epsilon}e^{\gamma_- z}\right)e^{-2az_0}}{\gamma_+\left(1 - \frac{1}{\epsilon}\right)\left[be^{\gamma_+ H_1} + \epsilon R_b e^{\gamma_- H_1}\right]} \quad \text{(K.44)}$$

and for $-H_1 \leq z < z_0$,

$$g^{(2)}_\omega(k, z, z_0) = \frac{2\left[e^{\gamma_+(z+H_1)} + R_b e^{\gamma_-(z+H_1)}\right]\left(e^{\gamma_+ z_0} - \frac{b}{\epsilon}e^{\gamma_- z_0}\right)e^{-2az_0}}{\gamma_+\left(1 - \frac{1}{\epsilon}\right)\left[be^{\gamma_+ H_1} + \epsilon R_b e^{\gamma_- H_1}\right]} \quad \text{(K.45)}$$

Substituting solutions (K.44) and (K.45) into (K.28) via (K.24) leads finally to

$$\Phi^{(2)}_{1a\omega}(\vec{R}) = -\frac{1}{4\pi}\int_S\left\{\int_0^\infty\left[\int_{-H_1}^0 \bar{S}_a(\omega, \vec{r}_0, z_0)g^{(2)}_\omega(k, z, z_0)\,dz_0\right]J_0(k\rho)k\,dk\right\}d\vec{r}_0 \quad \text{(K.46)}$$

The Green's function for the surface source, $G^{(2)}_b(\vec{R}, \vec{R}_0)$, can be established in a similar way. Writing expression (K.22) in the form

$$\Phi^{(2)}_{1b\omega}(\vec{R}_0) = \frac{1}{4\pi}\int_S\left\{\hat{G}^{(2)}_b(\vec{R}, \vec{R}_0)\left(\frac{\partial}{\partial z} - \frac{\omega^2}{g}\right)\Phi^{(2)}_{1b\omega}(\vec{R}) \right.$$

$$\left. - \Phi^{(2)}_{1b\omega}(\vec{R})\left[\frac{\partial}{\partial z} + (2a - \frac{\omega^2}{g})\right]\hat{G}^{(2)}_b(\vec{R}, \vec{R}_0)\right\}d\vec{R} \quad \text{(K.47)}$$

and referring to the surface condition, (K.13), we see that the required Green's function should satisfy the relation

$$\left[\frac{\partial}{\partial z} + (2a - \frac{\omega^2}{g})\right]\hat{G}^{(2)}_b(\vec{R}, \vec{R}_0) = 0 \quad \text{(K.48)}$$

which is the same as that applying to $\hat{G}^{(2)}_a$, Eq.(K.27). Therefore we can establish the corresponding function, $g^{(2)}_\omega$, from Eq.(K.45) by letting $z_0 = 0$, i.e.,

$$g^{(2)}_{\omega b}(k, z, 0) = \frac{2\left[e^{\gamma_+(z+H_1)} + R_b e^{\gamma_-(z+H_1)}\right]}{\gamma_+\left[e^{\gamma_+ H_1} + \frac{\epsilon}{b}R_b e^{\gamma_- H_1}\right](1 - \omega^2/g\gamma_+)} \quad \text{for} \quad -H_1 \leq z < 0$$

(K.49)

Solution of Second Order Potential

and

$$G_b^{(2)}(\vec{R}, \vec{R}_0) = \int_0^\infty g_{\omega b}^{(2)}(k, z, z_0) J_o(k\rho) k \, dk$$

This leads to the solution

$$\Phi_{1b\omega}^{(2)}(\vec{R}) = \frac{1}{4\pi} \int_S G_b^{(2)}(\vec{R}, \vec{R}_0)(\frac{\partial}{\partial z_0} - \frac{\omega^2}{g}) \Phi_{1b\omega}^{(2)}(\vec{R}_0) d\vec{R}_0 \qquad (K.50)$$

which can be expressed by using condition (K.13) as

$$\Phi_{1b\omega}^{(2)}(\vec{R}) = \frac{1}{4\pi g} \int_S \left\{ \int_0^\infty \bar{S}_b(\omega, \vec{r}_0, z_0) g_{\omega b}^{(2)}(k, z, z_0) J_0(k\rho) k \, dk \right\} d\vec{r}_0 \qquad (K.51)$$

when $\hat{G}_b^{(2)}(\vec{R}_0, \vec{R})$ in (K.47) is replaced by $G_b^{(2)}(\vec{R}, \vec{R}_0)$ and the condition (K.48) is applied.

Appendix L

Concerning the Source Pressure Spectrum of the Wave-Wave Interaction Process in a Shallow-Water Environment

From the text of Chap.5 (Eqs.(5.42), (5.43), and (5.32)) we have

$$\nabla \tilde{\Phi}_1^{(1)}(\vec{R},t) = \frac{1}{(2\pi)^2} \cdot$$
$$\int A_0^{(1)}(q)\zeta^{(1)}(\vec{q})[(i\vec{q},\gamma_+)e^{\gamma_+(z+H_1)} + (i\vec{q},\gamma_-)R_b e^{\gamma_-(z+H_1)}]e^{i(\vec{q}\vec{r}-\sigma t)}d\vec{q}$$

$$[\nabla\tilde{\Phi}^{(1)}]^2\big|_{z=0} = \frac{1}{(2\pi)^4} \int\int A_{01}^{(1)}(\vec{q}_1)A_{02}^{(1)}(\vec{q}_2)\zeta^{(1)}(\vec{q}_1)\zeta^{(1)}(\vec{q}_2) \cdot$$

$$\cdot \{(\gamma_{+1}\gamma_{+2} - \vec{q}_1\vec{q}_2)e^{(\gamma_{+1}+\gamma_{+2})H_1} + R_{b1}R_{b2}(\gamma_{-1}\gamma_{-2} - \vec{q}_1\vec{q}_2)e^{(\gamma_{-1}+\gamma_{-2})H_1}$$

$$+ R_{b2}(\gamma_{+1}\gamma_{-2} - \vec{q}_1\vec{q}_2)e^{(\gamma_{+1}+\gamma_{-2})H_1} + R_{b1}(\gamma_{-1}\gamma_{+2} - \vec{q}_1\vec{q}_2)e^{(\gamma_{-1}+\gamma_{+2})H_1}\}$$

$$e^{i(\vec{q}_1+\vec{q}_2)\vec{r}-i(\sigma_1+\sigma_2)t}d\vec{q}_1\,d\vec{q}_2 \tag{L.1}$$

$$\bar{S}_b^{(2)}(\omega,\vec{r},0) = -\int\left\{\frac{\partial}{\partial t}[\nabla\tilde{\Phi}^{(1)}]^2\Big|_{z=0}\right\}e^{i\omega t}dt$$

$$= \frac{i}{(2\pi)^4}\int\int(\sigma_1+\sigma_2)A_{01}^{(1)}(q_1)A_{02}^{(1)}(q_2)\zeta^{(1)}(\vec{q}_1)\zeta^{(1)}(\vec{q}_2)\cdot$$

$$\cdot\{q_1,q_2\}e^{i(\vec{q}_1+\vec{q}_2)\vec{r}}\delta(\omega-\sigma_1-\sigma_2)d\vec{q}_1\,d\vec{q}_2 \tag{L.2}$$

where $\{q_1, q_2\}$ represents the terms in $\{..\}$ in (L.1). If, as an approximation, we set

$$\gamma_\pm = a \mp i\sqrt{\frac{\sigma^2}{\alpha_1^2} - q^2} \cong \pm q \qquad (L.3)$$

then

$$\bar{S}_b^{(2)}(\omega, \vec{r}, 0) = \frac{i}{(2\pi)^4} \int\int (\sigma_1 + \sigma_2) A_{01}^{(1)}(q_1) A_{02}^{(1)}(q_2) \zeta^{(1)}(\vec{q}_1)\zeta^{(1)}(\vec{q}_2) q_1 q_2 \cdot$$

$$\cdot \{C_-[e^{(q_1+q_2)H_1} + R_{b1}R_{b2}e^{-(q_1+q_2)H_1}] - C_+[R_{b2}e^{(q_1-q_2)H_1} + R_{b1}e^{-(q_1-q_2)H_1}]\} \cdot$$

$$\cdot \delta(\omega - \sigma_1 - \sigma_2) e^{i(\vec{q}_1+\vec{q}_2)\vec{r}} d\vec{q}_1 \, d\vec{q}_2 \qquad (L.4)$$

where $C_+ = (1 + \cos\theta_{12})$, $C_- = (1 - \cos\theta_{12})$, $R_{b1,2} = R_b(\omega, q_{1,2})$, and $\cos\theta_{12} = \vec{q}_1 \cdot \vec{q}_2 / q_1 q_2$.

Under this approximation it also follows that

$$A_{01}^{(1)}(q_1) A_{02}^{(1)}(q_2) = -\frac{\sigma_1 \sigma_2}{q_1 q_2} \frac{e^{-(q_1+q_2)H_1}}{(1 - R_{b1}e^{-2q_1 H_1})(1 - R_{b2}e^{-2q_2 H_1})} \qquad (L.5)$$

so that finally

$$\bar{S}_b^{(2)}(\omega, \vec{r}, 0) = \frac{-i}{(2\pi)^4} \int\int (\sigma_1 + \sigma_2) \sigma_1 \sigma_2 \cdot$$

$$\cdot \left\{ \frac{C_-[1 + R_{b1}R_{b2}e^{-2(q_1+q_2)H_1}] - C_+[R_{b2}e^{-2q_2 H_1} + R_{b1}e^{-2q_1 H_1}]}{(1 - R_{b1}e^{-2q_1 H_1})(1 - R_{b2}e^{-2q_2 H_1})} \right\} \cdot$$

$$\cdot \zeta^{(1)}(\vec{q}_1)\zeta^{(1)}(\vec{q}_2) \delta(\omega - \sigma_1 - \sigma_2) e^{i(\vec{q}_1+\vec{q}_2)\vec{r}} d\vec{q}_1 \, d\vec{q}_2 \qquad (L.6)$$

Appendix M

The Geometric Description of Wave-Wave Interactions and the Calculation of the Source Spectrum $\langle |B^{(2)}(\vec{k},\omega)|^2 \rangle$

M.1 Geometric description of wave-wave interactions

In an earlier contribution we presented a geometric description of the interaction (sum-frequency interaction) of two surface wave trains and the resulting pressure field in a three-dimensional ω-$\vec{k}$ space[38, 40]. To aid the understanding of the developments given in Chap.6 the salient features of that analysis are reproduced here.

In the deep ocean the dispersion relation takes the well known form, $\sigma_i^2 = gq_i$, and the wave-interaction conditions become (Eq.(6.7))

$$\left.\begin{aligned} \sigma_1 + \sigma_2 &= \omega \\ q_{1x} + q_{2x} &= k_x \\ q_{1y} + q_{2y} &= k_y \\ \sigma_1^2 &= gq_1 \\ \sigma_2^2 &= gq_2 \\ \tan^{-1}(q_{1y}/q_{1x}) &= \theta_1 \end{aligned}\right\}$$

The key aspects of the interaction process are summarised in Figs. M.1a to M.1d and M.2. To appreciate these figures it is helpful to first recognize the significance of the cone OBCDEB centered on the ω axis — see Fig. M.1a.

Any point within the surface of this cone defines an ordinary plane acoustic wave satisfying the dispersion relation

$$|\vec{k}| = \epsilon \omega/\alpha_1, \quad 0 \leq \epsilon \leq 1.$$

The cone is a representation of the relation $k = \omega/\alpha_1$ when $\epsilon = 1$.

The two horn-like surfaces whose apexes O and O' are also centered on the ω axis in Fig. M.1a, represent, respectively, the rotation of the two dispersion curves describing the relations $\sigma_1 = \sqrt{gq_1}$ and $\sigma_2 = \sqrt{gq_2}$ for the case $\vec{k} = 0$, with the additional constraint $\sigma_1 + \sigma_2 = \omega$ also applied. (If $\vec{q}_1 + \vec{q}_2 = \vec{k} \neq 0$ then O and O' do not both lie on the ω axis. The more general case of $k \neq 0$ is presented in Fig. M.1b and is discussed below.) The intersection of these two rotational surfaces is indicated by the spatial curve L, the projection of which on the $\vec{k}$ plane is the curve L'. Corresponding to each point on L', say G' in the lower part of Fig. M.1a, there are two diametrically opposed vectors, $\vec{q}_1$ and $\vec{q}_2$, which add to give $\vec{k} = 0$.

In the more general case where $\vec{k} \neq 0$, the induced acoustic wave is propagating in a direction other than normal to the surface of the sea. We first consider the case of a homogeneous wave (see Fig. M.1b), for which the end point of $\vec{k}$ lies inside the surface of the cone defined by $\omega = |\vec{k}|/\alpha_1$. In this situation the apex of the upper horn is displaced as shown and the projection of the intersection curve L' in the $\vec{k}$ plane is slightly different from a circle. The two interacting vectors $\vec{q}_1, \vec{q}_2$ associated with point G', now differ slightly in magnitude and direction.

When $|\vec{k}|$ equals the acoustic wavenumber ω/α_1, the induced pressure field becomes a travelling plane wave propagating horizontally. For $|\vec{k}| > \omega/\alpha_1$ the acoustic response degenerates to an inhomogeneous wave decaying with increasing depth (distance from the surface). The intersection curve L and its projection L' gradually change from the dumbbell-like shape of Fig. M.1c to the two closed curves, centered on O and O', presented in Fig. M.1d. In both situations it is clear that for each point G' on L', up to three pairs of interacting vectors $\vec{q}_1, \vec{q}_2$ can now lead to the resultant $\vec{k}$. This is indicated in exaggerated form in Fig. M.1d. For a given $\vec{k}$ and a defined direction for $\vec{q}_1$ the three possible pairs of vectors

The Source Spectrum

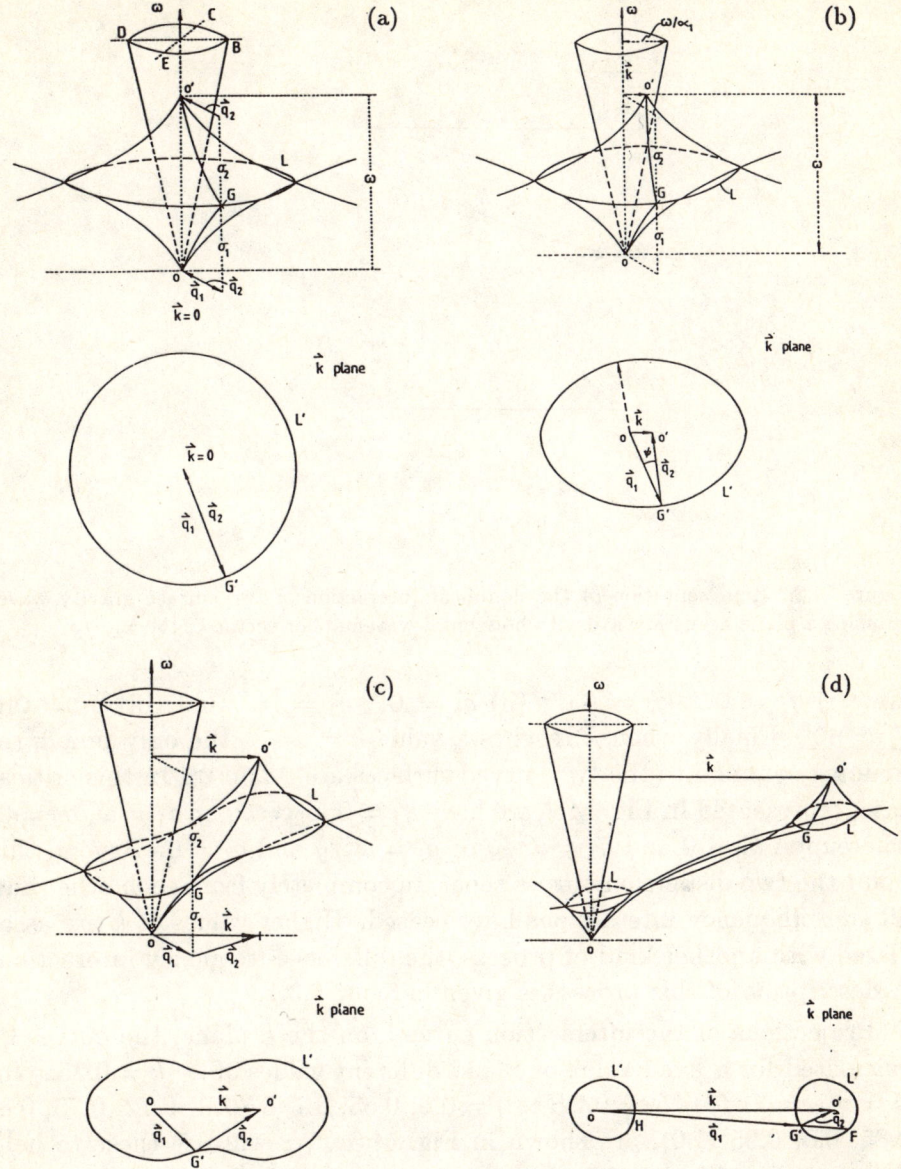

Figure M.1: Representation of the nonlinear interaction of two surface gravity waves inducing a plane acoustic wave with horizontal wavenumber vector $\vec{k}$: **a** for $|\vec{k}| = 0$; **b** for $0 \leq |\vec{k}| \leq \omega/\alpha_1$; **c** $\omega/\alpha_1 \leq |\vec{k}| < \omega^2/g$; **d** for $|\vec{k}| \geq \omega/\alpha_1$.

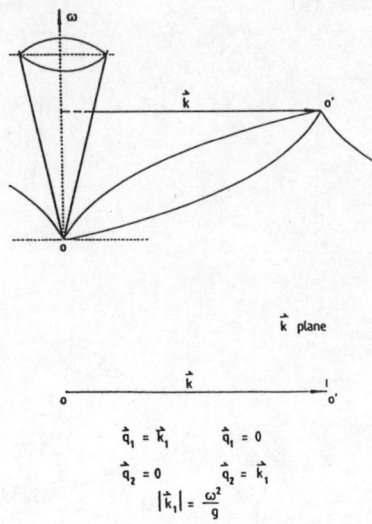

Figure M.2: Representation of the nonlinear interaction of two surface gravity waves inducing a plane acoustic wave with horizontal wavenumber vector $\vec{k}$: $|\vec{k}| = \omega^2/g$.

are: (i) $\vec{q}_1 = \text{OG}'$, $\vec{q}_2 = \text{G}'\text{O}'$; (ii) $\vec{q}_1 = \text{OF}$, $\vec{q}_2 = \text{FO}'$; and (iii) $\vec{q}_1 = \text{OH}$, $\vec{q}_2 = \text{HO}'$. Finally when $\vec{k}$ reaches a value $k = \omega^2/g$, the only points remaining common to the two curved surfaces are O and O'. In this critical case, represented in Fig. M.2, we have $\sigma_1 = 0$, $\sigma_2 = \omega$, or $\sigma_1 = \omega$, $\sigma_2 = 0$, whereupon $\vec{q}_1 = 0$ and $\vec{q}_2 = \omega^2/g$ or $\vec{q}_1 = \omega^2/g$ and $\vec{q}_2 = 0$. Beyond this point the two dispersion curves separate completely from each other and all sum-frequency interactions have ceased. Higher values of k are associated with another kind of process, the difference-frequency interaction. A description of this process is given in Sect. 6.3.1.

Projections of the intersection curve L on the $\vec{k}$ plane, the curves L', calculated for a fixed value of k and different values of ω ($k = 0.25\omega_0^2/g$, $\omega = \eta\omega_0 = 2\pi f_0 \eta$, $f_0 = 0.2 Hz$, $\eta = 0.6, 0.65, 0.7, 0.7071, 0.72, 0.75, 0.8, 0.85, 0.9, 0.95, 1.0$), are shown in Fig. M.3. We will use these to help define the qualitative descriptions given above. Before doing so, however, we note that Fig.M.3 is similar to the representation used in the oceanographic literature[56, 59, 83] to describe the resonant interaction of two gravity waves. It has been shown there that the second-order interactions of two gravity-wave trains cannot yield another gravity wave, but that resonance can occur in the third-order (four wave) interaction

The Source Spectrum

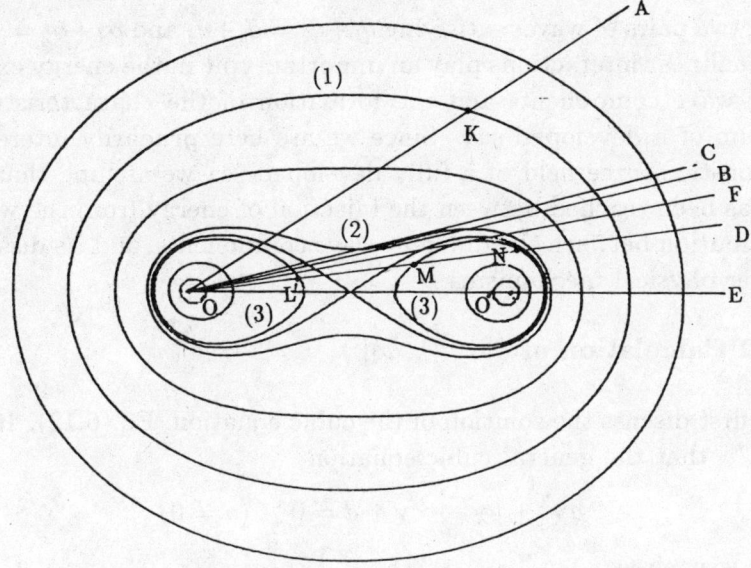

Figure M.3: Curves L' on the $\vec{k}$-plane for different values of $\hat{m}$

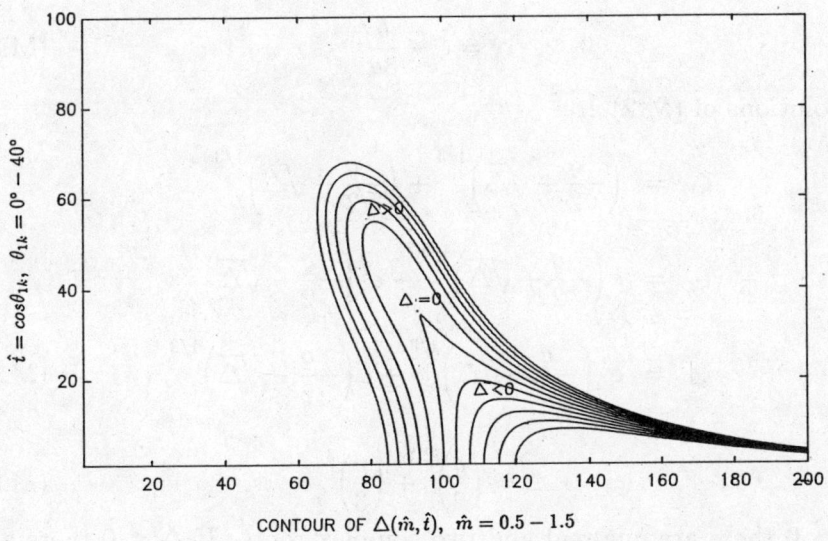

Figure M.4: Contour plot of the values of Δ

among two pairs of waves satisfying $\vec{q}_1+\vec{q}_2 = \vec{q}_3+\vec{q}_4$ and $\sigma_1+\sigma_2 = \sigma_3+\sigma_4$. Such nonlinear interactions play an important role in the energy exchange among wave components and the formation of the characteristic wave spectrum of a developed sea. Since we are here primarily interested in the acoustic source field of a fully developed sea we assume that a balance has been reached between the injection of energy from the wind, its redistribution between the different wave components, and its dissipation by other physical mechanisms.

M.2 Calculation of $\langle |B^{(2)}(\vec{k},\omega)|^2 \rangle$

We first discuss the solution of the cubic equation, Eq.(6.12). It is well known[86] that the general cubic equation

$$a\chi^3 + b\chi^2 + c\chi + d = 0 \quad (a \neq 0) \tag{M.1}$$

can be written as

$$\zeta^3 + p\zeta + q = 0 \tag{M.2}$$

through the transformation

$$\chi = \zeta - \frac{b}{3a} \tag{M.3}$$

The solutions of (M.2) are

$$\begin{aligned}
\zeta_1 &= \left(-\frac{q}{2}+\sqrt{\Delta}\right)^{1/3} + \left(-\frac{q}{2}-\sqrt{\Delta}\right)^{1/3} \\
\zeta_2 &= \epsilon\left(-\frac{q}{2}+\sqrt{\Delta}\right)^{1/3} + \epsilon^2\left(-\frac{q}{2}-\sqrt{\Delta}\right)^{1/3} \\
\zeta_3 &= \epsilon^2\left(-\frac{q}{2}+\sqrt{\Delta}\right)^{1/3} + \epsilon\left(-\frac{q}{2}-\sqrt{\Delta}\right)^{1/3}
\end{aligned} \tag{M.4}$$

with

$$\Delta = \left(\frac{q}{2}\right)^2 + \left(\frac{p}{3}\right)^3 \tag{M.5}$$

If $\Delta > 0$ there are one real and two complex roots. If $\Delta \leq 0$ there are three real roots.

For Eq.(6.12) it follows that

$$a = \pm\left(1 - \frac{\hat{m}^2}{4}\right)$$

$$b = -\left(\frac{3}{4}\hat{m}^2 + 1\right)$$

$$c = \pm\left(1 - \frac{3}{4}\hat{m}^2 + \hat{m}\hat{t}\right)$$

$$d = \left(\hat{m}\hat{t} - \frac{1}{4}\hat{m}^2 - 1\right)$$

$$p = (3a\rho^2 - 2b\rho + c)/a$$

$$q = (b\rho^2 - a\rho^3 - c\rho + d)/a$$

$$\rho = b/(3a)$$

where the signs '+' and '−' indicate respectively the sum- and difference-frequency interactions.

We first consider the sum-frequency interactions. Figure M.4 presents contours of the value of Δ as a function of $\hat{m}$ and $\hat{t}$ around the value $\Delta \sim 0$. We note that the common point of the two branches of the curve, $\Delta = 0$, is located around $\hat{m} = 0.96$. This confirms our previous prediction[38] that when $\hat{m}$ is less than 0.958 the ratio χ has a single root. This situation is represented in Fig. M.3 by the point K on the smooth single loop, curve (1). When $\hat{m}$ increases to lie between 0.96 and 1 there is a range of angles, θ_{1k}, in which three roots of χ can be found for a given $\hat{m}$. These are exemplified by the two radials OB and OC crossing L'(2) in Fig. M.3. Radials outside this interval intersect L' only once. With a further increase in $\hat{m}$ to a value greater than 1, L' splits into two separate loops, exemplified by curve (3) in Fig. M.3, and a radial such as OD in the plot, lying between OO'E($\theta_{1k} = 0$) and OF(θ_{1k})(the tangent to the right-hand-side loop of curve (3)) again crosses L'(3) at three points L, M, and N, resulting in three roots of χ. The angle θ_{1k} is bounded by a critical value, which is about 13.75° in this case.

We now turn to the calculation of the integral in Eq.(6.18). For convenience Eq.(6.18) can be written in the form

$$\langle |B^{(2)}(\vec{k},\omega)|^2\rangle = \frac{g^2}{2\omega V(k,\omega)}I(k,\omega) \qquad (M.6)$$

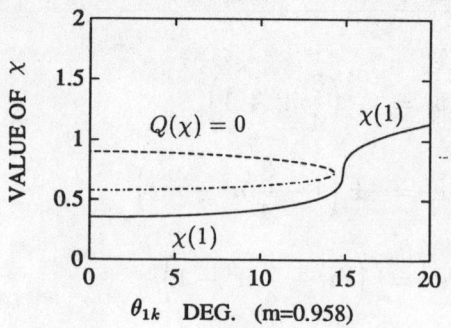

Figure M.5: The values of the roots of the cubic equation and $Q(\chi)$ as a function of θ_{1k} when $\hat{m} \simeq 0.958$.

where

$$I(k,\omega) = \sum_{i=1}^{3} \int_{\theta_{di}}^{\theta_{ui}} \frac{P(\chi^{(i)})}{Q(\chi^{(i)})} d\theta_{1k} \tag{M.7}$$

and

$$P(\chi) = \frac{(1-\cos\theta_{12})^2 \chi^2}{(1+\chi)^2} F_a\left(\frac{\omega}{1+\chi}\right) F_a\left(\frac{\omega\chi}{1+\chi}\right) H(\theta_1) H(\theta_2) \tag{M.8}$$

$$Q(\chi) = \left[\chi^2 - \left(1 + \frac{\hat{m}}{2}\hat{t}\right)\chi + \left(1 - \frac{\hat{m}}{2}\hat{t}\right)\right] \tag{M.9}$$

It is clear that the term in the denominator, $Q(\chi)$, can become zero at some values of χ when $\hat{m}$ is greater than the critical value $\hat{m}_c$. As was shown earlier, in this range three values of χ can exist for a given angle, θ_{1k} (there is only one value of χ when $\hat{m} < \hat{m}_c$). In Figs. M.5 to M.7a and b the value of all possible roots, $\chi^{(1)}$ to $\chi^{(3)}$, are plotted as a function of the angle, θ_{1k}, together with the location of the roots of $Q(\chi) = 0$, for values of $\hat{m}$ ranging from 0.958 to 1.5. It can be seen that when $\hat{m}$ is equal to (and less than) 0.958 (see Fig. M.5), there is only one root, $\chi^{(1)}$, and there is no point common to $Q(\chi) = 0$ and $\chi^{(1)}$.

When $\hat{m}_c < \hat{m} < 1$, the three branches of the solution, χ, form a continuous curve. The range of angles over which three roots exist lies between points A and B, say 11° to 13°, for the case $\hat{m} = 0.98$ (see Fig. M.6a). In this particular case there are two singularities along the continuous curve where $Q(\chi) = 0$; one at the common point of $\chi^{(1)}$ and $\chi^{(3)}$ (point A) and the other at that of $\chi^{(3)}$ and $\chi^{(2)}$ (point B). In Fig. M.3 this situation is represented by curve (2). As $\hat{m}$ increases further up to 1, the angle, θ_{1k}, associated with point A decreases to zero(Fig. M.6b and d). Beyond $\hat{m} = 1.0$ the branch $\chi^{(1)}$ separates from $\chi^{(3)}$, while the branch

The Source Spectrum

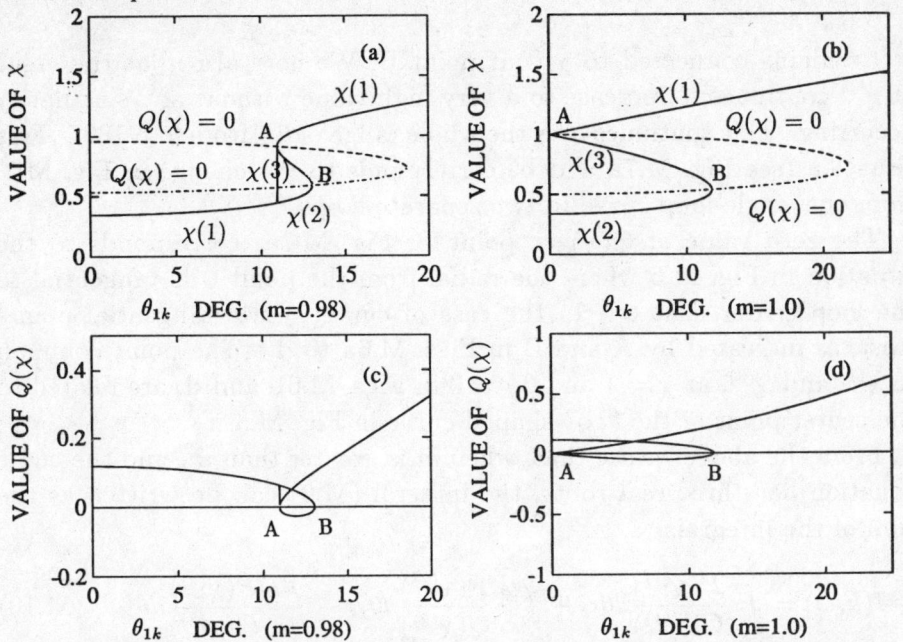

Figure M.6: The values of the roots of the cubic equation and $Q(\chi)$ as a function of θ_{1k} when $\hat{m}=0.98$ (**a** and **c**) and $\hat{m}=1.0$ (**b** and **d**).

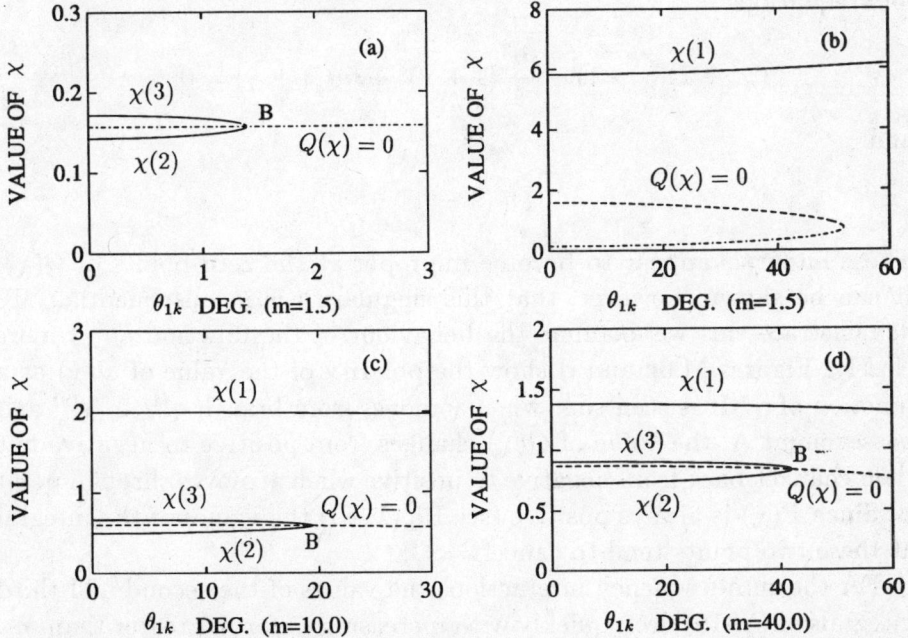

Figure M.7: The values of the roots of the cubic equation χ as a function of θ_{1k} when $\hat{m}=1.5$ (**a** and **b**), and $\hat{m}=10.0$ (**c**) and $\hat{m}=40.0$ (**d**).

$\chi^{(2)}$ remains connected to $\chi^{(3)}$ at point B. We note also that the value of $\chi^{(1)}$ continues to increase to a very high value without any singularity occurring. This continues over the whole range of θ_{1k} from 0 to 180°. This situation (see Fig. M.7a and b) corresponds to the change in Fig. M.3, from one single-loop curve to two separate loops.

The zero value of $Q(\chi)$ at point B, Fig. M.7a, corresponds to the situation in Fig. M.3 where the radial from the point O is tangential to the loop centered at O'. In the case of curve(2) two tangential points exist, as indicated by A and B in Figs. M.6a to d. The point common to $\chi^{(1)}$ and $\chi^{(3)}$, at $\chi = 1$ and $\theta_{1k} = 0$ in Figs. M.6b and d, are related to the center point of the "∞" shaped curve in Fig. M.3.

From the above we see that, when $\hat{m}$ is greater than $\hat{m}_c$ and the cubic equation has three real roots, the integral (M.9) can be written as the sum of the integrals,

$$I(k,\omega) = \int_0^\pi \frac{P(\chi^{(1)})}{Q(\chi^{(1)})} d\theta_{1k} + \int_{\theta_A}^{\theta_B} \frac{P(\chi^{(2)})}{Q(\chi^{(2)})} d\theta_{1k} + \int_{\theta_A}^{\theta_B} \frac{P(\chi^{(3)})}{Q(\chi^{(3)})} d\theta_{1k} \quad (M.10)$$

where θ_A and θ_B are found by eliminating the variable χ from the two joint equations

$$(\chi^2 + 1)(\chi - 1) - \frac{\hat{m}^2}{4}(1 + \chi)^3 + \hat{m}\hat{t}(1 + \chi) = 0$$

and

$$Q(\chi) = 0$$

These integrals appear to become improper at the zero points of $Q(\chi)$. It can be shown, however, that this singularity is inconsequential. To demonstrate this we examine the behaviour of the function $Q(\chi)$ more closely. Figures M.6c and d show the polarity of the value of $Q(\chi)$ as a function of χ. It is seen that when χ moves from branch $\chi^{(1)}$ to $\chi^{(3)}$ and passes point A, the value of $Q(\chi)$ changes from positive to negative but then changes back from negative to positive when χ moves through point B. Since $P(\chi)$ is always positive (see Eq.(M.8)) the values of the integral at these two points tend to cancel.

For the sum-frequency interactions the values of the second and third integrals of (M.10) drop quickly with increasing $\hat{m}$ for $\hat{m}$ greater than $\hat{m}_c$. Figures M.6b and d and M.7a and b show that with increasing $\hat{m}$ the values of $\chi^{(2)}$ and $\chi^{(3)}$ both decrease while that of $\chi^{(1)}$ increases. In both cases therefore the frequency difference of the two interacting waves is increasing. Since the surface-wave spectrum, $F_a(\sigma)$, decays as σ^{-5} for $\sigma > \sigma_0$, where σ_0 is the peak frequency, the product $F_a(\sigma_1)F_a(\sigma_2)$ drops quickly as $|\sigma_2 - \sigma_1|$ increases, and so too does the integral $I(k,\omega)$. Furthermore, when $\hat{m}$ increases the range of the angle, θ_{1k}, over which branches $\chi^{(2)}$ and $\chi^{(3)}$ exist, shrinks quickly to $\theta_{1k} \cong 0$. This effect also reduces the importance of the value of the integration upon θ_{1k}. Therefore, for the sum-frequency component we can restrict consideration to the range $0 \leq \hat{m} \leq \hat{m}_c (= 0.958)$ without introducing appreciable error.

A similar analysis can be made for the difference-frequency interaction. We recall that the sum-frequency interaction ceases when the value of k exceeds ($\hat{m} = 2$). The intersection of the two sets of dispersion curves ($\sigma_{1,2} = \pm\sqrt{gq_{1,2}}$) then involves the two branches of different sign, as is shown in Figs. 6.2c and d of Chap.6. From these figures it is clear that in this case three roots of χ always exist for a given angle θ_{1k}. The positions of the zeros of the function $Q(\chi)$ are indicated in Figs. M.7c and d by the dashed curves. We see that there is only one zero point and this is located at the common point of the branches $\chi^{(2)}$ and $\chi^{(3)}$. As before the effect of the singularity in the integrand is removed in the summation procedure.

Appendix N

Coherence and Spectral Density Functions of the First- and Second-Order Fields

N.1 First-Order Field

The coherence function for the first-order pressure field takes the form

$$C_p^{(1)}(\vec{R}_1, \vec{R}_2, \omega) \equiv \langle p_\omega^{(1)}(\vec{R}_1) p_\omega^{(1)*}(\vec{R}_2) \rangle \frac{d\omega}{(2\pi)^2} \qquad (N.1)$$

Assuming an active region of infinite size and a range-independent environment we can use Eq.(5.24) to establish

$$p_\omega^{(1)}(\vec{R}) = i\omega\rho_1 \Phi_{1\omega}^{(1)}(\vec{R}) =$$

$$\frac{i\omega\rho_1}{(2\pi)^2} \int S^{(1)}(\vec{k},\omega) \left[\frac{e^{\gamma_+(z+H_1)} + R_b e^{\gamma_-(z+H_1)}}{e^{\gamma_+ H_1} + \epsilon R_b e^{\gamma_- H_1}} \right] e^{i\vec{k}\vec{r}} d\vec{k} \qquad (N.2)$$

where

$$S^{(1)}(\vec{k},\omega) = \frac{-i\omega}{\gamma_+} \zeta_\omega^{(1)}(\vec{k}) \delta(\omega - \omega_k) \qquad (N.3)$$

with $\omega_k = \omega(k)$ satisfying the corresponding dispersion relation. Substituting for $p_\omega^{(1)}(\vec{R})$ then gives

$$C_p^{(1)}(\vec{R}_1, \vec{R}_2, \omega) = \omega^2 \rho_1^2 \frac{d\omega d\vec{k}}{(2\pi)^6} \int \langle |S^{(1)}(\vec{k},\omega)|^2 \rangle T^{(1)}(k, z_1) T^{(1)*}(k, z_2) e^{i\vec{k}(\vec{r}_1 - \vec{r}_2)} d\vec{k} \qquad (N.4)$$

where the function $T^{(1)}(k, z)$ is the term inside the square bracket in Eq.(N.2) and

$$\langle |S^{(1)}(\vec{k},\omega)|^2 \rangle = \frac{\omega^2}{|\gamma_+|^2} \langle |\zeta_\omega^{(1)}(\vec{k})|^2 \rangle \delta^2(\omega - \omega_k)$$

Referring to the definition in Chap.3 and incorporating the dispersion relation suggested by the surface condition, we can write the power-density spectrum of the surface displacement in $(\vec{k}, \omega)$ plane as

$$f_\zeta(\vec{k},\omega) = \langle |\zeta_\omega^{(1)}(\vec{k})|^2 \rangle \delta^2(\omega - \omega_k) \frac{d\vec{k}d\omega}{(2\pi)^6}.$$

The corresponding wave-number spectrum then takes the form

$$f_\zeta(\vec{k}) = \langle |\zeta_\omega^{(1)}(\vec{k})|^2 \rangle \delta(\omega - \omega_k) \frac{d\vec{k}d\omega}{(2\pi)^5} \qquad (N.5)$$

after completing one fold of integration upon ω, so that

$$\langle |S^{(1)}(\vec{k},\omega)|^2 \rangle = \frac{\omega^2}{|\gamma_+|^2} f_\zeta(\vec{k}) \delta(\omega - \omega_k) \frac{(2\pi)^5}{d\vec{k}d\omega}$$

Substituting this expression into Eq.(N.4) and completing another fold of integration upon ω leads to

$$C_p^{(1)}(\vec{R}_1, \vec{R}_2, \omega) = \frac{\omega^2 \rho_1^2}{2\pi} \int \frac{\omega^2}{|\gamma_+|^2} f_\zeta(\vec{k}) T^{(1)}(k, z_1) T^{(1)*}(k, z_2) \cdot$$

$$\cdot e^{i\vec{k}(\vec{r}_1 - \vec{r}_2)} d\vec{k} = \rho_1^2 g^2 \int f_\zeta(\vec{k}) T_p^{(1)}(k, z_1) T_p^{(1)*}(k, z_2) e^{i\vec{k}(\vec{r}_1 - \vec{r}_2)} \frac{d\vec{k}}{d\omega} \quad (N.6)$$

in which, for convenience, the dispersion relation

$$\omega^2 = g\gamma_+ \frac{1 + \epsilon R_b e^{(\gamma_- - \gamma_+)H_1}}{1 + R_b e^{(\gamma_- - \gamma_+)H_1}}, \qquad (N.7)$$

has been used to convert

$$T^{(1)}(k, z) = e^{\gamma_+ z} \frac{1 + R_b e^{(\gamma_- - \gamma_+)(z+H_1)}}{1 + \epsilon R_b e^{(\gamma_- - \gamma_+)H_1}}$$

to

$$T_p^{(1)}(k, z) = e^{\gamma_+ z} \frac{1 + R_b e^{(\gamma_- - \gamma_+)(z+H_1)}}{1 + R_b e^{(\gamma_- - \gamma_+)H_1}}. \qquad (N.8)$$

Coherence and Spectral Density Functions

Further, by invoking the relation,

$$f_\zeta^{(1)}(\vec{k}) k\, dk\, d\theta_k = F_a(\omega) H(\theta_k)\, d\omega\, d\theta_k \qquad (N.9)$$

where $F_a(\omega)$ is the surface-wave spectrum and $H(\theta_k)$ the normalised directional distribution function, Eq.(N.6) can be written as

$$C_p^{(1)}(\vec{R}_1, \vec{R}_2, \omega) = \rho_1^2 g^2 T_p^{(1)}(k, z_1) T_p^{(1)*}(k, z_2) F_a(\omega) \int_0^{2\pi} H(\theta_k) e^{ik\rho \cos(\theta_k - \theta_\rho)} d\theta_k \qquad (N.10)$$

where $\rho = |\vec{r}_1 - \vec{r}_2|$, $\theta_\rho = \arg(\vec{r}_1 - \vec{r}_2)$. The spectral-density function of the pressure field is then

$$F_p^{(1)}(\omega, z) \equiv \lim_{\vec{\rho} \to 0,\; z_1 \to z_2 \to 0} C_p^{(1)}(\vec{R}_1, \vec{R}_2, \omega) = \rho_1^2 g^2 |T_p^{(1)}(k, z)|^2 F_a(\omega) \qquad (N.11)$$

where

$$\gamma_\pm = a \mp \sqrt{a^2 - \gamma_1^2}, \qquad a \equiv g/2\alpha_1^2, \qquad \gamma_1^2 = \omega^2/\alpha_1^2 - k^2 \qquad (N.12)$$

When the water is regarded as incompressible and of infinite depth, we can set $H_1 \to \infty$, $R_b \to 0$, $\alpha_1 \to \infty$ and obtain

$$\gamma_\pm = \pm k, \qquad T_p^{(1)}(k, z) = e^{kz} = e^{-k|z|}, \qquad \omega^2 = gk$$

so that

$$F_p^{(1)}(\omega, z) = \rho_1^2 g^2 F_a(\omega) e^{-2\frac{\omega^2}{g}|z|} \qquad (N.13)$$

Equation (N.13) is recognised as the well known expression for the first-order, deep-water pressure spectrum associated with the ocean-wave field, $F_a(\omega)$.

N.2 Second-Order Field (Sum-Frequency Component)

The coherence function of the second-order pressure field is

$$C_p^{(2)}(\vec{R}_1, \vec{R}_2, \omega) = \langle p_\omega^{(2)}(\vec{R}_1) p_\omega^{*(2)}(\vec{R}_2) \rangle \frac{d\omega}{(2\pi)^2}$$

The dominant term of the second-order field in the reference wave guide is (Eq.(3.34))

$$p_\omega^{(2)}(\vec{R}) = i\omega \rho_1 \Phi_{1\omega}^{(2)}(\vec{R})$$

$$= \frac{i\omega \rho_1}{(2\pi)^2} \int S_b^{(2)}(\vec{k}, \omega) T_p^{(2)}(k, z) e^{i\vec{k}\vec{r}} d\vec{k}$$

where
$$T_p^{(2)}(k,z) = e^{\gamma_+ z}\frac{1+R_b e^{(\gamma_- -\gamma_+)(z+H_1)}}{1+\frac{\epsilon}{b}R_b e^{(\gamma_- -\gamma_+)H_1}} \quad (N.14)$$

and $S_b^{(2)}(\vec{k},\omega)$ is defined at the beginning of Chap. 6. Substituting for $p_\omega^{(2)}(\vec{R})$ gives

$$\begin{aligned}C_p^{(2)}(\vec{R}_1,\vec{R}_2,\omega) &= \frac{d\omega\omega^2\rho_1^2}{(2\pi)^6}\int\int \langle S_b^{(2)}(\vec{k}_1,\omega)S_b^{(2)*}(\vec{k}',\omega)\rangle \\ &\quad T_p^{(2)}(k,z_1)T_p^{*(2)}(k',z_2)e^{i(\vec{k}\vec{r}-\vec{k}'\vec{r}_2)}d\vec{k}\,d\vec{k}'\end{aligned} \quad (N.15)$$

where
$$b = \frac{(1-\omega^2/g\gamma_+)}{1-\omega^2/g\gamma_-}, \epsilon = \frac{\gamma_-}{\gamma_+} \quad (N.16)$$

and, referring to the definition Eq. (6.1),

$$\langle S_b^{(2)}(\vec{k}_1,\omega)S_b^{(2)*}(\vec{k}_1',\omega)\rangle = \frac{1}{(2\pi)^4|\omega^2-g\gamma_+|^2}\cdot$$

$$\cdot \int\int \sigma_1\sigma_2\sigma_1'\sigma_2'(\sigma_1+\sigma_2)(\sigma_1'+\sigma_2')\langle \zeta^{(1)}(\vec{q}_1)\zeta^{*(1)}(\vec{q}_1')\rangle\langle \zeta^{(1)}(\vec{q}_2)\zeta^{*(1)}(\vec{q}_2')\rangle \cdot$$

$$\cdot \{\vec{q}_1,\vec{q}_2\}\{\vec{q}_1',\vec{q}_2'\}^*\delta(\omega-\sigma_1-\sigma_2)\delta(\omega-\sigma_1'-\sigma_2')d\vec{q}_1\,d\vec{q}_1' \quad (N.17)$$

By virtue of the orthogonality of the process

$$\langle \zeta^{(1)}(\vec{q}_i)\zeta^{*(1)}(\vec{q}_j)\rangle = \begin{cases} 0 & i\neq j \\ \frac{(2\pi)^4}{d\vec{q}_i}f_\zeta(\vec{q}_i) & i=j \end{cases} \quad (N.18)$$

we can establish

$$\begin{aligned}\langle|S_b^{(2)}(\vec{k},\omega)|^2\rangle &= \frac{(2\pi)^6\omega^2}{|\omega^2-g\gamma_+|^2}\int\int \sigma_1^2\sigma_2^2 f_\zeta(\vec{q}_1)f_\zeta(\vec{q}_2)|\{\vec{q}_1\vec{q}_2\}|^2\frac{d\vec{q}_1d\vec{q}_2}{d\vec{q}_1d\vec{q}_2}\frac{1}{(d\omega)^2}\\ &= \frac{(2\pi)^6\omega^2}{|\omega^2-g\gamma_+|^2d\omega d\vec{q}_1}\int\int \sigma_1^2\sigma_2^2 F_a(\sigma_1)F_a(\sigma_2)H(\theta_1)H(\theta_2)|\{\vec{q}_1\vec{q}_2\}|^2\\ &\quad \frac{d\sigma_1d\sigma_2d\theta_1d\theta_2}{d\omega d\vec{q}_1d\vec{q}_2}d\vec{q}_1 = \frac{(2\pi)^6\omega^2}{|\omega^2-g\gamma_+|^2d\omega d\vec{q}_1}\int \sigma_1^2\sigma_2^2\cdots|\{\vec{q}_1\vec{q}_2\}|^2(\frac{\partial\sigma_1}{\partial q_1})(\frac{\partial\sigma_2}{\partial q_2})\\ &\quad \frac{\partial q_1}{q_2\partial(\sigma_1+\sigma_2)}d\theta_1 = \frac{(2\pi)^6}{d\vec{k}d\omega}\langle|B^{(2)}(k,\omega)|^2\rangle\end{aligned} \quad (N.19)$$

where

$$\langle |B^{(2)}(\vec{k},\omega)|^2\rangle = \frac{\omega^2}{|\omega^2 - g\gamma_+|^2}\int_0^{2\pi} F_a(\sigma_1)F_a(\sigma_2)H(\theta_1)H(\theta_2)$$
$$|\{q_1,q_2\}|^2\frac{\sigma_1^2\sigma_2^2}{q_2}\left(\frac{\partial\sigma_1}{\partial q_1}\right)\left(\frac{\partial\sigma_2}{\partial q_2}\right)\frac{\partial q_1}{\partial(\sigma_1+\sigma_2)}d\theta_1 \quad \text{(N.20)}$$

We recognise this as the source function given by Eq.(6.5). It follows that

$$C_p^{(2)}(\vec{R}_1,\vec{R}_2,\omega) = \omega^2\rho_1^2\int\langle |B^{(2)}(\vec{k},\omega)|^2\rangle T_p^{(2)}(k,z_1)T_p^{(2)*}(k,z_2)e^{i\vec{k}\cdot\vec{\rho}}d\vec{k} \quad \text{(N.21)}$$

where $\vec{\rho} = \vec{r}_1 - \vec{r}_2$ as before. Setting $\rho = 0$ and $z_1 = z_2$ in Eq.(N.21), the spectral-density function of the second-order pressure field is established as

$$F_p^{(2)}(\omega,z) = \omega^2\rho_1^2\int\langle |B^{(2)}(\vec{k},\omega)|^2\rangle |T_p^{(2)}(k,z)|^2 d\vec{k} \quad \text{(N.22)}$$

For convenience in numerical calculations we also develop (N.19) in the following normalised form, which proves invaluable in testing the numerical codes,

$$\langle |B^{(2)}(\vec{k},\omega)|^2\rangle = \frac{1}{\omega^2}F_a^2(\omega/2)(1-\cos\theta_{12}^{(0)})^2\frac{1}{q}\left(\frac{\omega}{2}\right)^4\left(\frac{g}{\omega}\right)^2\frac{2\omega}{4g}\frac{1}{|1-\frac{g\gamma_+}{\omega^2}|^2}$$
$$\int_0^{2\pi}\frac{F_a(\sigma_1)F_a(\sigma_2)}{F_a^2(\omega/2)}H(\theta_1)H(\theta_2)\frac{|\{q_1,q_2\}|^2}{(1-\cos\theta_{12}^{(0)})^2}$$
$$\left(\frac{2\sigma_1}{\omega}\right)^2\left(\frac{2\sigma_2}{\omega}\right)^2\left(\frac{q}{q_2}\right)\frac{\left(\frac{\partial\sigma_1}{\partial q_1}\right)\left(\frac{\partial\sigma_2}{\partial q_2}\right)}{g^2/\omega^2}\frac{\partial q_1}{\partial(\sigma_1+\sigma_2)}\frac{2g}{\omega}d\theta_1$$

In this normalisation, use has been made of the classical deep-ocean (standing-wave) approximation, i.e., $\sigma_1 = \sigma_2 = \frac{\omega}{2}$ and $\theta_{12}^{(0)} = \pi$,

$$\frac{\partial\sigma_i}{\partial q_i} = \frac{g}{2\sigma_i} = \frac{g}{\omega}$$

$$g = \frac{\sigma_i^2}{q_i} = \frac{\omega^2}{4q}$$

and

$$\frac{\partial q_1}{\partial(\sigma_1+\sigma_2)} = \frac{2\chi^3\omega}{g(1+\chi)^2[\chi^2-(1+\frac{1}{2}\hat{m}\hat{t})\chi+(1-\frac{1}{2}\hat{m}\hat{t})]} = \frac{\omega}{2g}$$

In this case we can write

$$\langle |B^{(2)}(\vec{k},\omega)|^2 \rangle = \frac{g^2}{2\omega} F_a^2(\omega/2) \frac{1}{|1 - \frac{g\gamma_\pm}{\omega^2}|^2} \int_0^{2\pi} \frac{F_a(\sigma_1) F_a(\sigma_2)}{F_a^2(\omega/2)} H(\theta_1) H(\theta_2)$$

$$\frac{1}{4} |\{q_1, q_2\}|^2 \frac{16\sigma_1^2 \sigma_2^2}{\omega^4} \frac{\omega^2}{4gq_2} \frac{\omega^2}{g^2} \left(\frac{\partial \sigma_1}{\partial q_1}\right) \left(\frac{\partial \sigma_2}{\partial q_2}\right) \frac{2g}{\omega} \frac{\partial q_1}{\partial(\sigma_1 + \sigma_2)} d\theta_1$$

$$= \frac{g^2}{2\omega} F_a^2(\omega/2) I_s \frac{1}{|1 - \frac{g\gamma_\pm}{\omega^2}|^2} \frac{1}{I_s} \int_0^{2\pi} \left[\frac{2}{\omega g^2} |\{q_1, q_2\}|^2 \frac{\sigma_1^2 \sigma_2^2}{q_2} \left(\frac{\partial \sigma_1}{\partial q_1}\right)\right.$$

$$\left. \cdot \left(\frac{\partial \sigma_2}{\partial q_2}\right) \frac{\partial q_1}{\partial(\sigma_1 + \sigma_2)} \frac{F_a(\sigma_1) F_a(\sigma_2)}{F_a^2(\omega/2)} H(\theta_1) H(\theta_2) \right] d\theta_1$$

with I_s defined below. Substituting this expression back into (N.22) leads to

$$F_p^{(2)}(\omega, z) = \frac{\omega^2 \rho_1^2 g^2}{2\omega} F_a^2(\omega/2) 2\pi I_s \int_0^{k_g} \frac{1}{|1 - \frac{g\gamma_\pm}{\omega^2}|^2} \left\{\frac{1}{2\pi I_s} \int_0^{2\pi} \int [\ ] d\theta_1 d\theta_k\right\} \cdot$$

$$\cdot |T_p^{(2)}(k, z)|^2 k \, dk = \pi \omega^3 \frac{\rho_1^2 g^2}{\alpha_1^2} F_a^2(\omega/2) I_s \int_0^{u_g} S_p^{(2)}(u, \omega) |T_p^{(2)}(u, z)|^2 u \, du$$

(N.23)

where

$$I_s = \int_0^{2\pi} H(\theta) H(\theta + \pi) d\theta$$

$$= \frac{\Gamma(s+1)}{\sqrt{\pi} 2^{(2s+1)} \Gamma(s + \frac{1}{2})}$$

Here $k_g(u_g = k_g \alpha_1/\omega)$ is the upper limit of the value k for the sum-frequency interactions and is determined by the dispersion relation involved. As was shown in Chap. 6 an effective upper limit, $k_e(u_e)$, can be used in place of $k_g(u_g)$. Under this approximation only the first branch of the roots of the interaction equation needs to be considered. For the difference-frequency interactions, on the other hand, $k(u)$ runs from $k_g(u_g)$ to ∞ and all three branches of the roots need to be taken into account.

For power density spectra of the pressure and ocean-wave field the following relationships apply (see Sect.6.3.1):

$$F_p^{(2)}(f, z) = (2\pi) F_p^{(2)}(\omega, z)$$

Using these we can also express the pressure spectrum as (by using $u = k\alpha_1/\omega$ and u_e as the upper limit)

$$F_p^{(2)}(f,z) = \frac{\omega^3 \rho_1^2 g^2}{2\alpha_1^2} F_a^2(f/2) I_s \int_0^{u_e} S_p^{(2)}(u,\omega)|T_p^{(2)}(u,z)|^2 u\,du \qquad (N.24)$$

where

$$S_p^{(2)}(u,\omega) = \frac{1}{\left|1 - \frac{g\gamma_+}{\omega^2}\right|^2 2\pi I_s} \int_0^{2\pi}\int_0^{2\pi} \frac{2}{\omega g^2}|\{q_1,q_2\}|^2 \frac{\sigma_1^2 \sigma_2^2}{q_2}\left(\frac{\partial \sigma_1}{\partial q_1}\right)\left(\frac{\partial \sigma_2}{\partial q_2}\right) \cdot$$
$$\cdot \frac{\partial q_1}{\partial(\sigma_1+\sigma_2)} \frac{F_a(\sigma_1)F_a(\sigma_2)}{F_a^2(\omega/2)} H(\theta_1)H(\theta_2)d\theta_1\,d\theta_k, \qquad (N.25)$$

$$|T_p^{(2)}(k,z)|^2 = \left|e^{\gamma_+ z}\frac{1 + R_b e^{(\gamma_- - \gamma_+)(z+H_1)}}{1 + \frac{\epsilon}{b}R_b e^{(\gamma_- - \gamma_+)H_1}}\right|^2$$

and

$$\gamma_\pm = a \mp i\gamma_1, \qquad \gamma_1 = \frac{\omega}{\alpha_1}\sqrt{1-u^2}, \qquad a = \frac{g}{2\alpha_1^2},$$

$$b = \frac{1 - \omega^2/g\gamma_+}{1 - \omega^2/g\gamma_-} = \left(\frac{g\gamma_+ - \omega^2}{g\gamma_- - \omega^2}\right)\frac{\gamma_-}{\gamma_+}, \qquad \epsilon = \frac{\gamma_-}{\gamma_+}$$

If we also assume that $a = 0$ so that $\gamma_\pm = \mp i\gamma_1$, then

$$\frac{\epsilon}{b} = \frac{g\gamma_- - \omega^2}{g\gamma_+ - \omega^2} = \frac{\omega^2 - ig\gamma_1}{\omega^2 + ig\gamma_1} \qquad (N.26)$$

and

$$|T_p^{(2)}(k,z)|^2 = \left|e^{-i\gamma_1 z}\right|^2 \left|\frac{1 + R_b e^{i2\gamma_1(z+H_1)}}{1 + \left(\frac{\epsilon}{b}\right)R_b e^{i2\gamma_1 H_1}}\right|^2 \qquad (N.27)$$

Further, we have seen that in the case of a deep and bottomless ocean $|\{q_1,q_2\}|^2 = (1 - \cos\theta_{12})^2$, and $\sigma_{1,2}^2 = gq_{1,2}$, so that

$$\frac{\sigma_1^2 \sigma_2^2}{q_2}\left(\frac{\partial \sigma_1}{\partial q_1}\right)\left(\frac{\partial \sigma_2}{\partial q_2}\right) = g^2 q_1 \frac{g^2}{4\sigma_1 \sigma_2} = \frac{g^3 \sigma_1^2}{4\sigma_1 \sigma_2} = \frac{g^3}{4\chi}$$

and
$$\frac{\partial q_1}{\partial(\sigma_1 + \sigma_2)} = \frac{2\omega\chi^3}{g(1+\chi)^2\left[\chi^2 - \left(\frac{1}{2}\hat{m}\hat{t} + 1\right)\chi + \left(1 - \frac{1}{2}\hat{m}\hat{t}\right)\right]}$$

In this situation

$$S_p^{(2)}(u,\omega) = \frac{1}{2\pi\left|1 - \frac{g\gamma_\pm}{\omega^2}\right|^2 I_s} \int_0^{2\pi}\int_0^{2\pi} \frac{(1-\cos\theta_{12})^2\chi^2}{(1+\chi)^2\left[\chi^2 - \left(\frac{1}{2}\hat{m}\hat{t} + 1\right)\chi + \left(1 - \frac{1}{2}\hat{m}\hat{t}\right)\right]} \\ \cdot \frac{F_a(f_1)F_a(f_2)}{F_a^2(f/2)} H(\theta_1)H(\theta_2) d\theta_1\, d\theta_k$$

N.3 The Total Pressure Field

By regarding the spectral components of the first- and second-order fields as independent the covariance and spectral functions of the total field (for a given frequency) can be written as

$$C_p(\vec{R}_1, \vec{R}_2, f) = C_p^{(1)}(\vec{R}_1, \vec{R}_2, f) + C_p^{(2)}(\vec{R}_1, \vec{R}_2, f) \tag{N.28}$$

$$F_p(f, z) = F_p^{(1)}(f, z) + F_p^{(2)}(f, z) \tag{N.29}$$

where, as has been shown above,

$$F_p^{(1)}(f, z) = \rho_1^2 g^2 F_a(f) \left|T_p^{(1)}(k, z)\right|^2 \tag{N.30}$$

$$F_p^{(2)}(f, z) = \frac{\rho_1^2 g^2 \omega^3 F_a^2(\frac{f}{2}) I_s}{2\alpha_1^2} \int_0^{u_e}\left[\int_0^{2\pi}\int_0^{2\pi} W(\vec{k}, \theta_1) d\theta_1\, d\theta_k\right] \left|T_p^{(2)}(k, f, z)\right|^2 u\, du \tag{N.31}$$

and

$$W(\vec{k}, \theta_1) = \frac{1}{\pi\omega g^2 I_s \left|1 - \frac{g\gamma_\pm}{\omega^2}\right|^2} |\{\vec{q}_1, \vec{q}_2\}|^2 \frac{\sigma_1^2\sigma_2^2}{q^2} \left(\frac{\partial\sigma_1}{\partial q_1}\right)\left(\frac{\partial\sigma_2}{\partial q_2}\right)\frac{\partial q_1}{\partial(\sigma_1+\sigma_2)} \\ \cdot \frac{F_a(f_1)F_a(f_2)}{F_a^2(\frac{f}{2})} H(\theta_1)H(\theta_2) \tag{N.32}$$

$$T_p^{(1)} = e^{\gamma_+ z}\frac{1 + R_b e^{(\gamma_- - \gamma_+)(z+H_1)}}{1 + R_b e^{(\gamma_- - \gamma_+)H_1}} \tag{N.33}$$

$$T_p^{(2)} = e^{\gamma_+ z}\frac{1 + R_b e^{(\gamma_- - \gamma_+)(z+H_1)}}{1 + (\epsilon/b)R_b e^{(\gamma_- - \gamma_+)H_1}} \tag{N.34}$$

The covariance functions then take the form

$$C_p^{(1)}(\vec{R}_1, \vec{R}_2, f) = \rho_1^2 g^2 T_p^{(1)}(k, z_1) T_p^{(1)*}(k, z_2) F_a(f) \int_0^{2\pi} H(\theta_k) e^{ik\rho \cos(\theta_k - \theta_\rho)} d\theta_k \quad (N.35)$$

and

$$C_p^{(2)}(\vec{R}_1, \vec{R}_2, f) = \frac{\rho_1^2 g^2 \omega}{2} F_a^2(\frac{f}{2}) I_s \int_0^\infty \int_0^\infty \left[\int_0^{2\pi} W(\vec{k}, \theta_1) d\theta_1 \right] \cdot$$
$$\cdot T_p^{(2)}(k, f, z_1) T_p^{*(2)}(k, f, z_2) e^{i(k_x \rho_x + k_y \rho_y)} dk_x \, dk_y \quad (N.36)$$

in which $\vec{\rho} = \vec{\rho}_2 - \vec{\rho}_1 \equiv \rho e^{i\theta_\rho}$ and $\vec{\rho}_1$ and $\vec{\rho}_2$ are position vectors in the horizontal plane.

Appendix O

Expression for the Reflection Coefficient in Terms of the Parameters of the Top Sedimentary Layer

In this Appendix we establish the relationship between the reflection coefficient, R_b, for a pressure wave incident at the water-sediment interface, and those for the fast- and slow-compressional waves and shear wave, R_{p1}, R_{p2} and R_s respectively, at the lower interface of a fluid-saturated, porous viscoelastic layer.

As in the previous sections, Φ_1 denotes the displacement potential in the water and Φ_{2p1}, Φ_{2p2} and Ψ_{2s} the displacement potentials of the fast- and slow-compressional waves and the shear wave in the sediment. In plane-wave form we can write

$$\left.\begin{aligned}
\Phi_1 &= e^{-i\gamma_1(z+H_1)} + R_b e^{i\gamma_1(z+H_1)} \\
\Phi_{2s1} &= T_1[e^{-i\gamma_{21}(z+H_2)} + \epsilon_{p1} e^{i\gamma_{21}(z+H_2)}] \\
\Phi_{2s2} &= T_2[e^{-i\gamma_{22}(z+H_2)} + \epsilon_{p2} e^{i\gamma_{22}(z+H_2)}] \\
\Psi_{2s} &= T_s[e^{-i\gamma_{23}(z+H_2)} + \epsilon_s e^{i\gamma_{23}(z+H_2)}]
\end{aligned}\right\} e^{ikx-i\omega t} \quad (O.1)$$

On the basis of the discussion in Appendix F, we can express the above equations as

$$\left.\begin{aligned}
\Phi_1 &= e^{-i\gamma_1(z+H_1)} + R_b e^{i\gamma_1(z+H_1)} \\
\Phi_{2s1} &= A_{21} e^{-i\gamma_{21}(z+H_1)} + B_{21} e^{i\gamma_{21}(z+H_2)} \\
\Phi_{2s2} &= A_{22} e^{-i\gamma_{22}(z+H_1)} + B_{22} e^{i\gamma_{22}(z+H_2)} \\
\Psi_{2s} &= A_{23} e^{-i\gamma_{23}(z+H_1)} + B_{23} e^{i\gamma_{23}(z+H_2)}
\end{aligned}\right\} e^{ikx-i\omega t} \quad (O.2)$$

and so establish the relations

$$\left.\begin{aligned}
T_1 &= A_{21} e_{21} & \epsilon_{p1} &= \tfrac{B_{21}}{A_{21}} e_{21}^{-1} \\
T_2 &= A_{22} e_{22} & \epsilon_{p2} &= \tfrac{B_{22}}{A_{22}} e_{22}^{-1} \\
T_s &= A_{23} e_{23} & \epsilon_s &= \tfrac{B_{23}}{A_{23}} e_{23}^{-1}
\end{aligned}\right\} \quad (O.3)$$

where

$$e_{21} = e^{i\gamma_{21}(H_2-H_1)}, \qquad e_{22} = e^{i\gamma_{22}(H_2-H_1)}, \qquad e_{23} = e^{i\gamma_{23}(H_2-H_1)} \quad (O.4)$$

Invoking the boundary conditions at the water-sediment interface, $z = -H_1$, (Appendix F, Eq.(F.11)),

$$\begin{aligned}
U_{1z} &= u_{2z} - V_{2z} \\
-p_w &= p_{2zz} \\
0 &= p_{2zx} \\
p_w &= p_{2f}
\end{aligned} \quad (O.5)$$

leads to the system of equations (Eqs.(F.20) to (F.23) of Appendix F),

$$\begin{aligned}
S_{11} R_b + S_{12} A_{21} + S_{13} A_{22} + S_{14} A_{23} &= \gamma_1 \\
S_{21} R_b + S_{22} A_{21} + S_{23} A_{22} + S_{24} A_{23} &= -\rho_1 \omega^2 \\
S_{32} A_{21} + S_{33} A_{22} + S_{34} A_{23} &= 0 \\
S_{41} R_b + S_{42} A_{21} + S_{43} A_{22} &= \rho_1 \omega^2
\end{aligned} \quad (O.6)$$

Expression for the Reflection Coefficient

where

$$\left.\begin{aligned}
S_{11} &= \gamma_1 \\
S_{12} &= \gamma_{21}(1-\delta_{12})(1-\bar{\epsilon}_{p1}) \\
S_{13} &= \gamma_{22}(1-\delta_{22})(1-\bar{\epsilon}_{p2}) \\
S_{14} &= k(1-\delta_{32})(1+\bar{\epsilon}_s) \\
\\
S_{21} &= \rho_1\omega^2 \\
S_{22} &= [\;]_{21}(1+\bar{\epsilon}_{p1}) \\
S_{23} &= [\;]_{22}(1+\bar{\epsilon}_{p2}) \\
S_{24} &= -2\mu_2 k\gamma_{23}(1-\bar{\epsilon}_s) \\
\\
S_{31} &= 0 \\
S_{32} &= 2k\gamma_{21}(1-\bar{\epsilon}_{p1}) \\
S_{33} &= 2k\gamma_{22}(1-\bar{\epsilon}_{p2}) \\
S_{34} &= (k^2 - \gamma_{23}^2)(1+\bar{\epsilon}_s) \\
\\
S_{41} &= -\rho_1\omega^2 \\
S_{42} &= (\;)_{21}(1+\bar{\epsilon}_{p1}) \\
S_{43} &= (\;)_{22}(1+\bar{\epsilon}_{p2}) \\
S_{44} &= 0
\end{aligned}\right\} \quad (O.7)$$

with

$$\left.\begin{aligned}
[\;]_{21} &= [(C_2\delta_{21} - \bar{H}_2)l_{c12}^2 + 2\mu_2 k^2] \\
[\;]_{22} &= [(C_2\delta_{22} - \bar{H}_2)l_{c22}^2 + 2\mu_2 k^2] \\
(\;)_{21} &= (C_2 - M_2\delta_{21})l_{c12}^2 \\
(\;)_{22} &= (C_2 - M_2\delta_{22})l_{c22}^2
\end{aligned}\right\} \quad (O.8)$$

and

$$\bar{\epsilon}_{p1} = \epsilon_{p1}e_{21}^2, \qquad \bar{\epsilon}_{p2} = \epsilon_{p2}e_{22}^2, \qquad \bar{\epsilon}_s = \epsilon_s e_{23}^2 \qquad (O.9)$$

The parameters C_2, $\bar{H}_2$, M_2, μ_2, δ_{21}, δ_{22}, δ_{23}, l_{c12}, l_{c22}, etc., in the above equations are defined in Appendix C. Solving the system of equations leads to

$$R_b = \frac{1}{\Delta}\begin{vmatrix} \gamma_1 & S_{12} & S_{13} & S_{14} \\ -\rho_1\omega^2 & S_{22} & S_{23} & S_{24} \\ 0 & S_{32} & S_{33} & S_{34} \\ \rho_1\omega^2 & S_{42} & S_{43} & 0 \end{vmatrix} \qquad \Delta = \begin{vmatrix} S_{11} & S_{12} & S_{13} & S_{14} \\ S_{21} & S_{22} & S_{23} & S_{24} \\ 0 & S_{32} & S_{33} & S_{34} \\ S_{41} & S_{42} & S_{43} & 0 \end{vmatrix} \quad (O.10)$$

These relations can be significantly simplified when the sedimentary layer is elastic. In this case, the fourth equation in Eq.(O.6) and all terms involving A_{22} disappear so that

$$R_b = \frac{1}{\Delta} \begin{vmatrix} \gamma_1 & S_{12} & S_{14} \\ -\rho_1\omega^2 & S_{22} & S_{24} \\ 0 & S_{32} & S_{34} \end{vmatrix} = \frac{\gamma_1[S_{22}S_{34} - S_{24}S_{32}] + \rho_1\omega^2[S_{12}S_{34} - S_{14}S_{32}]}{\gamma_1[S_{22}S_{34} - S_{24}S_{32}] - \rho_1\omega^2[S_{12}S_{34} - S_{14}S_{32}]} \quad (O.11)$$

where S_{ij} reduces to

$$\left. \begin{array}{l} S_{11} = \gamma_1 \\ S_{12} = \gamma_{21}(1 - \bar{\epsilon}_{p1}) \\ S_{14} = k(1 + \bar{\epsilon}_s) \\ \\ S_{21} = \rho_1\omega^2 \\ S_{22} = -\mu_2(k_{b2}^2 - 2k^2)(1 + \bar{\epsilon}_{p1}) \\ S_{24} = -2\mu_2 k \gamma_{23}(1 - \bar{\epsilon}_s) \\ \\ S_{31} = 0 \\ S_{32} = 2k\gamma_{21}(1 - \bar{\epsilon}_{p1}) \\ S_{34} = -(k_{b2}^2 - 2k^2)(1 + \bar{\epsilon}_s) \end{array} \right\} \quad (O.12)$$

in which $k_{b2} = \omega/\beta_2$, $\bar{\epsilon}_{p1} = \epsilon_{p1} e_{21}^2$, $\bar{\epsilon}_{p2} = \epsilon_{p2} e_{22}^2$ and $\bar{\epsilon}_s = \epsilon_s e_{23}^2$.

Using Eqs.(O.11) we establish

$$R_b = \frac{A + B}{A - B} \quad (O.13)$$

where $A = \rho_2 \gamma_1 [(k_{b2}^2 - 2k^2)^2 (1 + \bar{\epsilon}_{p1})(1 + \bar{\epsilon}_s) + 4k^2 \gamma_{21}\gamma_{23}(1 - \bar{\epsilon}_{p1})(1 - \bar{\epsilon}_s)]$ and $B = \rho_1 \gamma_{21} k_{b2}^4 (1 - \bar{\epsilon}_{p1})(1 + \bar{\epsilon}_s)$.

Another useful expression is

$$\frac{1 - R_b}{1 + R_b} = \frac{\gamma_{21}}{m_{21}\gamma_1} \frac{k_{b2}^4 (1 - \bar{\epsilon}_{p1})(1 + \bar{\epsilon}_s)}{(k_{b2}^2 - 2k^2)^2(1 + \bar{\epsilon}_{p1})(1 + \bar{\epsilon}_s) + 4k^2\gamma_{21}\gamma_{23}(1 - \bar{\epsilon}_{p1})(1 - \bar{\epsilon}_s)} \quad (O.14)$$

where $m_{21} = \rho_2/\rho_1$.

The corresponding expression for the porous layer is

$$R_b = \frac{\gamma_1 C_{11} + \rho_1\omega^2(C_{21} + C_{41})}{\gamma_1 C_{11} - \rho_1\omega^2(C_{21} + C_{41})} \quad (O.15)$$

so that

$$\frac{1 - R_b}{1 + R_b} = -\frac{\rho_1\omega^2}{\gamma_1} \frac{(C_{21} + C_{41})}{C_{11}} \quad (O.16)$$

Expression for the Reflection Coefficient

Here C_{11}, C_{21} and C_{41} are, respectively, the values of the sub-determinants of the elements S_{11}, S_{21} and S_{41}, ie.,

$$C_{11} = \begin{vmatrix} S_{22} & S_{23} & S_{24} \\ S_{32} & S_{33} & S_{34} \\ S_{42} & S_{43} & S_{44} \end{vmatrix}, \quad C_{21} = \begin{vmatrix} S_{12} & S_{13} & S_{14} \\ S_{32} & S_{33} & S_{34} \\ S_{42} & S_{43} & S_{44} \end{vmatrix}, \quad C_{41} = \begin{vmatrix} S_{12} & S_{13} & S_{14} \\ S_{22} & S_{23} & S_{24} \\ S_{32} & S_{33} & S_{34} \end{vmatrix}$$

(O.17)

Appendix P

Green's Functions Linking the Source Pressure Field and the Seismic Field on the Seafloor

In this Appendix we establish the Green's functions, $G_{v,h}(\vec{r},\vec{r_0})$, defined as the vertical and horizontal components of the displacement of the interface between the water- and a layered seabed- half-space, generated by a unit point-pressure source, $p_{00}\delta(\vec{r}-\vec{r_0})$ (assuming $p_{00}=1$ and the time factor $e^{-i\omega t}$ omitted), acting at the point $\vec{r_0}$ on the same interface. With these Green's functions available, the vertical and horizontal displacement field $u_{v,h}$, generated by any distributed pressure field, $p_b(\vec{r},\omega)$, acting at the seafloor (at $z=-H_1$), can then be expressed as

$$u_{v,h} = \int_{z=-H_1} p_b(\vec{r_0},\omega) G_{v,h}(\vec{r},\vec{r_0}) d\vec{r_0}$$

The required Green's functions can be derived in many ways. In one approach use is made of the formula,

$$\delta(\vec{r}-\vec{r_0}) = \frac{1}{(2\pi)^2} \int_{-\infty}^{\infty}\int_{-\infty}^{\infty} e^{ik_x(x-x_0)+ik_y(y-y_0)} dk_x dk_y,$$

which allows the point-pressure source to be replaced by a set of plane-wave components,

$$p_{00}\delta(\vec{r}-\vec{r_0}) = \frac{1}{(2\pi)^2} \int_{-\infty}^{\infty}\int_{-\infty}^{\infty} e^{ik_x(x-x_0)+ik_y(y-y_0)-i\gamma_1(z+H_1)} dk_x dk_y \Big|_{z=-H_1}$$

where $\gamma_1 = \sqrt{k_{\alpha_1}^2 - k_x^2 - k_y^2}$.

For convenience, we temporarily locate the point source at the origin, ie., $\vec{r_0}=0$. Since $p_1 = \omega^2 \rho_1 \Phi_1$, the displacement potential, Φ_1, generated by the source on the water side of the interface takes the form

$$\Phi_1(\vec{r}, z) = \frac{p_{00}}{(2\pi)^2 \omega^2 \rho_1} \int\int_{-\infty}^{\infty} \frac{1}{1+R_b} \left[e^{-i\gamma_1(z+H_1)} + R_b e^{i\gamma_1(z+H_1)}\right] e^{i(k_x x + k_y y)}$$

$$dk_x\, dk_y = A_0 \int_0^\infty \frac{1}{1+R_b} \left[e^{-i\gamma_1(z+H_1)} + R_b e^{i\gamma_1(z+H_1)}\right] J_0(kr) k\, dk \quad (P.1)$$

where

$$A_0 = \frac{p_{00}}{(2\pi)\omega^2 \rho_1} \quad (P.2)$$

and R_b is the reflection coefficient for the seafloor. Referring to Appendix O, we can write the three displacement potentials in the solid frame of the top sedimentary layer in the form

$$\Phi_{2s1}(\vec{r}, z) = \frac{A_0}{2\pi} \int\int_{-\infty}^{\infty} \frac{A_{21}}{1+R_b} \left[e^{-i\gamma_{21}(z+H_1)} + e_{21}\epsilon_{p1} e^{i\gamma_{21}(z+H_2)}\right] e^{i\vec{k}\vec{r}} d\vec{k}$$

$$(P.3)$$

$$\Phi_{2s2}(\vec{r}, z) = \frac{A_0}{2\pi} \int\int_{-\infty}^{\infty} \frac{A_{22}}{1+R_b} \left[e^{-i\gamma_{22}(z+H_1)} + e_{22}\epsilon_{p2} e^{i\gamma_{22}(z+H_2)}\right] e^{i\vec{k}\vec{r}} d\vec{k}$$

$$(P.4)$$

$$\vec{\Psi}_{2s}(\vec{r}, z) = \frac{A_0}{2\pi} \int\int_{-\infty}^{\infty} \frac{A_{23}}{1+R_b} \left[e^{-i\gamma_{23}(z+H_1)} + e_{23}\epsilon_s e^{i\gamma_{23}(z+H_2)}\right] (\vec{k}_0 \times \vec{z}_0) e^{i\vec{k}\vec{r}} d\vec{k}$$

$$(P.5)$$

where ϵ_{p1}, ϵ_{p2} and ϵ_s are the reflection coefficients of the three waves at the sub-interface of the upper layer at $z = -H_2$,

$$e_{2j} = e^{i\gamma_{2j}(H_2 - H_1)} \quad (P.6)$$

$j = 1, 2, 3$, and $\vec{k}_0$ and $\vec{z}_0$ are as before unit vectors in the $\vec{k}$ and $\vec{z}$ directions. All other parameters have the same meaning as in Appendix O.

For each plane-wave component, the displacement vector of the frame in the top sedimentary layer

$$\vec{u}_2 = [u_{2x}, u_{2y}, u_{2z}]$$

$$= \nabla(\Phi_{2s1} + \Phi_{2s2}) + \nabla \times \vec{\Psi}_{2s}$$

takes the form

$$u_{2x} = \frac{\partial}{\partial x}(\Phi_{2s1} + \Phi_{2s2}) + \frac{k_x}{k}\frac{\partial}{\partial z}\Psi_{2s}$$

$$u_{2y} = \frac{\partial}{\partial y}(\Phi_{2s1} + \Phi_{2s2}) + \frac{k_y}{k}\frac{\partial}{\partial z}\Psi_{2s}$$

$$u_{2z} = \frac{\partial}{\partial z}(\Phi_{2s1} + \Phi_{2s2}) - \frac{k_x}{k}\frac{\partial}{\partial x}\Psi_{2s} - \frac{k_y}{k}\frac{\partial}{\partial y}\Psi_{2s} \qquad (P.7)$$

where use has been made of the conventions

$$\vec{k}_0 \times \vec{z}_0 = \left[\frac{k_y}{k}, \frac{-k_x}{k}, 0\right]$$

$$\vec{\Psi}_{2s} = (\vec{k}_0 \times \vec{z}_0)\Psi_{2s}$$

$$\nabla \times (\vec{k}_0 \times \vec{z}_0)\Psi_{2s} = \left[\frac{k_x}{k}\frac{\partial}{\partial z}, \frac{k_y}{k}\frac{\partial}{\partial z}, -\left(\frac{k_x}{k}\frac{\partial}{\partial x} + \frac{k_y}{k}\frac{\partial}{\partial y}\right)\right]\Psi_{2s}$$

and Ψ_{2s1}, Ψ_{2s2} and $\vec{\Psi}_{2s}$ are the integrands of Eqs.(P.3) to (P.5) respectively.

Referring again to Appendix O, we can write the components u_{2x}, u_{2y} and u_{2z} at $z = -H_1$ as

$$u_{2x} = \frac{ik_x}{1+R_b}\left[A_{21}(1+\bar{\epsilon}_{p1}) + A_{22}(1+\bar{\epsilon}_{p2}) - A_{23}\frac{\gamma_{23}}{k}(1-\bar{\epsilon}_s)\right]e^{i\vec{k}\vec{r}}$$

$$u_{2y} = \frac{ik_y}{1+R_b}\left[A_{21}(1+\bar{\epsilon}_{p1}) + A_{22}(1+\bar{\epsilon}_{p2}) - A_{23}\frac{\gamma_{23}}{k}(1-\bar{\epsilon}_s)\right]e^{i\vec{k}\vec{r}}$$

$$u_{2z} = \frac{-i\gamma_{21}}{1+R_b}\left[A_{21}(1-\bar{\epsilon}_{p1}) + A_{22}\frac{\gamma_{22}}{\gamma_{21}}(1-\bar{\epsilon}_{p2}) + A_{23}\frac{k}{\gamma_{21}}(1+\bar{\epsilon}_s)\right]e^{i\vec{k}\vec{r}}$$

(P.8)

where

$$\bar{\epsilon}_{p1} = \epsilon_{p1}e_{21}^2, \qquad \bar{\epsilon}_{p2} = \epsilon_{p2}e_{22}^2, \qquad \bar{\epsilon}_s = \epsilon_s e_{23}^2 \qquad (P.9)$$

Since

$$\gamma_1(1-R_b) = \gamma_{21}A_{21}(1-\delta_{12})(1-\bar{\epsilon}_{p1}) + \gamma_{22}A_{22}(1-\delta_{22})(1-\bar{\epsilon}_{p2}) + kA_{23}(1-\delta_{32})(1+\bar{\epsilon}_s), \qquad (P.10)$$

from the boundary conditions (see Eq.(F.20) of Appendix F and Eq.(O.1) of Appendix O) we can establish the required Green's functions for the vertical and horizontal components of the seabed displacement, by integrating u_{2x}, u_{2y} and u_{2z} over the entire k-plane. For the vertical component this gives

$$G_v(\vec{r},\vec{r}_0) = \frac{-i}{2\pi\omega^2\rho_1} \int_0^\infty \left(\frac{1-R_b}{1+R_b}\right) R_v(k)\gamma_1 J_0(k\rho) k\,dk \qquad (P.11)$$

where the term $R_v(k)$ is expressed as

$$R_v(k) =$$
$$\frac{\gamma_{21}A_{21}(1-\bar{\epsilon}_{p1}) + \gamma_{22}A_{22}(1-\bar{\epsilon}_{p2}) + kA_{23}(1+\bar{\epsilon}_s)}{\gamma_{21}A_{21}(1-\delta_{12})(1-\bar{\epsilon}_{p1}) + \gamma_{22}A_{22}(1-\delta_{22})(1-\bar{\epsilon}_{p2}) + kA_{23}(1+\delta_{32})(1+\bar{\epsilon}_s)},$$

$$(P.12)$$

$\rho = |\vec{r}-\vec{r}_0|$ and for the two horizontal components

$$\begin{aligned}G_x(\vec{r},\vec{r}_0) &= \frac{-1}{(2\pi)^2\omega^2\rho_1}\int_0^\infty \frac{k_x}{k}\left(\frac{1-R_b}{1+R_b}\right) R_h(k)\gamma_1 e^{i\vec{k}\vec{\rho}} d\vec{k} \\ &= \frac{-i}{2\pi\omega^2\rho_1}\left(-\frac{\partial}{\partial x}\right)\int_0^\infty \left(\frac{1-R_b}{1+R_b}\right) R_h(k)\gamma_1 J_0(k\rho)\,dk \quad (P.13)\end{aligned}$$

or

$$G_x(\vec{r},\vec{r}_0) = \frac{-i}{2\pi\omega^2\rho_1}\cos\theta_\rho \int_0^\infty \left(\frac{1-R_b}{1+R_b}\right) R_h(k)\gamma_1 J_1(k\rho) k\,dk \qquad (P.14)$$

and

$$G_y(\vec{r},\vec{r}_0) = \frac{-i}{2\pi\omega^2\rho_1}\sin\theta_\rho \int_0^\infty \left(\frac{1-R_b}{1+R_b}\right) R_h(k)\gamma_1 J_1(k\rho) k\,dk \qquad (P.15)$$

where

$$R_h(k) = -i\cdot$$
$$\cdot\frac{kA_{21}(1+\bar{\epsilon}_{p1}) + kA_{22}(1+\bar{\epsilon}_{p2}) - \gamma_{23}A_{23}(1-\bar{\epsilon}_s)}{\gamma_{21}A_{21}(1-\delta_{12})(1-\bar{\epsilon}_{p1}) + \gamma_{22}A_{22}(1-\delta_{22})(1-\bar{\epsilon}_{p2}) + kA_{23}(1-\delta_{32})(1+\bar{\epsilon}_s)},$$

$$(P.16)$$

$$\cos\theta_\rho = \frac{x-x_0}{\rho} \quad \text{and} \quad \sin\theta_\rho = \frac{y-y_0}{\rho}$$

We can also define the Green's function of the horizontal component in the form

$$G_h(\vec{r},\vec{r}_0) = \frac{-i}{2\pi\omega^2\rho_1}\int_0^\infty \left(\frac{1-R_b}{1+R_b}\right) R_h(k)\gamma_1 J_1(k\rho) k\, dk \qquad (P.17)$$

For convenience we can then define the amplitude ratio of the response (vertical and horizontal) generated by the plane-wave component,

$$R_{hv} = R_h/R_v = -i\frac{kA_{21}(1+\bar{\epsilon}_{p1}) + kA_{22}(1+\bar{\epsilon}_{p2}) - \gamma_{23}A_{23}(1-\bar{\epsilon}_s)}{\gamma_{21}A_{21}(1-\bar{\epsilon}_{p1}) + \gamma_{22}A_{22}(1-\bar{\epsilon}_{p2}) + kA_{23}(1+\bar{\epsilon}_s)}$$
$$(P.18)$$

and examine characteristics of R_v, R_h and R_{hv} in special cases.

1. **When the seabed can be regarded as a porous halfspace.**
   In this case no energy returns to the interface from below so that $\bar{\epsilon}_{p1} = \bar{\epsilon}_{p2} = \bar{\epsilon}_s = 0$ and

$$R_v = \frac{\gamma_{21}A_{21} + \gamma_{22}A_{22} + kA_{23}}{\gamma_{21}A_{21}(1-\delta_{12}) + \gamma_{22}A_{22}(1-\delta_{22}) + kA_{23}(1+\delta_{32})} \qquad (P.19)$$

$$R_h = -i\frac{kA_{21} + kA_{22} - \gamma_{23}A_{23}}{\gamma_{21}A_{21}(1-\delta_{12}) + \gamma_{22}A_{22}(1-\delta_{22}) + kA_{23}(1-\delta_{32})} \qquad (P.20)$$

$$R_{hv} = -i\frac{kA_{21} + kA_{22} - \gamma_{23}A_{23}}{\gamma_{21}A_{21} + \gamma_{22}A_{22} + kA_{23}} \qquad (P.21)$$

Eq.(P.19) is the same as Eq.(4) of [41].

2. **When the seabed can be regarded as a layered elastic halfspace.**
   In this case all terms relating to the slow wave disappear, i.e., $A_{22} = 0$ and $\delta_{j2} = 0$ in (P.12) to (P.16), so that

$$R_v = 1 \qquad (P.22)$$

$$R_h = R_{hv} = -i\left[\frac{kA_{21}(1+\bar{\epsilon}_{p1}) - \gamma_{23}A_{23}(1-\bar{\epsilon}_s)}{\gamma_{21}A_{21}(1-\bar{\epsilon}_{p1}) + kA_{23}(1+\bar{\epsilon}_s)}\right] \qquad (P.23)$$

According to Appendix O, Eq.(O.6) degenerates in this case to

$$S_{11}R_b + S_{12}A_{21} + S_{14}A_{23} = \gamma_1$$

$$S_{21}R_b + S_{22}A_{21} + S_{24}A_{23} = -\rho_1\omega^2$$

$$S_{32}A_{21} + S_{34}A_{23} = 0 \qquad (P.24)$$

so that A_{21} and A_{23} can be expressed explicitly as

$$A_{21} = \frac{1}{\Delta}(-S_{34})(\gamma_1 S_{21} + \rho_1\omega^2 S_{11})$$

$$A_{23} = \frac{1}{\Delta}S_{32}(\gamma_1 S_{21} + \rho_1\omega^2 S_{11})$$

where Δ is the determinant of the coefficients S_{ij} ($i = 1,3$ and $j = 1,3$) of the left-hand-side of Eq.(P.18). This leads to

$$R_{hv} = i\frac{S_{34}k(1+\bar{\epsilon}_{p1}) + \gamma_{23}S_{32}(1-\bar{\epsilon}_s)}{-S_{34}\gamma_{21}(1-\bar{\epsilon}_{p1}) + kS_{32}(1+\bar{\epsilon}_s)}$$

$$= i\frac{k}{\gamma_{21}}\frac{(k^2 - \gamma_{23}^2)(1+\bar{\epsilon}_s)(1+\bar{\epsilon}_{p1}) + 2\gamma_{21}\gamma_{23}(1-\bar{\epsilon}_s)(1-\bar{\epsilon}_{p1})}{(\omega^2/\beta_2^2)(1+\bar{\epsilon}_s)(1-\bar{\epsilon}_{p1})}$$

If we denote $\frac{\alpha_1}{\beta_2} = n_b$ and $\frac{\alpha_1}{\alpha_2} = n_a$ so that $\gamma_{23} = \frac{\omega}{\alpha_1}\sqrt{n_b^2 - u^2}$, $\gamma_{21} = \frac{\omega}{\alpha_1}\sqrt{n_a^2 - u^2}$ then

$$R_h = R_{hv} = -i\frac{u}{n_b^2}\left[\frac{(n_b^2 - 2u^2)}{\sqrt{n_a^2 - u^2}}\frac{(1+\bar{\epsilon}_{p1})}{(1-\bar{\epsilon}_{p1})} - 2\sqrt{n_b^2 - u^2}\frac{(1-\bar{\epsilon}_s)}{(1+\bar{\epsilon}_s)}\right]$$
(P.25)

Finally the required Green's functions can also be written in a normalised form as

$$G_{v,h}(\vec{r},\vec{r}_0) = \frac{-i\omega}{2\pi\rho_1\alpha_1^3}\int_0^\infty \frac{1-R_b}{1+R_b}\sqrt{1-u^2}R_{v,h}J_{0,1}(u\xi)u\,du \qquad (P.26)$$

where $\xi = \frac{\omega}{\alpha_1}|\vec{r}-\vec{r}_0|$.

Bibliography

[1] V.O. Knudsen, R.S. Alford and J.W. Emling, 'Underwater ambient noise', J. Mar. Res., **7** 410 (1948).

[2] G.M. Wenz, 'Acoustic ambient noise in the ocean: Spectra and sources', J. Acoust. Soc. Am. **34**(12) 1936-1956 (1962).

[3] C.L. Piggot, 'Ambient sea noise at low frequencies in shallow water off the Scotian Shelf', J. Acoust. Soc. Am. **36**(11) 2152-2163 (1964).

[4] E.M. Arase and T. Arase, 'Ambient sea noise in the deep and shallow ocean', J. Acoust. Soc. Am. **42**(1) 73-77 (1967).

[5] A.J. Perrone, 'Deep-ocean ambient noise spectra in the Northwest Atlantic', J. Acoust. Soc. Am. **46** 762-770 (1969).

[6] A.C. Kibblewhite, 'The acoustic detection and location of an underwater volcano', N.Z. Jl. Sci. **9** 178-199 (1966).

[7] A.C. Kibblewhite, R.N. Denham, and D.J. Barnes, 'Unusual low frequency signals observed in New Zealand waters', J. Acoust. Soc. Am. **41** 644-655 (1967).

[8] A.C. Kibblewhite and R.N. Denham, 'Hydroacoustic signals from the CHASE V explosion', J. Acoust. Soc. Am. **45** 944-956 (1969).

[9] G.V. Latham, R.S. Anderson and M. Ewing, 'Pressure variations produced at the ocean bottom by hurricanes', J. Geophys. Res. **2** 5693-5704 (1967).

[10] R.J. Urick, 'Sea-bed motion as a source of the ambient noise background of the sea', J. Acoust. Soc. Am. **56** 1010-1011 (1974).

[11] A.J. Perrone, 'Infrasonic and low-frequency ambient noise measurements on the Grand Banks', J. Acoust. Soc. Am. **55**(2) 754-758 (1974).

[12] D.H. Cato, 'Ambient sea noise in waters near Australia', J. Acoust. Soc. Am. **60**(2) 320-328 (1976).

[13] A.C. Kibblewhite, J.A. Shooter and W.L. Watkins, 'Examination of attenuation at very low frequencies using the deep-water ambient-noise field', J. Acoust. Soc. Am. **60**(5) 1040-1047 (1976).

[14] J.R. McGrath, 'Infrasonic sea noise at the Mid-Atlantic Ridge near 37degN', J. Acoust. Soc. Am. **60**(6) 1290-1299 (1976).

[15] A.C. Kibblewhite, P.R. Bergquist, B.A. Foster, M.R. Gregory, and M.C. Miller, 'Maui Development Environmental Study - Report on Phase Two, 1977-1981', University of Auckland, Auckland, New Zealand October(1982).

[16] G.J.M. Copeland, 'Low frequency ambient noise- generalised spectra', in *Natural Physical Sources of Underwater Sound*, Ed. B.R. Kerman, Kluwer Academic Publishers, pp 17-30 (1993).

[17] R.H. Nichols, 'Infrasonic ocean noise sources: Wind versus waves', J. Acoust. Soc. Am. **82**(4) 1395-1402 (1987).

[18] T.E. Talpey and R.D. Worley, 'Infrasonic ambient noise measurements in deep Atlantic water', J. Acoust. Soc. Am. **75** 621-622 (1984).

[19] A.C. Kibblewhite and K.C. Ewans, 'Wave-wave interactions, microseisms and infrasonic ambient noise in the ocean', J. Acoust. Soc. Am. **78**(3) 981-994 (1985).

[20] A.C. Kibblewhite 'Ocean noise spectrum below 10 Hz - mechanisms and measurements', in *Sea Surface Sound, Natural Mechanisms of Surface Generated Noise in the Ocean*, Ed. B.R. Kerman, Kluwer Academic Publishers, pp 337-359(1988).

[21] R.G. Adair, J.A. Orcutt and T.H. Jordan, 'Low frequency noise observations in the deep ocean', J. Acoust. Soc. Am. **80**(2) 633-645 (1986).

[22] R.H. Nichols, 'Infrasonic ambient ocean noise measurements: Eleuthera', J. Acoust. Soc. Am. **69**(4) 974-981 (1981).

[23] F.D. Cotaras, I.A. Fraser, H.M. Merklinger, 'Near-surface ocean ambient noise measurements at very low frequencies', J. Acoust. Soc. Am. **83**(4) 1345-1359 (1988).

[24] M.S. Longuet-Higgins, 'Can sea waves cause microseisms?', Proc. Symposium on Microseisms, Harriman, N.Y., 74-86, (Sept. 1952).

[25] M.S. Longuet-Higgins, 'A theory of the origin of microseisms', Philos. Trans. Roy. Soc., London Ser. **A243** 1-35 (1950).

[26] K. Hasselmann, 'A statistical analysis of the generation of microseisms', Rev. Geophys. **1** 177-210 (1963).

[27] M.A. Isakovich and B.F. Kur'yanov, 'Theory of low-frequency noise in the ocean', Soviet Physics - Acoustics, **16**(1) 49-58 (1970).

[28] L.M. Brekhovskikh, 'Underwater sound waves generated by surface waves in the ocean', Izv. Atmos. Ocean. Phys. **2**(9) 582-587 (1966, translated by J. Gollob).

[29] B. Hughes, 'Estimates of underwater sound (and infrasound) produced by nonlinearly interacting ocean waves', J. Acoust. Soc. Am. **60**(5) 1032-1039 (1976).

[30] S.P. Lloyd, 'Underwater sound from surface waves according to the Lighthill-Ribner theory', J. Acoust. Soc. Am. **69**(2) 425-435 (1981).

[31] M. Miche, 'Mouvements ondulatories de la mer en profondeur constante ou decroissante', Ann. Ponts Chauss. **114** 25-87, 131-164, 270-292, 396-406 (1944).

[32] S.C. Webb and C.S. Cox, 'Observations and modelling of seafloor microseisms', J. Geophys. Res. **91** 7343–7358 (1986).

[33] G.H. Sutton and N. Barstow, 'Ocean-bottom ultra low-frequency (ULF)seismoacoustic ambient noise: 0.002 to 0.4 Hz', J. Acoust. Soc. Am. **87** 2005-2012 (1990).

[34] H. Schmidt and W.A. Kuperman, 'Estimation of surface noise source level from low-frequency seismo-acoustic ambient noise measurements', J. Acoust. Soc. Am. **84** 2153–2162 (1988).

[35] A.C. Kibblewhite and C.Y. Wu, 'The generation of infrasonic ambient noise in the ocean by nonlinear interactions of ocean surface waves', J. Acoust. Soc. Am. **85** 1935-1945 (1989).

[36] A.C. Kibblewhite and C.Y. Wu, 'An analysis of the ULF acoustic wave field in the ocean', in *Natural Physical Sources of Underwater Sound − Sea Surface Sound (2)*, Ed. B.R. Kerman, Kluwer Academic Publishers, pp 189-202 (1993).

[37] A.E. Schreiner and L.M. Dorman, 'Coherence length of seafloor noise: Effect of ocean bottom structure', J. Acoust. Soc. Am. **88** 1503-1514 (1990).

[38] A.C. Kibblewhite and C.Y. Wu, 'The theoretical description of wave-wave interactions as a noise source in the ocean', J. Acoust. Soc. Am. **89** 2241–2252 (1991).

[39] G.V. Frisk, 'Inhomogeneous waves and the plane-wave reflection coefficient', J. Acoust. Soc. Am. **66** 219-234 (1979).

[40] J.A. Orcutt, C.S. Cox, A.C. Kibblewhite, W.A. Kuperman, H. Schmidt, 'Observations and causes of ocean and seafloor noise at ultra-low and very-low frequencies', in *Natural Physical Sources of Underwater Sound — Sea Surface Sound (2)*, Ed. B.R. Kerman, Kluwer Academic Publishers, pp 203-232 (1993).

[41] A.C. Kibblewhite and C.Y. Wu, 'Acoustic source levels associated with the nonlinear interactions of ocean waves', J. Acoust. Soc. Am. **94** 3358-3378 (1993).

[42] A.C. Kibblewhite and C.Y. Wu, 'A solution for the seismo-acoustic effects induced by ocean waves applicable in shallow-water regions', Proceedings of the Conference — *Sea Surface Sound '94*, Lake Arrowhead, California, March 1994 (In press).

[43] M.V. Trevorrow and T. Yamamoto, 'Summary of marine sedimentary shear modulus and acoustic speed profile results using a gravity wave inversion technique', J. Acoust. Soc. Am. **90** 441-456 (1991).

[44] M.A. Biot, 'Theory of propagation of elastic waves in a fluid-saturated porous solid: I. low-frequency range', J. Acoust. Soc. Am. **28** 168-178 (1956).

[45] M.A. Biot, 'Theory of propagation of elastic waves in a fluid-saturated porous solid: II. high-frequency range', J. Acoust. Soc. Am. **28** 179-191 (1956).

[46] R.D. Stoll, 'Sediment Acoustics', *Lecture Notes in Earth Sciences*, edited by S. Bhattacharji, G.U. Friedman, H.J. Neugebauer and A.Seslacher, Springer-Verlag, (1989).

[47] R.D. Stoll, G.M. Bryan and E.O. Bautista, 'Measuring lateral variability of sediment geoacoustic properties', J. Acoust. Soc. Am. **96** 427–438 (1994).

[48] A.C. Kibblewhite and C.Y. Wu, 'A study of reflection loss: (I) a multilayered viscoelastic seabed at very low frequency' J. Acoust. Soc. Am. **96** 2965-2980 (1994).

[49] A.C. Kibblewhite and C.Y. Wu, 'A study of reflection loss: (II) involving a porous layer and a demonstration of the Biot slow wave' J. Acoust. Soc. Am. **96** 2981-2992 (1994).

[50] W. A. Kuperman and F. Ingentio, 'Spatial correlation of surface-generated noise in a stratified ocean', J. Acoust. Soc. Am. **67** 1988-1996 (1980).

[51] D.H. Cato, 'Sound generation in the vicinity of the sea surface: source mechanisms and the coupling to the received sound field', J. Acoust. Soc. Am. **89** 1076–1095 (1991).

[52] D.H. Cato, 'Theoretical and measured underwater noise from surface orbital motion', J. Acoust. Soc. Am. **89** 1096–1112 (1991).

[53] L.D. Landau and E.M. Lifshitz, *'Fluid Mechanics'*, Pergamon Press Ltd. (1959).

[54] Y.P. Guo, 'On sound generation by weakly non-linear interactions of surface gravity waves', J. Fluid Mech. **181** 311-328 (1987).

[55] W.M. Ewing, W.S. Jardetzky, F. Press, *'Elastic Waves in Layered Media'*, McGraw-Hill Book Company, (1957).

[56] O.M. Phillips, *'The Dynamics of the Upper Ocean'*, 2nd ed. Cambridge University Press (1977).

[57] M.B. Priestly, *'Spectral Analysis and Time Series'*, Academic Press, (1981).

[58] P.M. Morse and H. Feshbach, *'Methods of Theoretical Physics'*, Chap. 7, McGraw-Hill Publishing Company, (1953).

[59] M.S. Longuet-Higgins, 'Resonant interactions between two trains of gravity waves', J. Fluid Mech. **12** 321-322 (1962).

[60] D.S. Jones, *'The Theory of Generalised Functions'*, 2nd Edition, Cambridge Univ. Press, (1982).

[61] C.S. Cox, 'Low frequency pressure fluctuations at the deep sea floor', Proceedings of the Conference on *'Propagation and Noise in Underwater Acoustics'*,pp 170-179, University of Auckland, organized by C.T.Tindle and G.E.J.Bold, 7-9 Feb. (1990).

[62] H. Schmidt and F.B. Jensen, 'A full-wave solution for propagation in multilayered viscoelastic media with application to Gaussian beam reflection at a fluid-solid interface', J. Acoust. Soc. Am. **77** 813-825 (1985).

[63] F.B. Jensen, W.A. Kupermam, M.B. Porter and H. Schmidt, *'Computational Ocean Acoustics'*, AIP Press, New York, (1994).

[64] K. Aki and P. Richards, *'Quantitative Seismology'* Vol 1, W.H. Freeman and Company (1980).

[65] W.J. Pierson, Jr., 'A proposed spectral form for fully developed wind seas based on the similarity theory of S.A. Kitaigorodskii', J. Geophys. Res., **69** 5181-5190 (1964).

[66] K.C. Ewans and A.C. Kibblewhite, 'An examination of fetch-limited wave growth off the west coast of New Zealand by a comparison with the JONSWAP results', J. Phys. Oceanogr., **20** 1278-1296 (1990).

[67] K. Hasselmann, T.P. Barnett, E. Bouws, H. Carlson, D.E. Cartwright, K. Enke, J.A. Ewing, H. Gienapp, D.E. Hasselmann, P. Kruseman, A. Mearbug, P. Muller, D.J. Olbers, K. Richter, W. Sell and H. Walden, 'Measurements of wind-wave growth and swell decay during the joint North Sea wave project (JONSWAP)', Dent. Hydrogr. **2**. Suppl. A8, No 12 (1973).

[68] W.A. Kuperman, 'Propagation effects associated with ambient noise', in *Sea Surface Sound, Natural Mechanisms of Surface Generated Noise in the Ocean*, Ed. B.R. Kerman, Kluwer Academic Publishers, pp 253-272(1988).

[69] P. Spudich and J. Orcutt, 'A new look at the seismic velocity structure of the oceanic crust', Rev. of Geophysics and Space Physics, **18** 627-645 (1980).

[70] J.A. Collins, M.G. Purdy and T.M. Brocher, 'Seismic velocity structure at deep sea drilling project site 504B, PANAMA Basin: evidence for a thin oceanic crust', J. Geophy. Res., **94**(B7) 9283-9302 (1989).

[71] E.L. Hamilton, 'Geoacoustic modeling of the sea floor', J. Acoust. Soc. Am. **68** 1313-1340 (1980).

[72] P.L. Vidmar, R.A. Koch, Jo. B. Lindberg, D.W. Oakley and R. Pitre, 'Acoustic bottom interaction: A review of recent work at ARL:UT', Applied Research Laboratories, University of Texas at Austin, Report: ARL-TR-86-4 (1983).

[73] M.D. Max, 'Gas hydrate and acoustically laminated sediments: potential environmental cause of anomalously low acoustic bottom loss in deep-ocean sediments', Naval Research Laboratory, Washington DC, Report NRL 9235 (1990).

[74] D. Rauch, 'On the role of bottom interface waves in ocean seismo-acoustics, a review', in *Ocean Seismo-Acoustics*, Eds. T. Akal and J.M. Berkson, Plenum Press, New York, pp 625-642 (1986).

[75] H. Deresiewicz and R. Skalak, 'On uniqueness in dynamic poro-elasticity', Bull. Seism. Soc. Am. **53** 783-788 (1963).

[76] A.B. Duxbury and A.C. Duxbury, *'Fundamentals of oceanography'*, Wm. C. Brown Communications Inc., (1993).

[77] C.S. Cox and D.C. Jacobs, 'Cartesian diver observations of double frequency pressure fluctuations in the upper levels of the ocean', Geo. Res. Lett. **16** 807-810 (1989).

[78] D.C. Jacobs, C.S. Cox and X. Zhang,' Observations of the near-field double frequency pressure spectrum in the upper ocean using the Cartesian Diver Profiling Instrument', in *Natural Physical Sources of Underwater Sound-Sea Surface Sound(2)*, Ed. B.R. Kerman, Kluwer Academic Publishers, pp 175-188 (1993).

[79] L.M. Dorman, A.E. Schreiner, L.D. Bibee and J.A. Hildebrand, ' Deep-water seafloor array observations of seismo-acoustic noise in the Eastern Pacific and comparisons with wind and swell', in *Natural Physical Sources of Underwater Sound-Sea Surface Sound(2)*, Ed. B.R. Kerman, Kluwer Academic Publishers, pp 165-174(1993).

[80] S.C. Webb, ' Very low frequency sound studied using multielement surface arrays', in *Natural Physical Sources of Underwater Sound-Sea Surface Sound(2)*, Ed. B.R. Kerman, Kluwer Academic Publishers, pp 233-245 (1993).

[81] M.J. Lighthill, 'On sound generated aero dynamically I: general theory', Proc. R. Soc. London Ser. A **211** 564-587 (1952).

[82] P.E. Doak, 'Analysis of internally generated sound in continuous materials: 2', J. Sound Vib. **25** 263-335(1972).

[83] A.D.D. Craik, *'Wave Interactions and Fluid Flows'*, Cambridge University Press (1985).

[84] N. Curle, 'The influence of solid boundaries upon aerodynamic sound', Proc. Roy. Soc. London Ser. A **231** 505-514 (1955).

[85] J.A. Stratton, *'Electromagnetic Theory'*, McGraw-Hill Book Company, Inc. (1941).

[86] I.S and E.C. Sokolnikoff, *'Higher Mathematics for Engineers and Physicists'*, McGraw-Hill Book Company Inc. (1941)

[87] D.O. Zopf, H.C. Creech and V.H. Qinn, 'The wave meter - a land based system for measuring near shore ocean waves', Marine Technical Society Journal **10**(4) 19-25 (1976)

[88] L.B. Felsen and N.M. Marcuvitz, *'Radiation and Scattering of Waves'*, Prentice-Hall, Inc., (1973)

[89] G.V. Frisk, *'Ocean and Seabed Acoustics-A Theory of Wave Propagation'*, PTR Prentice Hall Englewood Cliffs, New Jersey 07632, (1994)

[90] A.C. Kibblewhite and E.P.M. Brown, 'The use of a shore-based seismometer for wave-energy resource assessment in New Zealand', Oceans '91, Proceedings, Vol.1, Oct. 1991, Hawaii, pp 375-379, (1991)

[91] A.C. Kibblewhite, P.L. Pearce, C.Y. Wu, 'Southland ocean wave studies, Phase III; Wave-energy statistics, 1992', Department of Physics, University of Auckland, (March 1993).

SUBJECT INDEX

A
Acoustic analogue equation[13,15]
Acoustic gravity equation[6]
Acoustic properties of the
 upper-sedimentary layers[105]
Acoustic wave equation[6,16,59]
Adiabatic sound velocity[23]
Adjoint
-Green's function[63]
-operator[63,67]
Ambient noise[ii,1,3,13]
Analysis based on
-pressure[56]
-the potential[57]
Atmospheric pressure[24]
Atmospheric turbulence[13]
Attenuation[108,141,170]

B
Bessel function[144,159]
Bilinear concomitant vector[63]
Biot/Stoll model[33] Biot waves[154]
Bottom
-hard[61,122,130]
-soft[61]
Boundary conditions[11-14,24,25,
 30,41,45,59,62]
-for the first- and second-order
 potential fields[38]
-homogeneous and inhomogeneous[15,37,39]
Boundary source[15,56,57,59,66]

C
Coherence functions[73,99,142,156]
-of the first-order pressure
 field[100,144,157]
-of the second-order field[101,145]
-of the seismic field[154]
-and power-density spectra[157]
Coherence properties of the
 pressure field[99]
Coherency spectra[138,140]
Comparison
-of analyses using the velocity
 potential and pressure as the field
 variable[7,26,38-49]
-of different theoretical analyses[50]
-of experimental and theoretical spectra
 [128-134]
-of plane-wave and Green's
 function analyses[44]
Conversion from the second-order
 potential to pressure[92]
Continental shelf[124]
Critical values of k or $(\hat{m})$[82]
Critical angles, θ_{c1} and θ_{c2}[86]

D
Deep-ocean
-analysis[55,61,129]
-assumptions[5,99,181]
-dispersion relation[159]
-source field[37]
-source pressure spectrum[138]
 see also Source pressure spectrum-
 deep water
-spectrum[121]
Definition of the generalised source
 spectrum(wavenumber-frequency)[74]
Delta functions[54]
Density as the field variable[15]
Depth-dependent properties
-dispersion relation[135]
-pressure spectra[119,121,122]
 see also Pressure spectra-theoretical
Difference-frequency
-component[8,19,74,80,86,88-90,131,132]
-interactions[5,74,82,89,99,132,174,182]
-spectra[90]
Dispersion
-equations[33]
-relation[5,7,19,61,62,64,69,73,77,
 80,89,92-94,102,158,162,177]
Displacement
-of the seafloor[33,152]
-of water particles[21]
-potentials[32,153]
-vector[153]
-vector of the frame and of the pore fluid[23]
Double-frequency pressure field
 see Second-order field
Elastic layer[34,165]

Elastic media[26]
Elastic moduli[24,25]
Environmental model[7,112]
  see also Geoacoustic model,
  Models of seabed
Equivalent source
-second-order field[15,37]
-spectra[134-136,155]
Experimental data[3,128,129,131-134
  175-177]
Evanescent field[134]

F

Fast-wave[33,34,153]
Fetch[103,104]
Field variable see
  Potential and Pressure
First-order field
-component[12-14,28-31,37-40,46-61,
  100,139,141,182]
-source functions[64,73]
-surface displacement[39]
Fourier-Stjeltjes integration[12]
Fourier transform[34]
Full-wave analysis
-of multilayered seabeds[18]
-of the wave-interaction process
  [4,5,19,73,99]

G

General dispersion relation[93,95,152]
Generalised solution of the potential
  field
-first-order potential field[59]
-second-order potential field[65]
Generalised source pressure spectrum
  of the second-order field
-definition[74]
-deep water case[77]
  sum-frequency component[84]
  difference-frequency component[88]
  second-order pressure spectrum
    (frequency)[96]
  total second-order spectrum[90]
-shallow water case[92]
Geoacoustic model[74,105,136,169]
  see also Models of seabed

Geometric description
-of integration limits[93]
-of interacting gravity waves[77]
-of the wave interaction process
  [77,80,90,181]
Gravity
-acoustic wave equation[14]
-wave trains[18]
-waves[4,81,94]
Green's function[17,153,160]
-analyses[66,169]
-for the reference waveguide[64]
-of the gravity-acoustic wave operator[15]
-solution[5,6,8,15,17,41,42,48,49,63]
-$G_v(\vec{r},\vec{r}_0)$ and $G_h(\vec{r},\vec{r}_0)$[152]
Green's theorem[62,63]

H

HDF
component[37,116,134,136,137,142,172,179]
  see also Second-order pressure field
Helmholtz equation[37]
Hydrates[105,108,111,119,121,165]

I

IDF
component[4,116,134,136,137,142,172,179]
  see also Second-order pressure field
Interface-wave modes[151]
Interface conditions[26,28,33]
Intrinsic attenuation[110]
Inversion procedures[138,164]

J

Joint North Sea Wave Project(JONSWAP)
  [103]
JONSWAP
-spectral levels[104]
-form of the ocean-wave spectrum[162]

L

Land-based system (LBS)[136,137,172]
Laplace equation[37]
Lighthill's equation[14,16,45]
Linearisation[11,26]

Index

M
Microseisms[3,11,13,152]
Models of seabed
-MDL1[105,110,111,116,119,122,123,161, 162,164]
-MDL2[106,110,111,122,123,164,165,175, 179,180]
-MDL3(1)[105,108,119,122,165]
-MDL3(2)[105,108,109,119,121,165]
-MDL3(3)[105,108,110,111,116,119,121, 122,165]
-MDL(0)[119]
-MDL(1)[119]
-MDK1[169,172]
-MDK2[169,172]
-MD1(1)[170]
-MD2[169]
Multilayered seabed[110,116]

N
Noise-notch[92,132,133]
Non self-adjoint equation[8]
Nonhomogeneous
-acoustic-wave equation[15]
-equation[12,14,15]
-boundary condition[15,39]
Normal modes in the water layer[131]
Notations of the Fourier transform[34]

O
Ocean-bottom seismometer(OBS) [136,137,172]
Ocean-wave field
-angular distribution function[100]
-dispersion relation[37]
-frequency spectral-density function[100]
-spectral forms[102,123,135]

P
Parameter values of the porous sediment[106]
Particle velocity (or displacement)[24]
Perturbation
-analysis[6,7]
-expansion of the Green's function[15]
-procedure[11,13,14,17,21]
-series[26,30]
-series for the velocity potential[12]

PF component-*see* First-order field
Pierson-Moskowitz (PM) spectrum[102,104]
Plane-wave -analysis[7,13,17,38,41,49,50,71]
-comparison with the Green's function analysis[38,50,71]
-reflection coefficient[33,60,110,152]
   see also Reflection coefficient
-solutions of the first- and second-order equations[14]
Point source[12,73,135]
Point-pressure source[152,153]
Porous, viscoelastic medium[23,106]
Porous elastic media[25,32,33,152,179]
Potential
-as the field variable[26,57,59]
Pressure
-as the field variable[14,16,29,56]
-in lieu of density as the field variable[16]
-in the pore fluid[24]
-relation to the velocity potential[28,29]
Pressure spectra-theoretical
-vs observation depth[112]
-vs ocean-wave spectrum[122]
-vs seabed structure[121]
-vs water depth[124]
-vs wind speed[123]
-source (frequency)[55]
-source (frequency-wavenumber)[53]
Pressure-rate spectra[133,134]
Primary-frequency pressure field
   see First-order field
Propagation[6,99,165]
Properties
-of the seabed *see* Geoacoustic models
-of the wave-induced pressure field[6,99]
   see also Pressure spectra, *and* Coherence functions

R
Range dependent environment[165]
Range-dependent model[151]
Range-independent model[21,100,151,165]
Ratio H/V[165,174]
Rayleigh wave[170]
Reference wave guide[59,60,61,69]
Reflection coefficient[8,19,61,93,100,

110,111,119,121,153,179]
-as a function of both frequency
and wavenumber[110]
-of multilayered seabeds[182]
Reflection loss[111]

S
Seabed
-displacement[137,161,162,164,165][169]
-exploration[121]
-model see Models of seabed
-sediments[152]
-structure see Models of seabed
Second-order pressure field[28-31,37,49-54,
87,90,116,133-
135,136,137,142,172,177,182]
-difference-frequency component
see Difference-frequency component
-"exact" description of homogeneous
component[37-57]
-generalised description[73] see also
Generalised source pressure spectrum
-homogeneous (HDF) component
see HDF component
-inhomogeneous (IDF) component
see IDF component
-source spectra
see Equivalent source,
Source pressure spectra,
Generalised source pressure spectrum
-standing-wave approximation
see Standing-wave approximation
-sum-frequency component
see Sum-frequency component
-theoretical pressure spectra
see Pressure spectra-theoretical
Seismo-acoustic response[9,17,19,26,36,157]
see also Wave-induced seismic field
and Wave-induced seismic spectra
Shear wave[33,34,164]
Slow wave[33,153]
Source
-equivalent see Equivalent source
-point see Point source
-surface and volume distributions
of wave-induced field[15,38,49,54,70]
Source pressure spectra

-deep-water
based on the potential[52-55]
based on the pressure[52-55]
frequency[55]
frequency-wavenumber[53]
-equivalent seeEquivalent source
-generalised[74-92]
see also Generalised source pressure spectra
-standing-wave approximation
see Standing-wave approximation
Source pressure spectrum -generalised
second-order form see Generalised
source pressure spectrum
Southern Ocean[132,174]
Spectra
-coherence and coherency spectra
see Coherence functions
-ocean waves see Ocean-wave field
-pressure(wave-induced)
see Pressure spectra -theoretical
-seismic(wave-induced)
see Wave-induced seismic spectra
Standing-wave approximation[4,13,18,37,
50,54,82,87,102,116,142,144]
Sum-frequency
component[5,84-88,90,175,182]
see Generalised source pressure spectrum
Surface-source distribution[49,70]

T
Theoretical pressure spectra
see Pressure spectra -theoretical
Typical ULF spectrum[96-99]

U
Unconsolidated porous layer (sand)[110]

V
Velocity potential[6,27,38,66]
Vertical displacement of seabed
-and horizontal[12,152]
-coherence function[141,142]
Viscoelastic seabed[33]

W
Wave

Index

-fast, shear and slow compressional
  [33,34,153,164]
-evanescent
-homogeneous
  see Homogeneous component
-inhomogeneous
  see Inhomogeneous component
Wave-induced pressure field (first-order)
  see First-order field
Wave-induced pressure field (second-order)
  see Second-order pressure field
Wave-induced seismic field[151-177]
-coherence[154-159]
-propagation[165-171]
-ratio H/V[161-171]
-spectra *see* Wave-induced seismic spectra
-use as a wave recording system[173]
Wave-induced seismic spectra
-dependence on layer thickness[165]
-dependence on range[165]
-dependence on seabed structure[164]
-dependence on water depth[162]
-dependence on wind speed[164]
Wavenumber
-acoustic[73]
-frequency presentation of
  reflection coefficient[121]
-gravity-wave[73]
-horizontal[4,9,13,18,19,33,36,153]
-relative[87,110,182]

Lecture Notes in Earth Sciences

Vol. 1: Sedimentary and Evolutionary Cycles. Edited by U. Bayer and A. Seilacher. VI, 465 pages. 1985. (out of print).

Vol. 2: U. Bayer, Pattern Recognition Problems in Geology and Paleontology. VII, 229 pages. 1985. (out of print).

Vol. 3: Th. Aigner, Storm Depositional Systems. VIII, 174 pages. 1985.

Vol. 4: Aspects of Fluvial Sedimentation in the Lower Triassic Buntsandstein of Europe. Edited by D. Mader. VIII, 626 pages. 1985. (out of print).

Vol. 5: Paleogeothermics. Edited by G. Buntebarth and L. Stegena. II, 234 pages. 1986.

Vol. 6: W. Ricken, Diagenetic Bedding. X, 210 pages. 1986.

Vol. 7: Mathematical and Numerical Techniques in Physical Geodesy. Edited by H. Sünkel. IX, 548 pages. 1986.

Vol. 8: Global Bio-Events. Edited by O. H. Walliser. IX, 442 pages. 1986.

Vol. 9: G. Gerdes, W. E. Krumbein, Biolaminated Deposits. IX, 183 pages. 1987.

Vol. 10: T.M. Peryt (Ed.), The Zechstein Facies in Europe. V, 272 pages. 1987.

Vol. 11: L. Landner (Ed.), Contamination of the Environment. Proceedings, 1986. VII, 190 pages. 1987.

Vol. 12: S. Turner (Ed.), Applied Geodesy. VIII, 393 pages. 1987.

Vol. 13: T. M. Peryt (Ed.), Evaporite Basins. V, 188 pages. 1987.

Vol. 14: N. Cristescu, H. I. Ene (Eds.), Rock and Soil Rheology. VIII, 289 pages. 1988.

Vol. 15: V. H. Jacobshagen (Ed.), The Atlas System of Morocco. VI, 499 pages. 1988.

Vol. 16: H. Wanner, U. Siegenthaler (Eds.), Long and Short Term Variability of Climate. VII, 175 pages. 1988.

Vol. 17: H. Bahlburg, Ch. Breitkreuz, P. Giese (Eds.), The Southern Central Andes. VIII, 261 pages. 1988.

Vol. 18: N.M.S. Rock, Numerical Geology. XI, 427 pages. 1988.

Vol. 19: E. Groten, R. Strauß (Eds.), GPS-Techniques Applied to Geodesy and Surveying. XVII, 532 pages. 1988.

Vol. 20: P. Baccini (Ed.), The Landfill. IX, 439 pages. 1989.

Vol. 21: U. Förstner, Contaminated Sediments. V, 157 pages. 1989.

Vol. 22: I. I. Mueller, S. Zerbini (Eds.), The Interdisciplinary Role of Space Geodesy. XV, 300 pages. 1989.

Vol. 23: K. B. Föllmi, Evolution of the Mid-Cretaceous Triad. VII, 153 pages. 1989.

Vol. 24: B. Knipping, Basalt Intrusions in Evaporites. VI, 132 pages. 1989.

Vol. 25: F. Sansò, R. Rummel (Eds.), Theory of Satellite Geodesy and Gravity Field Theory. XII, 491 pages. 1989.

Vol. 26: R. D. Stoll, Sediment Acoustics. V, 155 pages. 1989.

Vol. 27: G.-P. Merkler, H. Militzer, H. Hötzl, H. Armbruster, J. Brauns (Eds.), Detection of Subsurface Flow Phenomena. IX, 514 pages. 1989.

Vol. 28: V. Mosbrugger, The Tree Habit in Land Plants. V, 161 pages. 1990.

Vol. 29: F. K. Brunner, C. Rizos (Eds.), Developments in Four-Dimensional Geodesy. X, 264 pages. 1990.

Vol. 30: E. G. Kauffman, O.H. Walliser (Eds.), Extinction Events in Earth History. VI, 432 pages. 1990.

Vol. 31: K.-R. Koch, Bayesian Inference with Geodetic Applications. IX, 198 pages. 1990.

Vol. 32: B. Lehmann, Metallogeny of Tin. VIII, 211 pages. 1990.

Vol. 33: B. Allard, H. Borén, A. Grimvall (Eds.), Humic Substances in the Aquatic and Terrestrial Environment. VIII, 514 pages. 1991.

Vol. 34: R. Stein, Accumulation of Organic Carbon in Marine Sediments. XIII, 217 pages. 1991.

Vol. 35: L. Håkanson, Ecometric and Dynamic Modelling. VI, 158 pages. 1991.

Vol. 36: D. Shangguan, Cellular Growth of Crystals. XV, 209 pages. 1991.

Vol. 37: A. Armanini, G. Di Silvio (Eds.), Fluvial Hydraulics of Mountain Regions. X, 468 pages. 1991.

Vol. 38: W. Smykatz-Kloss, S. St. J. Warne, Thermal Analysis in the Geosciences. XII, 379 pages. 1991.

Vol. 39: S.-E. Hjelt, Pragmatic Inversion of Geophysical Data. IX, 262 pages. 1992.

Vol. 40: S. W. Petters, Regional Geology of Africa. XXIII, 722 pages. 1991.

Vol. 41: R. Pflug, J. W. Harbaugh (Eds.), Computer Graphics in Geology. XVII, 298 pages. 1992.

Vol. 42: A. Cendrero, G. Lüttig, F. Chr. Wolff (Eds.), Planning the Use of the Earth's Surface. IX, 556 pages. 1992.

Vol. 43: N. Clauer, S. Chaudhuri (Eds.), Isotopic Signatures and Sedimentary Records. VIII, 529 pages. 1992.

Vol. 44: D. A. Edwards, Turbidity Currents: Dynamics, Deposits and Reversals. XIII, 175 pages. 1993.

Vol. 45: A. G. Herrmann, B. Knipping, Waste Disposal and Evaporites. XII, 193 pages. 1993.

Vol. 46: G. Galli, Temporal and Spatial Patterns in Carbonate Platforms. IX, 325 pages. 1993.

Vol. 47: R. L. Littke, Deposition, Diagenesis and Weathering of Organic Matter-Rich Sediments. IX, 216 pages. 1993.

Vol. 48: B. R. Roberts, Water Management in Desert Environments. XVII, 337 pages. 1993.

Vol. 49: J. F. W. Negendank, B. Zolitschka (Eds.), Paleolimnology of European Maar Lakes. IX, 513 pages. 1993.

Vol. 50: R. Rummel, F. Sansò (Eds.), Satellite Altimetry in Geodesy and Oceanography. XII, 479 pages. 1993.

Vol. 51: W. Ricken, Sedimentation as a Three-Component System. XII, 211 pages. 1993.

Vol. 52: P. Ergenzinger, K.-H. Schmidt (Eds.), Dynamics and Geomorphology of Mountain Rivers. VIII, 326 pages. 1994.

Vol. 53: F. Scherbaum, Basic Concepts in Digital Signal Processing for Seismologists. X, 158 pages. 1994.

Vol. 54: J. J. P. Zijlstra, The Sedimentology of Chalk. IX, 194 pages. 1995.

Vol. 55: J. A. Scales, Theory of Seismic Imaging. XV, 291 pages. 1995.

Vol. 56: D. Müller, D. I. Groves, Potassic Igneous Rocks and Associated Gold-Copper Mineralization. XIII, 210 pages. 1995.

Vol. 57: E. Lallier-Vergès, N.-P. Tribovillard, P. Bertrand (Eds.), Organic Matter Accumulation. VIII, 187 pages. 1995.

Vol. 58: G. Sarwar, G. M. Friedman, Post-Devonian Sediment Cover over New York State. VIII, 113 pages. 1995.

Vol. 59: A. C. Kibblewhite, C. Y. Wu, Wave Interactions As a Seismo-acoustic Source. XIX, 313 pages. 1996.